STUDY GUIDE

John Vincent Aliff
Georgia Perimeter College
Gwinnett University Center

to accompany

ENVIRONMENT

Fourth Edition

Peter H. Raven
Missouri Botanical Garden

Linda R. Berg
St. Petersburg College

John Wiley & Sons, Inc.

To order books or for customer service call 1-800-CALL-WILEY (225-5945).

ISBN 0-471-44494-4

Printed in the United States of America

10 9 8 7 6 5 4 3 2 1

Printed and bound Bradford & Bigelow, Inc.

Preface

"If you study to remember, you will forget, but, if you study to understand, you will remember."

Anonymous

Peter Raven and Linda Berg are committed educators and scientists who have devoted their time and resources to providing you a sound introduction to the study of the environment. Your authors intend to build the skills you will need to be a responsible citizen with an appreciation of the effects that human activities have on the Earth. Science is an intellectual activity that lights a fire in the human spirit to explain what we observe in nature. There is no final or complete knowledge in human experience—knowledge is infinite. We can forever search the expanses of the Universe or the microcosms of atoms. We search for the best answers that we can arrive at today. No claims are made that our conclusions are absolute, unalterable facts. Tomorrow, we will know more.

DEVELOPING YOUR KNOWLEDGE USING THIS STUDY GUIDE

Sections

Each chapter is divided into 6 sections as described below:

1. **Objectives** - these are taken verbatim from your Raven-Berg 4e textbook.
2. **Seeing the Forest** - this consists of one or more questions that help you organize a general overview of the related textbook chapter.
3. **Seeing the Trees** - this section explores specific concepts in the chapter. It is subdivided into these areas:
   a. **Vocab** - important vocabulary terms are defined using different words than the text. Further, you may reinforce learning by putting these definitions in your own words.
   b. **Questions to Ponder** - exploration questions allow for further research and review of major points made in the textbook chapter.
   c. **Pop Quiz** - matching and multiple choice questions are offered to help get you ready for a test!

4. **Cyber Surfin'** - current websites are listed for background information and further exploration. If a link dies, a tragedy of course, first try to access the server root of the URL, then click to the changed address, e.g., if the address was formerly **www.gpc.edu/~jaliff/scimen-vs.htm**, try **www.gpc.edu/~jaliff/** or **www.gpc.edu.** Please report dead links to me at **jaliff@gpc.edu**.
5. **Like Bookin' It** - these are generally many of the most current books available or recognized library resources. For other works or current scientific papers, see your text listings, do a resource search online, or visit your college library.
6. **Answers** - the study guide author's answers are included. The questions in *Seeing the Forest* and *Questions to Ponder* are to some extent open ended. Many responses are possible. These answers are provided to assist you if you get bogged down. They should not be memorized.
7. **Chemistry Appendix** - since environmental science may be offered by scientists whose background and course materials are biased towards the physical science, a chapter on elementary environmental chemisty will be found in the Appendix.

The University of North Carolina student counseling service has a good web page on improving study habits, check out **http://caps.unc.edu/TenTraps.html**. Virginia Tech student affairs also has a helpful website at **http://www.ucc.vt.edu/stdysk/stdyhlp.html**.

There is no such thing as common sense, there is only sense or nonsense. Education should empower us with the desire to distinguish reason from unreason. Overcome prejudice and the future is bright indeed.

John Vincent Aliff

"Let us think of education as the means of developing our greatest abilities, because in each of us there is a private hope and dream which, fulfilled, can be translated into benefit for everyone and greater strength for our nation."

President John F. Kennedy

ACKNOWLEGEMENTS

I want to thank John Wiley & Sons editorial staff Wendy Perez and Geraldine Osnato for their teamwork in producing this study guide. I want to thank my wife, Roxy Warren, and kids Elaine Opal, John Manley and William for sacrificing family time to complete it. Thanks are offered to Georgia Perimeter College for valuing writing for professional development.

Contents

1

Our Changing Environment

LEARNING OBJECTIVES

After you have studied this chapter you should be able to:

1. Define environmental science and explain why environmental sustainability is an important concern of environmental science.
2. Summarize human population issues, including population size and level of consumption.
3. Describe the three factors that are most important in determining human impact on the environment and solve a problem using the *IPAT* equation.
4. Briefly describe some of the data that suggest that certain chemicals used by humans may also function as endocrine disrupters in animals, including humans.
5. Provide an overview of how human activities have affected the following: the Georges Bank fishery, tropical migrant birds, wolf populations in Yellowstone National Park, and invasive species, such as comb jellies and zebra mussels.
6. Characterize human impacts on the global atmosphere, including stratospheric ozone depletion and climate warming.
7. Describe some of the consequences of tropical rain forest destruction.
8. Define environmental ethics and discuss distinguishing features of the Western and deep ecology worldviews.

SEEING THE FOREST

How has Oberlin College in Ohio put commitment to environmental quality into action?

Fill out the table below, applying the principles of the $I = P \times A \times T$ equation.

I = impact on the environment, P = the number of people, A = affluence (resource consumption per capita), and T = the environmental effects of the technologies used to obtain and consume resources.

Green architectural design compared to *traditional design*	**Number of *people* using the technology: Rate high, low, or moderate**	**Consumption per capita by *affluence:* Rate high, low, or moderate**	**Environmental effects of *technology:* Rate high, low, or moderate**	**Total *impact* on the environment: Rate high, low, or moderate (See ecological footprint, Ch. 9)**
Geothermal heat pump				
Open air heat pump				
Photovoltaic cells on roof				
All electricity provided by a local power company produced by coal burning (most)				
Triple-paned southern exposure windows				
Windows double paned and evenly distributed around the building				

Living machine wastewater treatment and recycling of "gray" water				
Sewage to outside pipes and conduction to municipal treatment plants to be released into rivers				
Recycling of wood, harvest of wood from sustainable sources, use of recycled metal, carpet				
Disposal of used materials in landfills				
Preservation of local marsh and forest				
Use of highly productive natural areas to build new dormatories				

SEEING THE TREES

VOCAB—If we can't agree on the meaning of terms, no issue is ever resolved.

1. Environmental Science—the interdisciplinary study of man's relationship with other organisms and the nonliving environment, combining information from biology, chemistry, geography, geology, physics, economics, sociology, cultural anthropology, agriculture, engineering, law, politics, and ethics.
2. Pollution—any alteration of air, water, or soil that harms the health, survival, or activities of living organisms.
3. Ecology—a branch of biology that studies the interactions of organisms and environment.
4. Environmental sustainability—a concept that the environment can function indefinitely without going into decline from the results of human activity.
5. Endocrine disrupters—chemicals that negate or enhance the effects of natural hormones by binding to their cellular receptors.

6. Green engineering—the design, commercialization and use of processes and products that are feasible and economical while minimizing the generation of pollution at its source and risks to human health and the environment (U.S. EPA).

Questions to Ponder

1. How is the interdisciplinary nature of Environmental Science reflected in the definition of ecology? Do you agree with the author's diagram below?

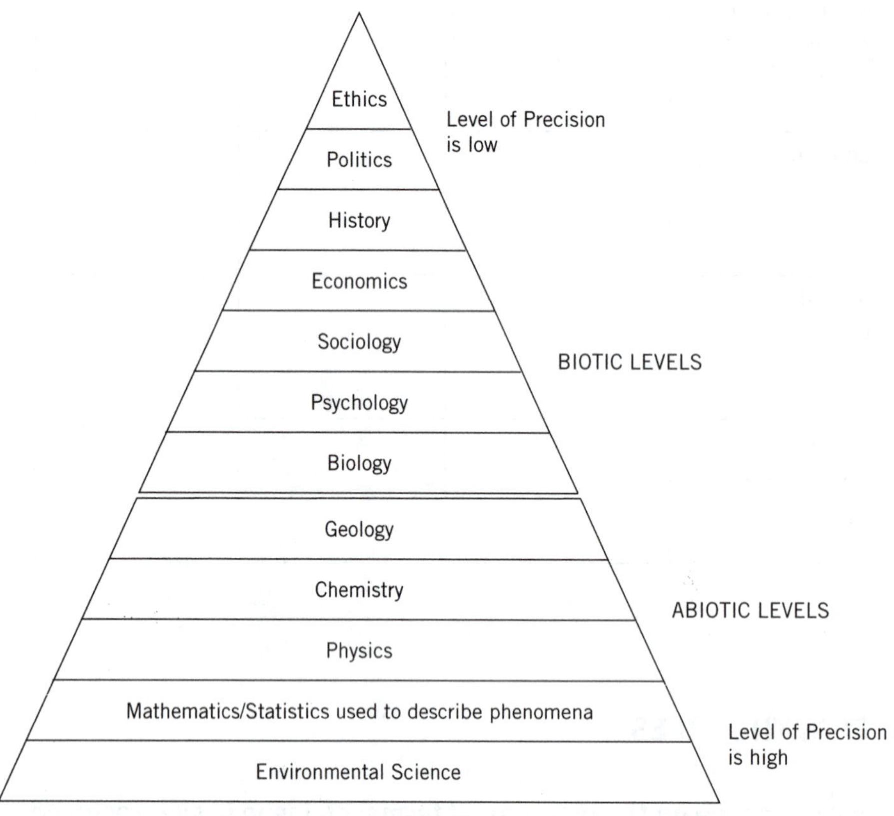

A-frame "House of Ecology."

2. List the pros and cons of a new policy governing forestry: "Tree biomass (dry weight) harvested = tree biomass replacement by growth." Fill-in the table below, using the perspectives of two different disciplines.

Discipline	Pros	Cons
Environmental science		
Stereotypic corporate economics		

3. Compare and contrast: attitudes of the early European explorers, and Native Americans at that time.

a. Fill-in the table below.

Group	Concept of nature	Nature must be nurtured or exploited?	Resources viewed as limited or unlimited?	Resources viewed as sustainable?
Native Americans	Nature as mother, or provider			
European settlers	Nature as resource to be exploited			

b. Now compare the economic attitudes of developed countries against the concept of environmental sustainability. Complete the table.

Group	Concept of nature	Costs to the health of organisms and people viewed as important or unimportant?	Natural resources viewed as limited or unlimited?
Developed countries economic policies	Nature as resource to be exploited		
Environmental science	Nature as a provider of sustainable resources		

4. Explain the concept of sustainability and human interaction by considering human activities that compromise environmental stability and our own survival.

a. Nonrenewable resources are being used up quickly. Describe four:

1. __________

2. __________

3. __________

4. __________

b. Renewable resources are being used up faster than they can be replenished naturally. Think of four:

1. __________

2.

3.

4.

5. In resource-rich developing/underdeveloped countries, how do the economic policies and consumption activities of developed countries perpetuate poverty there?
List four:

1.

2.

3.

4.

6. What are some of the possible effects of exotic chemical release into the biotic environment? Explain four.

1.

2.

3.

4.

7. List four invasive species in your locale.

1.

2.

3.

4.

8. Describe four reasons for the over 40% decline in eastern U.S. songbirds in the last 40 years.

1.

2.

3.

4.

9. How will the burning of tropical rain forest affect the problem of global warming? Research the processes of combustion and photosynthesis.

10. Refer to the graph below. Explain the seasonal increases and decreases in carbon dioxide in the northern hemisphere.

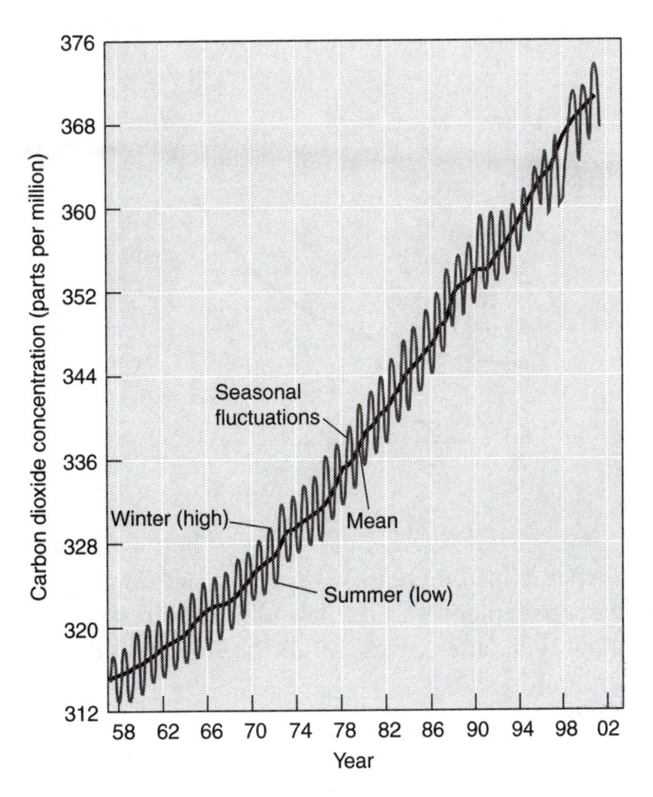

11. Explain two possible effects of exotic chemical release into the abiotic, atmospheric environment.

1. ______

2. ______

12. Propose four positive economic benefits of a sustained environment.

1. ______

2. ______

3. ______

4. ______

13. Describe four factors that most directly spurred the exponential growth in human population since the decline caused by the black death plague. See the graph below.

1. ______

2. ______

3. ______

4. ______

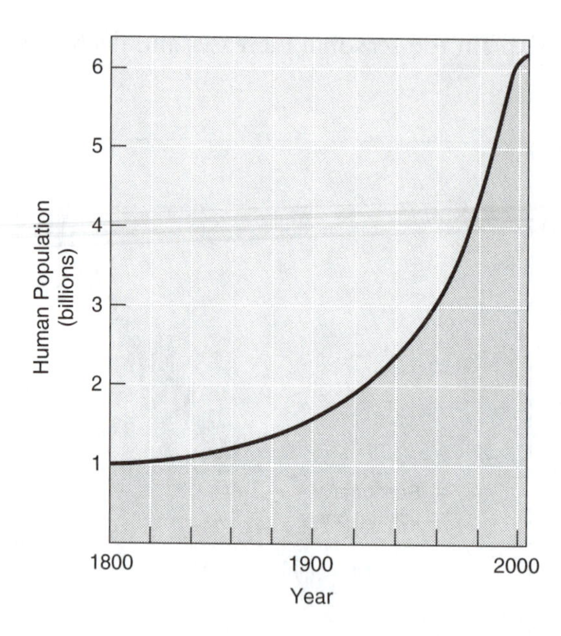

The agricultural discoveries that helped increase human population before the black death included crop rotation, the use of alfalfa to replenish nutrient and humus loss in soils, irrigation, and the invention of the scouring plow. The latter turns soil to the sides as it makes a furrow, instead of scratching the ground as early plows did.

POP QUIZ

Matching

Match the following terms and definitions. Use the best answer.

____1. Conserving species of plants and animals (biodiversity) in today's world	a. Great Lakes
____2. The burning of wood, coal, oil, and natural gas produces this "global warming" pollutant	b. Stratosphere
____3. The gray wolf, extinct in most of the U.S., was reintroduced here.	c. Synergism
____4. Many U.S. migratory birds overwinter here	d. Carbon dioxide
____5. The Georges Bank is closest to	e. Earth as Mother
____6. Zebra mussels plug up water pipes here	f. Yellowstone National Park
____7. Premise of 19th-century pioneer philosophy	g. Antarctica
____8. Philosophical premise of Native Americans	h. Environmental Science
____9. When various chemicals interact to have negative health effects	i. Environmental sustainability

____10. The “ozone hole” is found over	j. Unlimited resources
____11. CFCs release ozone-destroying chlorine gas here,	k. Massachusetts
____12. Interdisciplinary study of the interrelations of humans and other organisms to their physical environment.	l. Venezuela, Colombia

Multiple Choice

1. The most important and direct assumption made by scientists in concluding that there is a worldwide environmental crisis is:
   a. endangered species numbers are increasing
   b. renewable resources are being used up faster than they are replenished
   c. the human population is increasing
   d. global warming is happening
   e. toxic chemicals are being released into the environment

2. Approximately how many people in the world live in extreme poverty?
   a. 5.6 billion
   b. 600 million
   c. 1.2–2.8 billion
   d. less than 10%
   e. 828 million

3. From the perspective of Environmental Sciences, the reason why developing countries have more extreme poverty is:
   a. lack of proper education of the inhabitants that will allow better exploitation of their resources
   b. when developed countries extract resources from undeveloped countries, the damage done to nonrenewable resources is great
   c. there is not enough investment in providing water through irrigation and wells
   d. poor farming techniques lead to starvation
   e. all of the above

4. Which factor(s) represent the environmental effects of affluent populations in developed countries on the poorer people in undeveloped countries?
   a. higher emission of pollutants spread worldwide
   b. increased extraction of nonrenewable resources
   c. decreasing health of people in undeveloped countries
   d. more extraction of renewable resources than replacement
   e. all of the above

5. If human sperm counts are indeed decreasing due to the release of exotic chemicals into the environment, which statement best describes their effect?
   a. Synergism interferes with testosterone secretion and reception.
   b. Poisoning kills sperm cells.
   c. Antagonism destroys testicular cells.
   d. Agonism (enhanced action) makes estrogen overactive.
   e. All of the above are correct.

6. The overfishing done in the Georges Bank fishery, eliminated the commercial fishing of:
   a. Pacific salmon
   b. Atlantic salmon
   c. blue crabs in the Chesapeake Bay
   d. Atlantic cod
   e. Florida lobsters

7. Which of the following are major factors contributing to the loss of at least 40% of eastern U.S. migratory songbirds in the last 40 years?
   a. forest fragmentation
   b. increased nest parasitism by cowbirds
   c. loss of habitat and bird territory compression
   d. growing coffee in sun plantations rather than shade plantations
   e. all of the above

8. The Kyoto protocol that was renounced by the Bush administration for economic reasons would control ____________?
   a. combustion activities that produce sulfur dioxide.
   b. ozone layer destruction
   c. overfishing of commercial fisheries
   d. human population
   e. carbon dioxide emissions

9. The ozone layer protects life on Earth from:
   a. chlorine from CFC refrigerants
   b. ultraviolet light
   c. gamma and x-rays from the sun
   d. magnetic storms from the sun
   e. all of the above

10. According to environmental science, a positive effect of environmental preservation is:
    a. increased industrial productivity
    b. increased health of citizens
    c. contributions to the economy by ecotourism
    d. increased human population
    e. b and c

11. Tropical rain forest destruction and conversion to grazing or farmland has the following effect:
    a. increased carbon dioxide in the atmosphere
    b. increased global warming
    c. loss of species diversity
    d. loss of migratory birds in the U.S.
    e. all of the above

12. The body mass of the tree mostly comes from:
    a. minerals in the soil
    b. sunlight
    c. carbon dioxide
    d. oxygen
    e. chlorophyll

13. An area most closely approximate to the annual loss of tropical rain forest is ____________?
    a. Connecticut
    b. Georgia
    c. Texas
    d. Pennsylvania
    e. Greenland

14. According to graph on page 7, how much in percentage has carbon dioxide levels increased since 1960?
    a. 16%
    b. 14%
    c. 51%
    d. it has doubled—200%
    e. seasonal fluctuations prevent us from knowing

15. According to the graph on page 8, in what phase is human population trend?
    a. lag (pregrowth)
    b. decline
    c. plateau (stability)
    d. exponential growth
    e. we don't know

CYBER SURFIN'

1. Learn about the life and work of Rachel Carson: **www.rachelcarson.org/**
2. The National Wildlife Federation explains biodiversity and effects of global warming: **www.nwf.org**
3. Green Engineering: **www.epa.gov/oppt/greenengineering/**
4. The Sierra Club reflects on human population and environmental effects: **www.sierraclub.org/population/**
5. Invasive plants and animals and their environmental effects can be explored here:
   a. www.fsu.edu/~imsp/silent_invaders/new_weeds/main_html/
   b. biology.usgs.gov/s+t/SNT/noframe/ns112.htm
   c. short list of invasive species at **www.invasivespecies.gov/profiles/main.shtml**
6. See endangered species lists, U.S. Fish and Wildlife Service: **endangered.fws.gov/wildlife.html#Species**
7. Check out the depletion of commercial fisheries: **archive.greenpeace.org/~oceans/globaloverfishing/sinkingfast.html**
8. Learn about environmental ethics: **onlineethics.org/moral/carson/main.html**

'LIKE' BOOKING IT

1. Erlich P. ***The Population Bomb.*** New York: Ballantine Books, 1989.
2. Erlich P, Ehrlich A, and Holdren P. ***Ecoscience: Population, Resources, Environment.*** San Francisco, Freeman, 1977.
3. Carson RL. ***Silent Spring.*** New York: Houghton Mifflin Co., 1962.
4. Marco GJ, Hollingworth RM, and Durham W, eds. ***Silent Spring Revisited.*** Washington, D.C.: American Chemical Society, 1987.

5. Raven PH, and Williams T, eds. ***Nature and Human Society: The Quest for a Sustainable World*****, Committee for the Second Forum on Biodiversity,** Washington, D.C.: National Academy of Sciences and National Research Council, 2000.

ANSWERS

Questions to Ponder (These answers are some of many possible, please don't memorize them.)

The table responses relate to the example at Oberlin College.

Green architectural design compared to traditional design	Number of people using the technology: Rate high, low, or moderate	Consumption per capita by affluence: Rate high, low, or moderate	Environmental effects of technology: Rate high, low, or moderate	Total impact on the environment: Rate high, low, or moderate
Geothermal heat pump	*Moderate: Oberlin College is not New York City.*	*Moderate: The geothermal heat pump is more efficient than an air-mounted heat exchanger, particularly at lower temperatures. Per-capita consumption of electricity is reduced.*	*Moderate: Less surface land is used for the equipment, therefore displacing fewer plants and animals. Less coal is used to generate electricity, therefore reducing nonrenewable resource extraction and related land destruction. For the small % of power generated by nuclear energy, less nuclear waste has to be dispersed into the environment. Moderate as compared to traditional methods*	*of heating and cooling, such as "all electric" or natural gas.*
Open air heat pump	*Moderate*	*High: At low temperatures, air-mounted heat pumps must switch to total electrical heat, which is costly in economic and environmental terms. More heat is released into the air of urban "heat islands." Affluence increases consumption of resources and related pollution.*	*High: More coal has to be mined increasing environmental disruption and more radioactive materials are dispersed into the environment. Coal burning produces carbon dioxide that enhances global warming as well as acid rain gases that kill trees and lakes in higher elevations.*	*High but not as high as electric.*

Photovoltaic cells on roof	*Moderate*	*Low: No ecological communities are affected as compared to a bare roof. However, some are experimenting with roof gardens as a way of reducing urban heating. Excess power is sold to conventional power distributors.*	*Low: Generation and local transmission space requirements are low. See below.*	*Low*
All electricity provided by a local power company produced by coal burning (most)	*Moderate*	*High: See above. Affluence increases resource use and related pollution.*	*High: Consider the environmental disruption of mining for coal, the area occupied by power plants, and the transmission lines.*	*High*
Triple-paned southern exposure windows	*Moderate*	*Low*	*Low: A simple design that increases heat available in the Winter.*	*Low*
Windows double paned and evenly distributed around the building	*Moderate*	*Moderate: Heat loss from windows is higher than walls, even when the windows are insulated. More affluent households or occupants of buildings have more windows. Affluence increases loss of resources and pollution.*	*High: Heat loss in the winter increases.*	*Moderate*
Living machine wastewater treatment and recycling of "gray" water	*Moderate*	*Moderate: Affluence increases renewable resource extraction and use.*	*Low: As compared to traditional, methods. Gray water's best use is probably not to irrigate grass and decorative plants in the long run.*	*Moderate*
Sewage to outside pipes and conduction to municipal treatment plants to be released into rivers	*Moderate*	*High: Affluence increases renewable resource extraction and use.*	*High: Consider the economic and environmental costs of wastewater transmission, treatment and release (e.g., land disruption for treatment plants and sewer lines).*	*High*

Recycling of wood, harvest of wood from sustainable sources, use of recycled metal, carpet	*Moderate: The principle of sustainability prevents increases in supply due to greater numbers of people.*	*Low: The principle of sustainability prevents increases in supply due to affluence.*	*Low: As compared to traditional methods.*	*Low in this case, but moderate in other locations where extraction is greater than renewal.*
Disposal of used materials in landfills	*Moderate*	*High: Affluence greatly increases renewable resource extraction and use.*	*High: A "throwaway" mentality magnifies renewable resource extraction and use.*	*High*
Preservation of local marsh and forest	*Moderate*	*Low: Preservation and sustainability principles prevent loss of the natural resources.*	*Low: Natural biodiversity, water purification, ground water charging and aesthetics are maintained.*	*Low*
Use of highly productive natural areas to build new dormatories	*Moderate: Increased numbers of people would multiply the effects.*	*High: Permanent loses of these areas increase as affluence increases.*	*High: Permanent losses of natural biodiversity, water purification, groundwater charging and aesthetics.*	*High*

1. How is the interdisciplinary nature of Environmental Science reflected in the definition of ecology? *Ecology is defined as the science dealing the relationship of organisms and their environment. It necessarily involves the basic sciences—biology, chemistry, geology and physics, and areas such as mathematics, sociology, psychology, and history.*
2. List the pros and cons of a new policy governing forestry: "Tree biomass (dry weight) harvested = tree biomass replacement by growth." Fill-in the table below, using the perspectives of two different disciplines:

Discipline	Pros	Cons
Environmental science	*The area covered by forest is stable. "Standing biomass" is maintained. Carbon dioxide is taken from the air by trees. Groundwater is charged by forests.*	*Some forests may still be "clear-cut" with resulting water pollution, loss of plants, animals, and soil.*
Stereotypical corporate economics	*Ecotourism increases with resulting economic benefits to local residents.*	*No further growth in forestry industries occurs. Some jobs lost.*

3. Compare and contrast: attitudes of the Early European Explorers, and Native Americans at that time.

a.

Group	Concept of Nature	Nature must be nurtured or exploited?	Resources viewed as limited or unlimited?	Resources viewed as sustainable?
Native Americans	Nature as mother or provider	*Nurtured*	*Limited*	*Sustainable*
European settlers	Nature as resource to be exploited	*Exploited*	*Unlimited*	*Not relevant*

b.

Group	Concept of Nature	Costs to the health of organisms and people viewed as important or unimportant?	Natural resources viewed as limited or unlimited?
Developed countries economic policies	Nature as a resource to be exploited	*Unimportant*	*Unlimited*
Environmental science	Nature as a provider of sustainable resources	*Important*	*Limited*

4. Explain the concept of sustainability and human interaction by considering the human activities that compromise environmental stability and our own survival.

a. Nonrenewable resources are being used up quickly.

1. Coal
2. Oil
3. Natural gas
4. Uranium

b. Renewable resources are being used up faster than they can be replenished naturally.

1. Water
2. Wood
3. Soil
4. Whales

5. In resource-rich undeveloped countries, how do the economic policies and consumption activities of developed countries perpetuate poverty there?
List four:

   1. Strip mining for coal and metal ores destroys and pollutes land that can be used for farming.

   2. Land destruction and loss of biodiversity reduces tourism and its economic benefits.

   3. Pollution and poor nutrition impacts health.

   4. To keep profit margins high, low wages are paid to local residents for resource extraction.

6. What are some of the possible effects of exotic chemical release into the biotic environment? Explain four.

   1. Disruption of reproductive cycles by reason of endocrine disruption

   2. Neurological and behavioral effects

   3. Loss of biodiversity. Not only do communities of plants and animals change, they decrease in numbers of species

   4. Increased diseases such as cancer

7. List four invasive species in your locale.

   1. Fire ants

   2. Argentine ants

   3. Dutch elm fungus

   4. Japanese beetles

Where do I live?

8. Describe four reasons for the over 40% decline in eastern U.S. songbirds in the last 40 years.

   1. Compression of bird territories increases stress and conflict and reduces feeding area

   2. Increasing edge effect of field and forest junctions that allows increased nest parasitism by cowbirds and cuckoos

   3. Reduction of the total area of bird habitat

   4. Loss of habitat due to the conversion of coffee plantations from shade forest growth to sun grown

9. How will the burning of tropical rain forest affect the problem of global warming? Research the processes of combustion and photosynthesis. *When hydrocarbon compounds in wood, oil, gasoline, and natural gas are burned (combusted) the carbon combines with oxygen to produce carbon dioxide*

and water. Conversely, the carbon dioxide and water can be combined in plants, using sunlight energy, to make carbon compounds such as sugars, starches, proteins, fats and cellulose. Theoretically, in order to maintain a constant level of carbon dioxide would require that plants take in or "fix" carbon into their bodies and fast as it is being produced by natural and human combustion processes. If plant life decreases, how will carbon dioxide levels be stabilized?

10. Refer to the graph. Explain the seasonal increases and decreases in carbon dioxide in the northern hemisphere. *Due to increased growth in the sunny days of Summer, plants in the northern hemisphere are using up carbon dioxide at higher rates than in the Winter.*

11. Explain two possible effects of exotic chemical release into the abiotic, atmospheric environment? Explain two.

    1. In the atmosphere, certain compounds like CFCs destroy the layer of ozone that protects life from ultraviolet radiation.
    2. If the chemicals are particulates, they can screen out sunlight, creating temperature inversions that trap pollution at the surface.

12. Propose four positive economic benefits of a sustained environment.

    1. Increased human physical health and longevity
    2. Increased natural beauty that contributes to mental health
    3. Sustained biodiversity with the economic benefits of ecotourism and sources for new drugs
    4. Better physical and mental health contributes to increased human productivity

13. Describe four factors that most directly spurred the exponential growth in human population since the decline caused by the black death plague.

    1. Improved agricultural quality and production through selection of superior strains of crops
    2. Mechanization of agriculture
    3. Improvement in health care
    4. Increased protection and nurturing of children

Pop Quiz Answers

Matching

1. i
2. d
3. f
4. l
5. a
6. k
7. j

8. e
9. c
10. g
11. b
12. h

Multiple Choice

1. b
2. c
3. b
4. e
5. a
6. d
7. e
8. e
9. b
10. e
11. e
12. c
13. a
14. a
15. d

2

Using Science to Address Environmental Problems

LEARNING OBJECTIVES

After you have studied this chapter you should be able to:

1. Outline the steps of the Scientific Method.
2. Distinguish between deductive and inductive reasoning.
3. Define risk assessment and explain how it helps determine adverse health effects.
4. Describe how a dose-response curve is used in determining the health effects of environmental pollutants.
5. Discuss the precautionary principle as it relates to the introduction of new technologies or products.
6. Explain how policy makers use cost-benefit analyses to help formulate and evaluate environmental legislation.
7. List and briefly describe the five stages of solving environmental problems.
8. Briefly describe the history of the Lake Washington pollution problem of the 1950s and how it was resolved.
9. Relate Garrett Hardin's description of the tragedy of the commons in medieval Europe to the global commons today.

SEEING THE FOREST

The Nature of Science

Environmental Science is a problem-solving method that observes happenings in nature and attempts to explain them; the method can identify environmental problems and evaluate solutions.

The evening news features an interview with a man who reports that he has seen a flying saucer.

Compare two approaches to how the public might evaluate his claim, on the one hand, without using Science, and on the other hand, using the Scientific Method.

Statement of explanation (hypothesis) includes a belief, a prejudiced, or a biased conclusion. Yes or No?	Claim is testable using the Scientific Method including experiments. Yes or No?	Conclusion is reported as an anecdote or as mathematical data. Which?	Conclusion is repeatable as an accurate description of other sightings. Yes or No?	Evidence allows for falsifying the conclusion. Yes or No?
Nonscientific				
Scientific				

SEEING THE TREES

VOCAB

Let the mind be enlarged ... to the grandeur of the mysteries, and not the mysteries contracted to the narrowness of the mind.—Francis Bacon, originator of the formal scientific method.

1. Science—using the Scientific Method to explain happenings in nature.
2. Scientific Method—a rigorous procedure for explaining occurrences in nature. It is characterized by the following: observations of natural phenomena, proposal of explanations of the observations and mathematical descriptions thereof, construction of unbiased experiments to determine the accuracy of the explanations, and comparison of those data with similar observations and explanations.

3. Scientific Hypothesis—a tentative, testable, and falsifiable explanation of the observations of a natural phenomenon.
4. Control—a group used in an experiment that has all characteristics of the experimental group except one—the experimental variable. The control variable remains unchanged during the experiment.
5. Scientific Theory—a general explanation of related phenomena that will predict additional observations and the outcome of new experiments. The results of many experiments and their supported hypotheses are collected to form a theory.

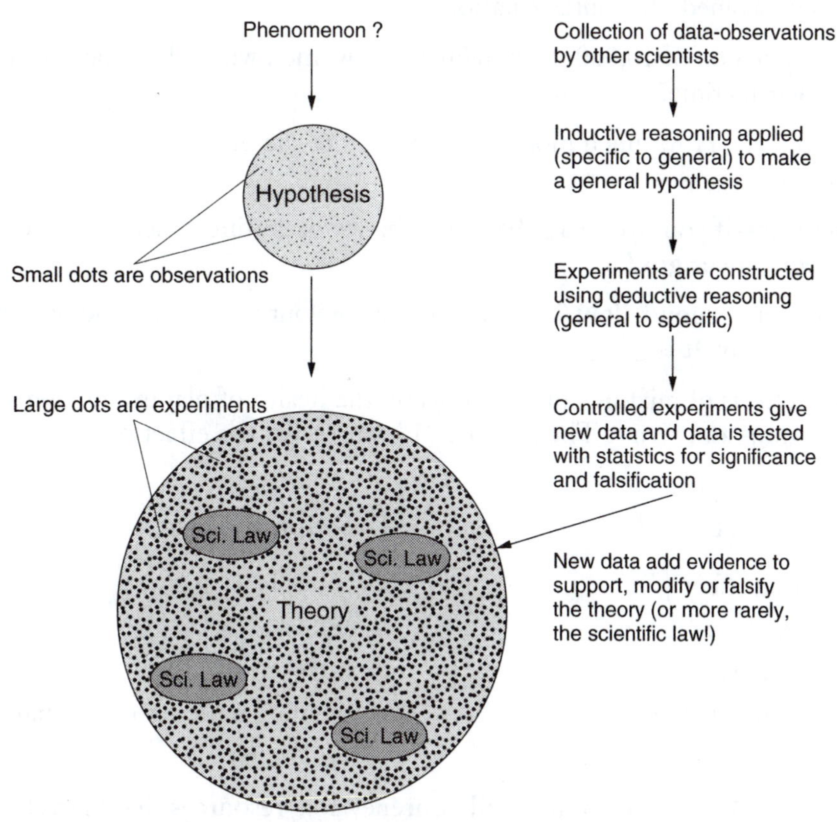

Scientific theories are large, general concepts. Scientific laws are smaller, mathematically precise concepts.

6. Scientific Principle—a theory promoted to the level of near certainty by a preponderance of evidence from experiments and mathematical descriptions.
7. Scientific Law—a mathematically precise principle of great importance.
8. Deduction—in the Scientific Method, a logical process of arriving at specific predictions derived from a general explanation.
9. Induction—in the Scientific Method, a logical process of arriving at a general conclusion based on specific observations.
10. Risk—the possibility of harm occurring due to certain environmental circumstances.
11. Risk Management—the development of procedures and regulations to limit risk.

Questions to Ponder

1. Reflect on this statement, "There is no final certainty in science, all statements and explanations are provisional."
2. Nuclear bomb testing in Nevada spread a cloud of radioactive fallout over the U.S. in the 1950s. Groundwater contamination from the Hanford, WA, bomb processing plant added to the contamination, particularly in Washington and Idaho. Iodine is incorporated into the thyroid gland to make its hormones, thyroxines. The average lifetime risk of thyroid cancer in the U.S. due to exposure to radioactive iodine is 0.64% for women, about two-thirds of 1% of developing thyroid cancer. The average lifetime risk for men is 0.25%. See rex.nci.nih.gov/massmedia/falloutcalculation.html
   a. In a U.S. population of 10,000, how many more women would be expected to get thyroid cancer in their lifetime?
   b. Consult table 2.1. How much more likely are you to die of cardiovascular disease compared to all cancers?

   Interesting fact—If you live to age 95, your chance of getting cancer is 43.26%. See **canques.seer.cancer.gov/**

   c. Go to spike.nci.nih.gov/fallout/html/ and calculate your radioactive iodine exposure based on age and locality. It is ______?
3. Regarding the effects of mixtures of chemicals on the health of plants and animals, match the following equations and terms. The number "1" represents the effect of one chemical acting alone.

Expression of Effect	**Descriptive Term**
___ 1a + 1b = 1.3x effect	a. synergism
___ 1a + 1b = 3x effect	b. additive
___ 1a + 1b = 2x effect	c. antagonism
Examples: 1a + 1b = 0 effect	"Corporate antienvironmentalist"
1a + 1b = 20x	"Tree-hugger"

4. List in the tables below, six renewable and nonrenewable resources that Garret Hardin and you regard as common to all humanity.

Renewable Resources	Nonrenewable Resources
1.	
2.	
3.	
4.	
5.	
6.	

5. Complete this table of "five stages in addressing an environmental problem."

Stage No.	Process	Activities
1.	(a.) ________________	Edmondson's hypothesis that nutrients from sewage was causing algal blooms in Lake Washington
2.	Risk analysis	(b.) ________________
3.	(c.) ________________	Results of scientific studies are published.
4.	Political action	(d.) ________________
5.	(e.) ________________	Monitoring the effects of reducing sewage nutrient inflow into the lake.

POP QUIZ

Matching

Match the following terms and definitions.

____1. A tentative, testable explanation of a natural phenomenon:	a. Theory
____2. Any hypothesis, theory, or principle must retain this characteristic:	b. Phosphates, nitrates
____3. A general principle that is confirmed by experiments and predicts new discoveries is _______.	c. ED_{50}
____4. If one makes many observations, then proposes an explanation or hypothesis, the logical process is called:	d. Global commons
____5. A single variable that has all the characteristics of the experimental group except one is called a _______?	e. Eutrophication
____6. If one starts with an explanation and constructs experiments to test predictions derived from it, the logical process is _______?	f. Threshold level
____7. If 140 female brown pelicans were exposed to a given dose of endocrine disrupter and 70 did not reproduce, the dosage term used is _______?	g. Deduction
____8. If 140 pelicans were exposed to a given dose of endocrine disrupter and 70 died, the dosage term used is _______?	h. Scientific hypothesis
____9. The maximum dose that has no measurable effect or the minimum dose that has a measurable effect.	i. Oxygen
____10. The 1950s drug thalidomide caused major birth defects. Today, a similar drug would not be allowed on the U.S. market until its risks had been determined to be small, demonstrating the _______?	j. LD_{50}
____11. The increase of nutrients in an ecosystem is termed _______.	k. Induction

____12. Increased sewage pollution released ________ into the water of Lake Washington.	l. Falsification
____13. When cyanobacteria in thick algal mats rot, this decreases.	m. Control
____14. The atmosphere and oceans are ________?	n. Precautionary principle

Multiple Choice

1. Which statement is correct regarding the scientific method?
   a. Once an experiment is constructed and data observed, the hypothesis cannot be altered.
   b. Experiments always confirm the hypothesis.
   c. The experiment must have a group that has no change in the experimental variable.
   d. The formulation of a hypothesis from observations represents deductive reasoning.
   e. The formulation of predictions to be tested in an experiment represents inductive reasoning.

2. Which statement is *incorrect* in defining the term theory?
   a. Theories are the results of many tested hypotheses.
   b. Theories are tentative, testable explanations of observations.
   c. Theories predict new discoveries.
   d. A theory, widely held and confirmed, may become a scientific principle.
   e. Theories are falsifiable in the Scientific Method.

3. The reason why society regards the Scientific Method valuable is:
   a. Popular ideas are debunked.
   b. Repeatability of experiments and conclusions make them reliable.
   c. It is highly conservative, so new ideas are slowly accepted or rejected.
   d. It has led to atomic weaponry.
   e. All of the above are correct.

4. Malaria has killed or debilitated more people in the history of civilization than any other organism (not including viruses). It is now on the increase for this reason:
   a. Global warming has extended ranges into higher altitude and formerly cooler areas.
   b. Deforestation has created new breeding areas for mosquito larvae.
   c. It is easily passed from one person to another through sneezes and contact.
   d. a and b only
   e. a, b, and c

5. For years ecologists and agricultural entomologists (insect study) debated the issue of the cause of spruce tree death in the higher elevations of the Appalachian Mountains. Respectively it was explained due to the acids in rain and beetle infestations. The outcome of this debate was that acid rain debilitated the trees allowing for the insect attack. In this case the case, acid rain is a(n):
   a. carcinogen
   b. agent of eutrophication
   c. toxicant with an established LD_{50}
   d. environmental stressor
   e. all of the above

6. The (rat) LD_{50} for nicotine is 60mg/kg which makes it fit the definition of a toxic chemical (range = 25–200, < 25 is very toxic and > 200 is harmful). This means that:

a. A dose of 6000 mg will likely kill half of the human subjects that weigh 100 kg.
b. A 60 gram dose will kill half of the subjects.
c. A 60 mg dose will make half of the subjects sick.
d. Nicotine is 60 times more toxic than standard water at 1.
e. A dose of 60 mg will kill 50% of the time.

7. A certain mixed alcohol beverage delivers 30g of ethyl alcohol per 200 ml container. If the LD_{50} for ethanol is 10,000 mg/kg, how many drinks must a 100-lb woman chug-a-lug to leave her with a 50% chance of survival? Hint—convert kg to lbs (2.2 lbs/kg) and mg to g (1000 mg/1g).
a. about 15
b. approximately 30
c. nearly 10
d. She is petite, therefore about 6.
e. No amount! The drinks are too weak.

8. A dose of 0.1 g/deciliter in blood alcohol level will produce intoxication in 50% of the population with accompanying loss of reaction time and judgment. This dose is best termed a(n):
a. LD_{50}
b. ED_{50}
c. ineffective dose for 100% of the subjects
d. Effective Dose $_{0.1}$
e. All of the above are applicable

9. When mats of algal cyanobacteria rot, this happens next:
a. fish kills due to declining oxygen
b. foul smells result
c. eutrophication
d. water quality improves
e. a and b only

10. At the mouth of the Mississippi River off the coast of Louisiana, a "dead zone" of eutrophication and algal bloom is growing in the Gulf of Mexico. The source of the pollution is:
a. untreated sewage
b. treated sewage
c. chemical fertilizer runoff from the Mississippi River watershed
d. manure from animal farms
e. all of the above

11. Communities surrounding Lake Washington improved the aesthetics and water quality by:
a. not allowing raw sewage to flow into the lake
b. adding copper sulfate to the water to kill the cyanobacteria
c. preventing industries from releasing toxicants into the lake
d. stocking algae-eating fishes
e. all of the above

12. A cancer-causing chemical found to cause lung cancer in German industrial exposures and in Vietnam war era "Agent Orange" exposures is:
a. nicotine

b. benzpyrenes
c. dioxin
d. asbestos
e. cyclamate

13. Which statement is *incorrect* regarding eutrophication?
    a. Increases in nutrients can lead to algal blooms
    b. It only happens in cases of sewage pollution
    c. The bacteria that decompose algae consume oxygen
    d. Dumping a bag of fertilizer in a farm pond would cause an algal bloom
    e. Low oxygen levels cause the death of oxygen-consuming organisms

14. Pollution in Lake Washington directly affected this body of water:
    a. Portland Bay
    b. Puget Sound
    c. San Francisco Bay
    d. Washington, D.C.
    e. Prince Edward Sound

CYBER SURFIN'

1. For an interactive exercises on the use of the Scientific Method, from the QUIA Corporation, see **www.quia.com/jg/65726.html**; and San Diego City Schools, "Finding the Lighthouse Diamond Thief Using the Scientific Method," **projects.edtech.sandi.net/kroc/scimethod/t-index.htm**
2. For a detailed description of the Scientific Method as used for volcano research by **PunaRidge.org**, see **www.punaridge.org/doc/teacher/method/**
3. Information on harmful algal blooms from Woods Hole Oceanographic Institution and NOAA can be explored at **www.whoi.edu/redtide/**
4. The University of Edinburgh, U.K., Occupational and Environmental Health website tutorial "Introduction to Toxicology," **www.link.med.ed.ac.uk/HEW/tox/default.htm**
5. From the King County (WA) Natural Resources and Parks, "The Lake Washington Story," see **dnr.metrokc.gov/wlr/waterres/lakes/biolake.htm**

'LIKE' BOOKING IT

1. Popper K. ***The Logic of Scientific Discovery.*** London: Routledge, 14th Printing, 1977. First English Ed., Hutchinson, 1959. First published as ***Logik Der Forschung*** in Vienna, Springer, 1934.
2. Popper K. ***The Open Society and Its Enemies.*** London: Routledge & Kegan Paul, 1945.
3. Hardin G. ***Living Within Limits: Ecology, Economics, and Population Taboos.*** New York: Oxford University Press, 1993.
4. Ayres E. ***God's Last Offer – Negotiating for a Sustainable Future.*** New York: Four Walls Eight Windows, 1999.
5. Sagan C. ***The Demon-Haunted World: Science as a Candle in the Dark.*** New York: Ballantine Books, 1997.

ANSWERS

Seeing the Forest

The evening news features an interview with a man who reports that he has seen a flying saucer.

Compare two approaches to how the public might evaluate his claim without using Science and using the Scientific Method.

	Statement of explanation (hypothesis) includes a belief, a prejudiced, or a biased conclusion. Yes or No?	Claim is testable using the Scientific Method including experiments. Yes or No?	Conclusion is reported as an anecdote or as mathematical data. Which?	Conclusion is repeatable as an accurate description of other sightings. Yes or No?	Evidence allows for falsifying the conclusion. Yes or No?
Nonscientific	Yes	*No*	*Anecdotal*	*No*	*No*
Scientific	No	*Yes*	*Mathematical data*	*Yes*	*Yes*

Questions to Ponder

1. Reflect on this statement, "There is no final certainty in science, all statements and explanations are provisional." *All explanations in science depend on the state of knowledge at a certain point in time. Experiments and the state of technology will improve over time (we hope). All explanatory statements in science must be falsifiable. Even those historically accepted will be tested again, confirmed, refined or refuted. All explanations are therefore provisional. Science can disprove an explanation or prediction more certainly than prove one. Therefore, one can disprove the idea that plants have feelings more certainly than you can prove that they do have feelings.*

2. Nuclear bomb testing in Nevada spread a cloud of radioactive fallout over the U.S. in the 1950s. Groundwater contamination from the Hanford, WA, bomb processing plant added to the contamination, particularly in Washington and Idaho. Iodine is incorporated into the thyroid gland to make its hormones, thyroxines. The average lifetime risk of thyroid cancer in the United States due to exposure to radioactive iodine is 0.64% for women, about two-thirds of 1% of developing thyroid cancer. The average lifetime risk for men is 0.25%. See rex.nci.nih.gov/massmedia/falloutcalculation.html

   a. In a U.S. population of 10,000, how many more women would be expected to get thyroid cancer in their lifetime? *10,000 (x) 0.64/100 = 64 for women; 10,000 (x) 0.25/100 = 25, 64–25 = 39 more women.*

   b. Consult table 2.1. How much more likely are you to die of cardiovascular disease compared to all cancers? *940,600/541,500 = 1.83.*
   Interesting fact—If you live to age 95, your chance of getting cancer is 43.26%. See canques.seer.cancer.gov/

   c. Go to spike.nci.nih.gov/fallout/html/ and calculate your radioactive iodine exposure based on age and locality. It is______? *Answer varies according to sex, age, diet, and location.*

3. Regarding the effects of mixtures of chemical on the health of plants and animals, match the following equations and terms. The number "1" represents the effect of one chemical acting alone.

a. 1a + 1b = 1.3x effect	*antagonism*
b. 1a + 1b = 3x effect	*synergism*
c. 1a + 1b = 2x effect	*additive*
Examples: 1a + 1b = 0 effect	"Corporate antienvironmentalist"
1a + 1b = 20x	"Tree-hugger"

4. List in the tables below, six renewable and nonrenewable resources that Garret Hardin and you regard as common to all humanity.

Renewable Resources	Nonrenewable Resources
Water	Coal
Air	Oil
Soil? Top soil is lost faster than it is replaced in agricultural areas.	Metals
Crops	Nuclear energy
Wood	Natural gas
Detritus for humification	Phosphate-bearing rock, limestone

5. Complete this table of "five stages in addressing an environmental problem."

Stage No.	Process	Activities
1.	(a.) *Scientific assessment*	Edmondson's hypothesis that nutrients from sewage was causing algal blooms in Lake Washington
2.	Risk analysis	(b.) *Analyze the potential effects of continuation of the problem and its remediation.*
3.	(c.) *Public education and involvement*	Results of scientific studies are published.
4.	Political action	(d.) *In a political forum, the environmental, health and economic interests are debated and resolved.*
5.	(e.) *Evaluation*	Monitoring the effects of reducing sewage nutrient inflow into the lake.

Pop Quiz Answers

Matching

1. h
2. l
3. a
4. k
5. m
6. g
7. c
8. j
9. f
10. n
11. e
12. b
13. i
14. d

Multiple Choice

1. c
2. b
3. d
4. d
5. a
6. e
7. e
8. a
9. c
10. b
11. e
12. c
13. b
14. b

3

Environmental History, Legislation, and Economics

LEARNING OBJECTIVES

After you have studied this chapter you should be able to:

1. Define conservation and distinguish between conservation and preservation.
2. Briefly outline the environmental history of the U.S.
3. Describe the environmental contributions of the following people: John James Audubon, Henry David Thoreau, George Perkins Marsh, Theodore Roosevelt, Gifford Pinchot, John Muir, Franklin Roosevelt, Aldo Leopold, Wallace Stegner, Rachel Carson, and Paul Ehrlich.
4. Explain why the National Environmental Policy Act is the cornerstone of U.S. environmental law.
5. Relate how Environmental Impact Statements provide such powerful protection of the environment.
6. Sketch a simple diagram that shows how economics is related to natural capital. Make sure you include sources and sinks.
7. Describe various approaches to pollution control, including command and control regulation and incentive-based regulation (i.e., emissions charges and marketable waste-discharge permits).
8. Give two reasons why the national income accounts are incomplete estimates of national economic performance.
9. Distinguish among the following economic terms: marginal cost of pollution, marginal cost of pollution abatement, optimum amount of pollution.
10. Discuss some of the complexities of the "jobs versus the environment" issue in the Pacific Northwest.
11. Describe some of the environmental problems facing formerly communist governments in Central and Eastern Europe.

SEEING THE FOREST

Multiple Choice Question

Which problem did municipalities have in implementing the Safe Water Act prior to 1995?

a. assaying for pesticides that were not used in their area
b. state and local funding had to be provided for a federal mandate
c. removing 90% of a pollutant costs less than removing the final 10%
d. determining whether pesticides were within limits specified by the Safe Water Act
e. all of the above

SEEING THE TREES

VOCAB

> ***Worldwide practice of Conservation and the fair and continued access by all nations to the resources they need are the two indispensable foundations of continuous plenty and of permanent peace.***
> ***—Gifford Pinchot***

1. Conservation—management of natural resources that allows for use while maintaining stability.
2. Preservation—setting aside and protecting natural resources in a pristine state.
3. Frontier/Pioneer attitude—nature is something to be subdued and exploited.
4. Environmentalists—people concerned that the environment may be negatively affected by human activity.
5. Economics—the study of the production, distribution, and consumption of products and resources.
6. Natural economic source—the physical areas from which raw materials for production are extracted.
7. Sink—the environment that receives an input of waste or nonwaste materials.
8. Natural capital—the resources of the Earth that sustain life.
9. External cost or negative externality—the cost borne by people not directly involved in the production or distribution of a product.
10. Marginal costs—added costs borne by present and future organisms due to environmental damage and pollution abatement.
11. The economic optimum amount of pollution—the data point at which the graphed curves of the marginal cost of pollution and the marginal cost of pollution abatement intersect. The resulting graph is called a cost-benefit diagram.
12. Command and control regulation—pollution is controlled by laws and rules that set limits for amounts of pollutants.

13. Incentive-based regulation—a market-oriented strategy to use economic forces to reduce pollution and minimize the costs of abatement.
14. Emission charge—a tax on the amount of pollution.
15. Marketable waste-discharge permit—regulation accomplished by setting an absolute amount of pollution.
16. Emission reduction credit (ERC)—a pollution production credit that can be traded by companies producing emissions.

Questions to Ponder

1. Match the person and his or her contribution to Environmentalism.

Person	Contribution
____1. John James Audubon	a. U.S. President who preserved 23.9 million hectares (59 million acres) of national forest.
____2. Henry David Thoreau	b. warned of the dangers of pesticides and exotic chemicals in the environment, wrote *Silent Spring.*
____3. George Perkins Marsh	c. led efforts to preserve Yosemite National Park and founded the Sierra Club.
____4. Theodore Roosevelt	d. wildlife biologist who promoted wildlife preserves, wrote *Sand County Almanac.*
____5. Gifford Pinchot	e. painted the beauty of the North American birds and plants.
____6. John Muir	f. wrote *Man and Nature* explaining how humans can cause environmental damage.
____7. Franklin Roosevelt	g. promoted conservation of America's forest as first head of the U.S. Forest Service.
____8. Aldo Leopold	h. promoted philosophy of living simply in harmony with nature
____9. Rachel Carson	i. established the 1930s Civilian Conservation Corps that took unemployed men to work on nature conservation.
____10. Wallace Stegner	j. warned about the dangers of overpopulation and made proposals for dealing with the problem.
____11. Paul Ehrlich	k. Pulitzer Prize winning author, professor and environmentalist, 1909–1993.

2. Since the first Earth Day, April 1 of 1970, environmental laws have improved air and water quality in which specific areas.

   a. List six and explain how they were improved.

   1. ______________________________

   2. ______________________________

   3. ______________________________

4.

5.

6.

b. List six areas where you believe that much more improvement or control should be implemented.

1.

2.

3.

4.

5.

6.

3. List the three steps of the Environmental Impact Statement.

1.

2.

3.

4. Now that environmental laws to protect air and water are in place, how can citizens ensure that the intent and specific provisions of the laws are enforced? List three ways.

1.

2.

3.

5. Provide four examples of the *negative externality principle of specific costs,* not entirely borne by the producer or distributors of a product, but instead borne by consumers of a product or local residents where natural resources are extracted.

1.

2.

3.

4.

POP QUIZ

Matching

Match the following terms and definitions

____1. Setting aside national forests and allowing for managed logging	a. U.S. National Environmental Policy Act
____2. Preventing development or logging in old growth forests	b. Acid rain gases, nitrogen and sulfur oxides
____3. This endangered species limited logging in Oregon.	c. National resource capital loss
____4. Required Environmental Impact Statements	d. Bald eagle
____5. U.S. agency that determines allowable amount of pollution	e. Spotted owl
____6. "Scrubbers" are devices that remove ______ from coal exhaust gases.	f. Communism
____7. Formerly endangered, this species has recovered somewhat	g. Command and control regulation
____8. Area that had significant sewage pollution and a river that caught on fire in the 1960s	h. Salvage logging
____9. The overuse of the Coconino National Forest near the Grand Canyon leads to	i. Conservation
____10. The measure that determines economic loss due to persisting pollution	j. U.S. E.P.A.
____11. The U.S. Clean Air Act of 1977 is an example of	k. Cleveland, OH, U.S.A.
____12. Production "demand control" by the state for determining industrial and agricultural quotas	l. Czech Republic
____13. A toxic environment's a leftover of Cold War politics here	m. Marginal cost
____14. Removal of dead and diseased trees	n. Preservation

Multiple Choice

1. This U.S. regulatory act was an "unfunded federal mandate that created local problems in determining reasonable amounts of pollutants and supporting the costs of testing."
   a. 1977 Clean Air Act
   b. 1969 National Environmental Policy Act
   c. 1974 Safe Water Act
   d. 1990 Clean Air Act Amendments
   e. all of the above

2. The corporate environmental philosophy reflected in the building of tall smoke stacks on coal-burning power generators is:
   a. Acid rain pollution must be controlled
   b. Sulfur dioxide that causes lung diseases must be reduced to insignificant levels
   c. The solution to pollution is dilution
   d. Reduction in the local effects of pollution
   e. c and d only

3. The worst nuclear power plant accident occurred here:
   a. Three Mile Island, PA, U.S.A.
   b. Bhopal, India
   c. Prince Edward Sound, Alaska
   d. Chernobyl, Ukraine
   e. Cleveland, OH, U.S.A.

4. Pollution can be defined as:
   a. the occurrence of exotic (unnatural) chemical into the environment
   b. an excess of natural waste products of animals
   c. the release of exotic genes into the environment that cause changes in biodiversity
   d. the overuse of natural "sinks"
   e. all of the above

5. From the standpoint of Environmental Science, the *Gross Domestic Product* and *Net Domestic Product* indexes that estimate national economic performance are deficient is this way:
   a. The benefits of pollution and land use regulation are not formulated.
   b. The depletion of natural renewable resources is not calculated.
   c. The depletion of nonrenewable resources is not included.
   d. Decreases in productivity due to diminished health of citizens are not included.
   e. All of the above are correct.

6. Which of the examples below illustrate the principle of external cost/negative externality?
   a. the cost of repairing county roads damaged by heavy coal trucks
   b. the cost of transporting natural resources to a processor
   c. the cost of labor to sell the goods
   d. the cost of building hospitals
   e. Trick question alert! All the above are valid external costs.

7. Which statement is incorrect about the cost-benefit diagram on the top of page 36?
   a. The value of ecotourism is included.
   b. As more pollution is eliminated, abatement costs rise exponentially.
   c. As the level of pollution rises, the social costs increase exponentially.
   d. As the level of pollution rises, the environmental costs increase exponentially.
   e. The costs of reducing the pollution can exceed the economic harm done by the pollutant.

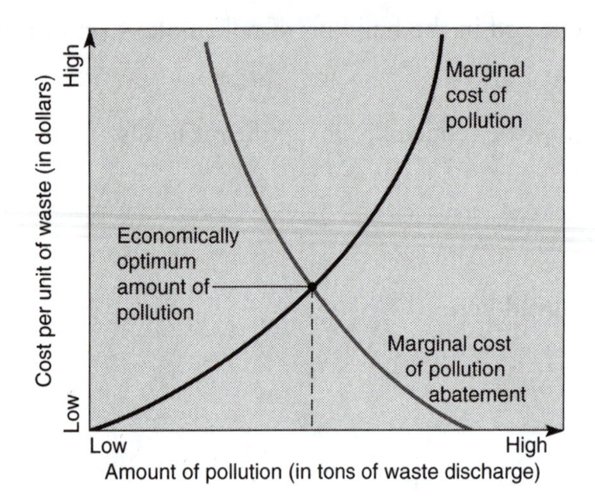

8. A cost-benefit diagram compares only:
   a. the marginal cost of pollution abatement with the projected costs of investment in pollution control equipment
   b. the marginal cost of each unit of pollution with pollution abatement cost
   c. the abatement cost is graphed with the environmental cost
   d. the benefits of pollution control are graphed with the costs of abatement
   e. all of the above

9. Which statement is correct about the graph below?
   a. The cost of the pollution exceeds the harm done by the pollution.
   b. The harm caused by the pollution exceeds the cost of reducing the pollution.
   c. The ideal economic optimum amount of pollution is located near the left y axis of the graph.
   d. The ideal economic optimum amount of pollution is located near the right y axis of the graph.
   e. All of the above are correct.

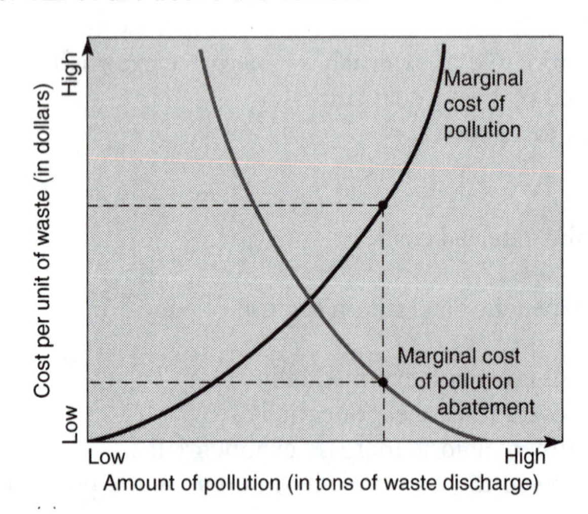

10. Which of the following are not examples of incentive-based pollution control?
    a. imposing emission charges or taxes on pollution
    b. issuing marketable waste-discharge permits allowing the holder to emit a specified amount of pollutant
    c. limiting the extraction of trees from a forest to sustainable yields

d. trading emission reduction credits
e. all of the above

11. Which of the following are major factor(s) contributing to the loss of at least 40% of eastern U.S. migratory songbirds in the last 40 years?
    a. forest fragmentation
    b. increased nest parasitism by cowbirds
    c. loss of habitat and bird territory compression
    d. growing coffee in sun plantations rather than shade plantations
    e. all of the above

12. Which creature or creatures illustrates the conflict between old growth forest management and economic interests?
    a. officially extinct (1940s) in the U.S., the ivory billed woodpecker required old hardwoods for food and nesting
    b. the endangered red cockaded woodpecker that requires old pines for nesting
    c. the spotted owl
    d. the loss of biodiversity of tree and animal species in replanted forests
    e. all of the above

13. Which statement is incorrect concerning the *demand production economy* of communist East Europe in years before the fall of the Berlin Wall?
    a. Pollution control was ignored so that production quotas could be met.
    b. Environmental quality was a natural concern of the proletariat bureaucracy.
    c. Respiratory diseases increased greatly over time.
    d. Political opposition to governmental policies on the environment was not tolerated.
    e. Life expectancies were approximately 10 years less there.

14. Turtle Exclusion Devices are required for the commercial fishing of ______ sold in the U.S.
    a. Atlantic cod
    b. tuna
    c. blue crabs
    d. shrimp
    e. lobsters

15. Old growth forests are valuable natural capital because:
    a. Older trees provide nesting and feeding habitat for certain endangered species.
    b. Forests charge ground water aquifers.
    c. Clean water flows from old growth forests.
    d. Biodiversity is higher there.
    e. All of the above are correct.

CYBER SURFIN'

1. The history of the U.S. conservation movement 1850–1920, Library of Congress: cweb2.loc.gov/ammem/amrvhtml/conshome.html
2. The National Wildlife Federation on biodiversity and effects of global warming: www.nwf.org

3. Design your own Index of Sustainable Economic Welfare with the help of the Friends of the Earth (U.K.), **http://www.foe.co.uk/campaigns/sustainable development/progress/**
4. A comprehensive set of environmental sustainability links can be seen at Ecosustainable Hub.com, **http://www.ecosustainable.com.au/links.htm#10**
5. From the U.N. Food and Agriculture Organization, various topics on environmental sustainability, biodiversity and women's issues, see **http://www.fao.org/waicent/faoinfo/sustdev/index_en.htm.**

'LIKE' BOOKING IT

1. Guha, R. ***Environmentalism: A Global History.*** San Francisco: Longman, 2000.
2. Nattrass, B and Altomare, M. ***Dancing With the Tiger: Learning Sustainability Step by Natural Step (Conscientious Commerce).*** Gabriola Island, British Columbia: New Society Publishers, 2002.
3. Lee, K. N. ***Compass and Gyroscope: Integrating Science and Politics for the Environment.*** New York: Island Press, 1995.
4. Ostrom, E. ***Governing the Commons: The Evolution of Institutions for Collective Action.*** Oxford, U.K.: Cambridge University, 1991.
5. Stegner, W. ***The Sound of Mountain Water (includes his renown "Coda: Wilderness letter").*** New York: Doubleday & Company, 1980.

ANSWERS

Seeing the Forest

Questions to Ponder (Many of these answers are examples of many possible responses, please don't memorize them.)

1. Match the person and his or her contribution to Environmentalism.

    1. e
    2. h
    3. f
    4. a
    5. g
    6. c
    7. i
    8. j
    9. b
    10. k
    11. d

2. Since the first Earth Day, April 1 of 1970, environmental laws have improved air and water quality in which specific areas.

    a. List six and explain how they were improved.

    1. Lead levels in the air have declined by 98%. It has also declined in water. Lead is no longer used in U.S. gasolines to prevent pre-ignition and knocking.

2. Hydrocarbon emissions, unburned gasoline and combustion products, have declined by nearly 50%. Engines are more efficient today.

3. Emissions of sulfur dioxide and carbon monoxide have declined by more than 30%. Lower sulfur fuels and catalytic converters have made a difference.

4. Ozone-destroying CFCs have been reduced 70%. New refrigerant gases like R-134A have replaced R-12 (used in pre-1993 automobiles) and R-22.

5. Little raw sewage leaks into water supplies today because cities were required to build secondary treatment facilities and are fined if sewage is released into streams. Hepatitis A and polio have declined as a result.

6. Certain materials like DDT and asbestos were banned from building products. The bald eagle and brown pelican recovered from their population drop due to DDT alteration of their eggs—they were shell-less. Mesothelioma, cancer of the outer lining of the lungs, has declined as a result of asbestos reduction.

b. List six areas where you believe that much more improvement or control needs to be implemented.

1. Sulfur dioxide and acid rain can be reduced with cleaner fuels technology and scrubbers.

2. Carbon dioxide emissions (which also contribute to acid rain) can be reduced by having higher efficiency combustion engines and hydrogen fuels.

3. Solar, wind, and geothermal energy can be increasingly developed in appropriate regions, such as the desert southwest, mountains, and near areas of geothermal activity, respectively.

4. Energy use can be reduced by inline water heaters so that water heating is not continuous, but only on demand immediately before its use.

5. Asbestos can be completely eliminated from automobile brake shoes and discs.

6. The precautionary principle can be more adequately applied to the release of exotic chemical and genes into the environment.

3. List the three elements of the Environmental Impact Statement.

   1. The project is described and its necessity established.

   2. The short- and long-term effects on the environment are delineated and measured.

   3. Ways to reduce the adverse effects of the project are determined.

4. Now that environmental laws to protect air and water are in place, how can citizens ensure that the intent and specific provisions of the laws are enforced? List three ways:

   1. Citizens and groups can seek legal action against the offending company or governmental agency.

   2. Citizens can persuade their political representatives to make regulatory agencies respond to the problem.

3. If the offender is a commercial establishment, the products or services of that company can be boycotted.

5. Provide four examples of the *negative externality principle of specific costs*, not entirely borne by the producer or distributors of a product, but instead borne by consumers of a product or local residents where natural resources are extracted.

   1. The health costs of caring for people that have lung cancer or emphysema from cigarette smoking.
   2. The costs of cleaning water polluted by strip mining waste or siltation.
   3. The costs of cleaning air polluted by coal-burning power plants.
   4. The cost of cleaning soil polluted by coal tars or dioxins.

Pop Quiz Answers

Matching

1. i
2. n
3. e
4. a
5. j
6. b
7. d
8. k
9. c
10. m
11. g
12. f
13. l
14. h

Multiple Choice

1. e
2. c
3. d
4. e
5. e
6. d
7. a
8. b
9. b
10. c
11. e
12. e
13. a
14. d
15. e

4

Ecosystems and Energy

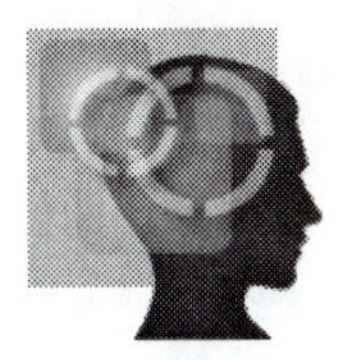

LEARNING OBJECTIVES

After you have studied this chapter you should be able to

1. Define ecology and distinguish among the following ecological levels: population, community, ecosystem, landscape, and biosphere.
2. Define energy and explain how it is related to work and to heat.
3. Use examples to contrast potential and kinetic energy.
4. State the first and second laws of thermodynamics and discuss the implications of these laws as they relate to organisms.
5. Write summary reactions for photosynthesis and respiration and contrast these two biological processes.
6. Describe the communities around hydrothermal vents and explain the source of energy that sustains them.
7. Summarize how energy flows through a food web, using the terms *producer*, *consumer*, and *decomposer*.
8. Explain some of the impacts humans have had on the Antarctic food web.
9. Draw and explain typical pyramids of numbers, biomass, and energy.
10. Distinguish between gross primary productivity and net primary productivity and discuss human impact on NPP.

SEEING THE FOREST

In *The Marshes of Glynn* and *Hymns to Marshes*, poet Sidney Lanier extolled their beauty.

> *"But the air and my heart and the earth are a-thrill,*
> *And look where the wild duck sails round the bend of the river,*
> *And look where a passionate shiver*
> *Expectant is bending the blades*
> *Of the marsh-grass in serial shimmers and shades,*
> *And invisible wings, fast fleeting, fast fleeting,*
> *Are beating."*
>
> University of North Carolina at Chapel Hill Libraries, *Documenting the American South, The Poems of Sidney Lanier*, Edited by his wife [Mary D. Lanier], electronic edition, docsouth.unc.edu/lanier/lanier.html.

1. Estuarine salt marshes, like those in the Chesapeake Bay, are "natural sewage treatment plants" that recycle nutrients into living organisms. Which statement below is incorrect?
   a. Algae and *Spartina* marsh grass use phosphate and nitrate nutrients in sewage and agricultural runoff.
   b. Because algae and marsh grass are not food for humans, salt marshes are regarded as economically unproductive.
   c. Salt marsh lands serve as the basis of food chains that include commercial fishes and shrimps.
   d. Frogs and salamanders can't live in salt marshes.
   e. A few mammals and reptiles are found there.

SEEING THE TREES

VOCAB

There are no passengers on Spaceship Earth, we are all crew.
Marshall McLuhan

1. Ecology—the science of the interrelationships of living organisms with their biotic (living) and abiotic (physical) environments.
2. Species—organisms of similar genetic makeup, anatomy, and physiology that breed successfully with their own kind and unsuccessfully with other species.
3. Population—all the members of a certain species living in a given geographical area.
4. Community—the members of all species living in a given geographical area.
5. Biosphere—the relatively thin layer of the atmosphere and the Earth's surface that contains all living organisms.
6. Physical environment—the atmosphere, hydrosphere, and lithosphere (soils and rocks).
7. Energy—a force that is capable of doing work—atomic, electromagnetic, chemical, kinetic (motion), electrical, and gravitational—that is expressed either in units of work and heat.
8. Isolated thermodynamic system—neither mass nor energy to cross the system boundary. The universe is an isolated system.
9. Closed thermodynamic system—energy, in the form of work or heat, can cross the boundary of the system, but no matter can cross. The Earth is such a system, but remember it is open to energy exchange.
10. Open thermodynamic system—matter and energy can cross the system boundary. An ecosystem is a good example.

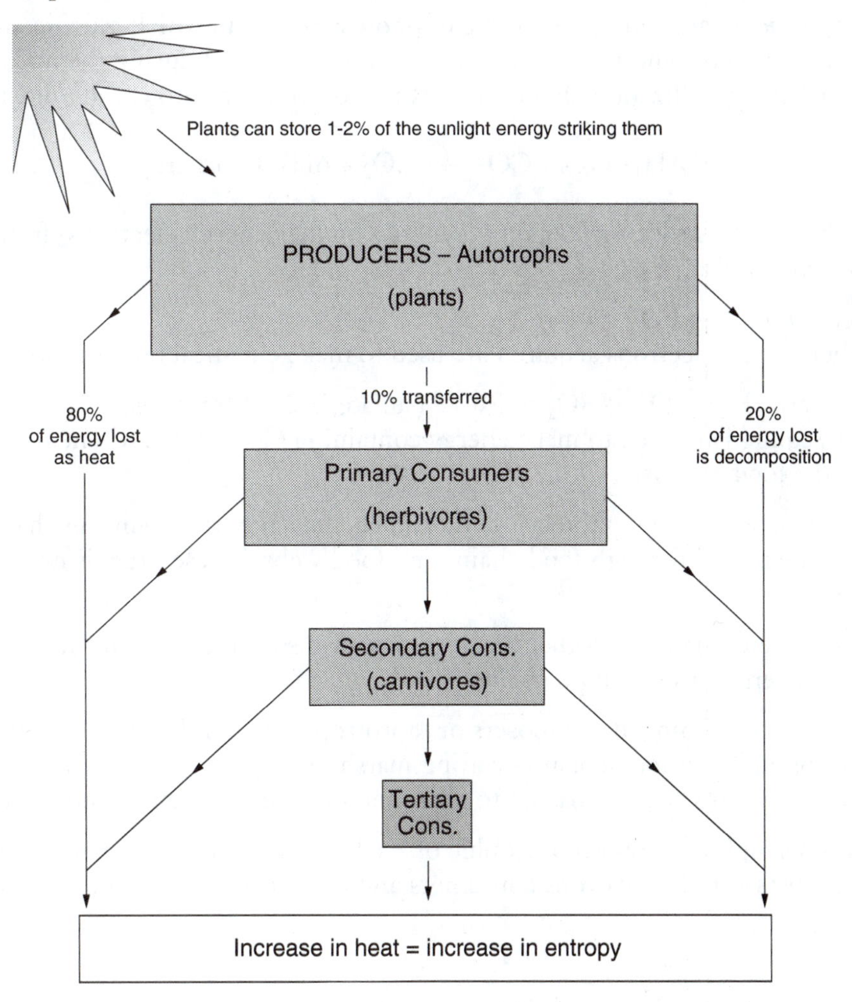

11. First law of thermodynamics—energy cannot be created or destroyed.
12. Energy transformations—energy can be changed in form. Starting with the sun, atomic energy is changed to electromagnetic/radiant energy, which is transformed by plants into chemical energy or food.
13. Second law of thermodynamics—all energy passing through a thermodynamic system is ultimately turned into infrared photons/heat.
14. Entropy—An increase in heat is equivalent to an increase in entropy. Heat is a form of energy that has limited use in thermodynamics. It is considered to be "disorganized," although it does activate chemical reactions. It can be said that over time, the universe as a thermodynamic system is transforming its "organized, concentrated" energy such as atomic energy to "disorganized" heat.
15. Photosynthesis—a process occurring in cyanobacteria, algae, and plants that converts electromagnetic/radiant energy from the sun into chemical energy that is used as food. Carbon dioxide, water, and the energy captured by the antenna molecules of chlorophyll are combined to form food such as glucose sugar and the by-product of oxygen:

$$6CO_2 + 6H_2O + \text{Energy} \rightarrow C_6 H_{12} O_6 + 6O_2$$

The oxygen comes from the water that serves as a source of higher energy electrons for making chemical bonds in the glucose molecule. See the appendix.

16. Cellular respiration—generally the reverse of photosynthesis in which glucose, assisted by oxygen, that soaks up lower energy electrons and protons produced during the making of the universal cellular energy utilization chemical ATP, is broken down into carbon dioxide and water:

$$C_6 H_{12} O_6 + 6CO_2 \rightarrow 6CO_2 + 6H_2O + \text{Energy}$$

17. Chemosynthesis—the process of obtaining energy (higher energy electrons) from the oxidation of inorganic chemicals:

    Iron bacteria: $Fe^{+2} \rightarrow Fe^{+3} + 1\ e-$
    (the higher energy electrons acquired are used to make chemical bonds in food for cells)

    Sulfur bacteria: $H_2S + O_2 \rightarrow SO_2$ or SO_4^{-2} (sulfate) + 2 H^+ (protons) + 2e– (the higher energy electrons acquired are used to make energy containing C–H and C–C chemical/covalent bonds in the food of cells)

18. Producers—in an ecosystem, photosynthetic and chemosynthetic organisms that can produce food that can be passed through food chains and food webs. These organisms are "self-feeding" or autotrophs.
19. Consumers—heterotrophic or "other feeding" organisms acquire food from the producers and pass chemical energy through food chains.
20. Decomposition food chain—decomposers or saprotrophs (rotten feeding) acquire energy from dead organisms and start food chains—rotting marsh grass gives energy to bacteria that make nitrates and phosphates that are passed to algae and on to consumers in the salt marsh.
21. Food chain—a simple illustration of trophic or feeding levels of an ecosystem that captures energy that is passed to herbivorous consumers and on to carnivorous consumers.

22. Food web—a more complex representation of feeding levels within an ecosystem that illustrates that some herbivores can also be carnivores, and that carnivores can feed on both herbivores or other carnivores.
23. Pyramid of numbers—the numbers of individual organisms are counted for each feeding level, with the bottom of the pyramid representing the producers and the consumers on higher levels:

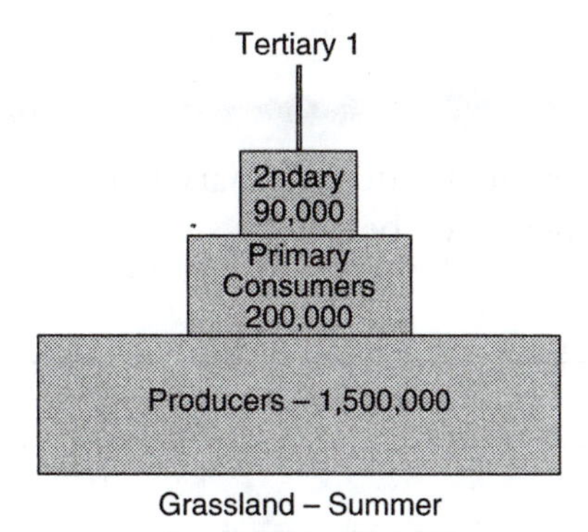

24. Pyramid of biomass—a pyramid of feeding levels that is determined by the dry weight or biomass of the organism:

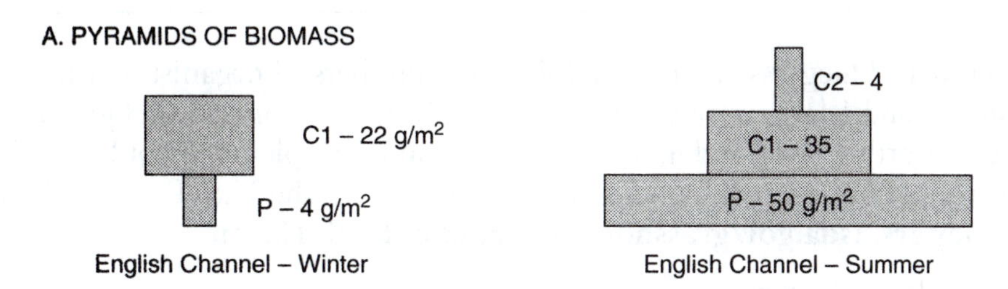

The left pyramid is inverted because the algae reproduce much faster than the copepods that eat them in the winter in the English Channel.

25. Pyramid of energy—the energy content of the organisms in a simple food chain are represented in terms of kilocalories, the amount of heat required to raise the temperature of 1 kg of water, 1°C.

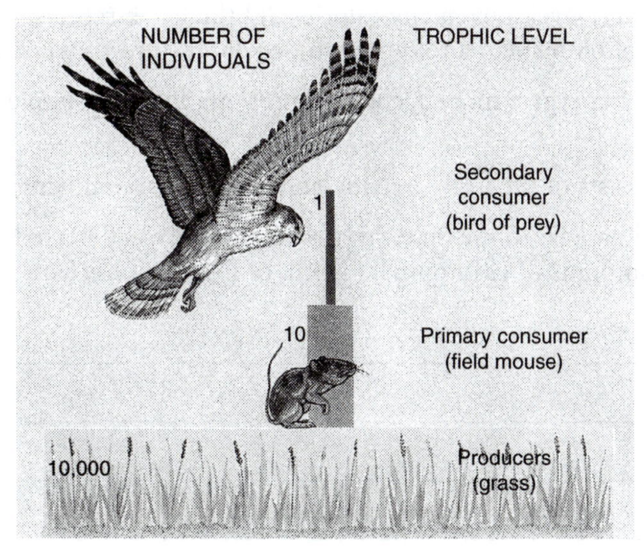

26. Gross primary productivity—the caloric or biomass value of the amount of energy in a given amount of time transformed into chemical energy or biomass by plants doing photosynthesis.
27. Net primary productivity—the amount of biomass or energy that exceeds that used to maintain its survival or the rate at which this excess energy is stored by plants in the process of growth:gross primary productivity – plants cellular respiration = net primary productivity.

Questions to Ponder

1. After reading Chapter 4 consider the multidisciplinary nature of environmental science by listing the contributions of the disciplines listed below.

   1. Biology ______
   2. Chemistry ______
   3. Engineering ______
   4. Mathematics ______
   5. Physics ______

2. An acre of Kentucky bluegrass yielded the following numbers of organisms: 1 hawk, 4 insectivorous garter snakes and 1 large copperhead, 22 mice, 6 meadow larks, 4,608 grasshoppers and other herbivorous arthropods, and 4,500,003 grasses and forb plants (dicot "weeds"). For a primer of the role of birds in controlling grasshoppers, go to the U.S. Dept. of Agriculture website: **www.sidney.ars.usda.gov/grasshopper/Handbook/I/i_10.htm**

   a. Outline a simple food chain.
   b. Outline a food web.
   c. Draw a pyramid of numbers. Make the size of each trophic level proportional to the numbers represented.
   d. Why is the pyramid of numbers of an old-growth forest inverted?
   e. Applying the assumptions that in a given time period the plants in c (above) 1) captured 2% of 10 million calories of sunlight energy available and 2) an average transfer rate of 7% for each trophic (feeding) level; how many calories did the hawk receive?
   f. Why is a pyramid of biomass of a told growth deciduous forest inverted?

3. Why is the pyramid of energy a more accurate representation of a simple food chain than a pyramid of biomass?
4. The environment consists of the ______ living and the ______ nonliving or physical.
5. Consider what would happen to life on earth if satellite analysis detected a consistent worldwide trend of decreasing amounts of chlorophyll. Think of four major effects.

   1. ______
   2. ______
   3. ______
   4. ______

6. Speculate on the relative biological productivity of the following ecosystems by rating an equal unit of each: salt marsh, open (pelagic ocean, outside the continental shelf), a Saskatchewan wheat field, a farm in south Florida and a tropical rain forest. Things to estimate are the annual primary productivity, biodiversity, standing biomass, nutrient inflow, and energy inflow.

Rank	**Primary productivity: estimate High, Low, Moderate**	**Biodiversity: estimate High, Low, Moderate**	**Standing biomass: estimate High, Low, Moderate**	**Natural nutrient inflow: estimate High, Low, Moderate**	**Energy inflow: estimate High, Low, Moderate**
1. Brackish marsh estuary					
2. Open ocean					
3. Saskatchewan wheat field					
4. South Florida farm					
5. Tropical rain forest					

POP QUIZ

Matching

Match the following terms and definitions

____1. A highly productive ecosystem that supports recreation and commercial fishing of oysters, blue crabs, flounder, and shrimps:	a. primary productivity
____2. As the atmosphere and hydrosphere break down the lithosphere, this is formed:	b. community
____3. A biologist studied a 10-square hectare plot of Nebraska marsh land and recorded 22 yellow headed blackbirds. This group of birds is called a ______	c. decrease in solar entropy
____4. A biologist recorded 208 Spartina marsh grass plants, 10 snails, and 1 blue crab in a 1 m^2 plot of salt marsh. The organisms represent a ______	d. saprotrophs

____5. A community and its physical environment is a(n):	e. producers
____6. In photosynthesis, the function of chlorophyll:	f. soil
____7. When herbivores eat plants, they can obtain and store about 5–10% of the calories available. This represents a(n):	g. pyramid of biomass
____8. These organisms do chemosynthesis or photosynthesis:	h. population
____9. Shelf fungus can be seen growing on the trunks of old trees. These organisms are:	i. antenna molecule for sunlight
____10. The second law of thermodynamics makes this impossible:	j. pyramid of energy
____11. A hawk has the smallest amount of biomass or caloric value in a certain ecosystem, it is called a(n):	k. biodiversity
____12. Trophic levels expressed as gram/m^2:	l. ecosystem
____13. Trophic levels expressed in kilocalories/m^2:	m. top carnivore
____14. An increase in the biomass of plants per m^2 in a growing season:	n. energy flow
____15. Typically, as worldwide productivity of crops increase, this decreases:	o. brackish estuary

Multiple Choice

1. A high-energy ultraviolet photon (packet of electromagnetic energy) strikes an ozone molecule in the stratosphere, eventually releasing red and infrared photons (heat) of lower energy, but of equal total energy as the excitation U-V photon. This would best illustrate the:
   a. first law of thermodynamics
   b. second law of thermodynamics
   c. a decrease in entropy
   d. an impossibility! The energy content of this system must decrease as the U-V photons are degraded in energy content.
   e. third law of thermodynamics

2. Which statement is *incorrect* concerning the second law of thermodynamics?
   a. All sunlight energy and chemical energy is degraded to heat.
   b. The Earth is a closed system that prevents the passage of energy into the biosphere.
   c. The universe is an isolated system as to energy content.
   d. A salt marsh is an open system that can exchange matter and energy with its surroundings.
   e. There is an increase in entropy in the universe, but very little here on Earth.

3. Ginny watches a twig that her Dad planted in the yard grow into a large maple tree. The increase in biomass of the tree is mostly due to:
   a. the water soaked up by the tree
   b. the minerals incorporated from the soil
   c. the carbon from carbon dioxide in the air
   d. the hydrogen soaked up from the air that is incorporated into carbohydrates
   e. all of the above

4. Which statement describes the process of photosynthesis?
   a. Chlorophyll molecules trap photons of sunlight.
   b. Electromagnetic energy is transformed to chemical energy.
   c. Sunlight energizes electrons in chlorophyll that are used to make chemical bonds.
   d. Carbon from the air becomes carbon in the body of the plant.
   e. All of the above

5. The oxygen produced in photosynthesis comes from:
   a. the air
   b. carbon dioxide
   c. water
   d. the breakdown of food by cellular respiration
   e. minerals in the soil

6. An ecosystem is discovered near volcanic hydrothermal vents at a depth of 2,800 m below the surface. Which statement below best describes it?
   a. Photosynthetic organisms make up the producers.
   b. Certain bacteria can use sulfur compounds to obtain energy to synthesize food.
   c. Crabs, barnacles, and mussels are the producers and consumers.
   d. Red tubeworms have a special chlorophyll that can capture the low amount of light here.
   e. This is a decomposition food chain.

7. The rate at which plants in a grassland capture energy in photosynthesis is called:
   a. net primary productivity
   b. gross cellular respiration
   c. gross primary productivity
   d. It is measured in g/m^2 per year.
   e. net cellular respiration

8. The blue whale is the largest mammal on Earth. It gets its large biomass from eating:
   a. algae and kelp
   b. squid and octupus
   c. porpoises
   d. tiny shrimp-like krill
   e. all of the above

9. The function of bacteria and fungi to ecosystems is:
   a. removal of the dead bodies or indigestible remains of organisms
   b. create disease in organisms that are weakened
   c. return minerals stored in dead creatures' bodies to the soil
   d. break down salt marsh grass to provide nutrients for algae
   e. all of the above

10. Algal productivity in Antarctic ocean waters has declined because of:
    a. overfishing of krill
    b. toxic water pollution from ships
    c. increased ultraviolet radiation
    d. decreased sunlight from increased cloud formation due to global warming
    e. all of the above

11. Which phenomenon is probably responsible for the decline in krill in Antarctic waters?
    a. global warming due to increased methane from herbivore guts
    b. ozone-layer destruction from CFCs and other gases
    c. pollution from fishing trawlers
    d. toxins washed by rain or snow into the water
    e. too many baleen whales

12. One could argue that eating a vegetarian diet is more efficient than eating a diet of meat because:
    a. there is less energy lost as heat during energy flow to higher trophic levels
    b. a carnivore has less energy available to eat than a herbivore
    c. about 90% of the energy available to carnivores in the bodies of herbivores is lost in transfer.
    d. when herbivores, such as cattle breathe, walk, run, munch, and moo, their cellular respiration produces energy that is lost in transfer to carnivores.
    e. all of the above

13. Regarding the destruction of tropical rain forests, which statement is *incorrect*?
    a. Their conversion to grazing and farmlands will increase their net primary productivity.
    b. One effect would be enhanced carbon dioxide in the atmosphere and therefore global warming.
    c. A long-term effect could be a reduction in atmospheric oxygen.
    d. The burning of the forests can produce smoke that screens out sunlight, somewhat reversing the trend of global warming.
    e. The biodiversity of the world would decrease.

14. If fossil fuel combustion continues to increase, what theoretical mechanism would the Earth (or Gaia) use to homeostatically correct for the increased levels of carbon dioxide and global warming?
    a. increased formation of carbonate/limestone rocks
    b. increased algal populations
    c. increased plant populations
    d. Higher temperatures will lead to increased cloud formation and rain that will screen out sunlight and wash carbon dioxide into the ocean.
    e. all of the above

15. Some ecologists have asserted that there is a *grand balance of life*, that is chemosynthesis + photosynthesis = cellular respiration. Therefore, which statement would be *incorrect* regarding this balance?
    a. A satellite assay of worldwide chlorophyll would tell us whether photosynthesis could keep up with increased carbon dioxide emissions.
    b. Chemosynthesis is of minor importance in maintaining forest ecosystems.
    c. Only a certain number of animals can be supported by a certain biomass of plants.
    d. There is no place for predation in ecosystems.
    e. If plant populations decline, so will animal populations.

16. Elks are herbivores, and ravens are omnivores that will eat berries and songbird eggs and scavenge elk carcasses. Which of the following happened as a result of feeding elks in the National Elk Refuge, Jackson, WY?
    a. Ravens switched from scavenging elk carcasses to eating songbird eggs.
    b. Songbird populations declined.
    c. Raven populations increased.

d. There is now justification for importing gray wolves.
e. All of the above are correct.

CYBER SURFIN'

1. To see a movie that summarizes photosynthesis, go to the BrainPop.com site: **www.brainpop.com/science/plantsandanimals/photosynthesis/**
2. From the U.S. National Air and Space Administration, see an explanation of a hydrothermal volcanic vent and related community: **liftoff.msfc.nasa.gov/news/2001/news-thermalvent.asp**
3. What killed the dinosaurs? Two competing theories are an asteroid collision with Earth 65 million years ago, **web.ukonline.co.uk/a.buckley/dino.htm**, and a global warming theory from geologist Dewey McLean at the University of Vermont, **filebox.vt.edu/artsci/geology/mclean/Dinosaur_Volcano_Extinction/.**
4. At Discovery.com, play a game of energy budgets and survival, go to **dsc.discovery.com/stories/dinos/dinos.html**.
5. To learn more about Antarctic whales and reasons for their decline, see the WWF (formerly World Wildlife Fund), **www.nwf.org/climate/whales.html**; and Whales Online at **www.whales-online.org/**.
6. On salt marsh ecology, see **www.ens-news.com/ens/aug2002/2002-08-07-06.asp.**

'LIKE' BOOKING IT

1. Hinrichs RA. ***Energy: Its Use and the Environment,*** **Third ed.** Philadelphia: Harcourt College Publishers, 2002.
2. Odum EP. ***Fundamentals of Ecology.*** Philadelphia: Saunders, 1971.
3. Odum EP. ***Ecology: A Bridge Between Science and Society.*** New Zealand: MacMillan, 2000.

ANSWERS

Seeing the Forest

1. e

Questions to Ponder (Many of these answers are examples of many possible responses, please don't memorize them.)

2. After reading Chapter 4 consider the multidisciplinary nature of environmental science by listing the contributions of the disciplines listed below.

   1. Biology—identifications of plants and animals, habitat and nutrition descriptions
   2. Chemistry—chemical reactions of photosynthesis, cellular respiration, and acid rain formation are described.
   3. Engineering—sewage treatment plants are constructed to use natural processes to break down organic waste.

4. Mathematics—the science of mathematics contributes statistical analysis of the data of all observations and experiments.

5. Physics—forms of energy and their transformations are described.

3. An acre of Kentucky bluegrass yielded the following numbers of organisms: 1 hawk, 4 insectivorous garter snakes and 1 large copperhead, 22 mice, 6 meadow larks, 4,608 grasshoppers and other herbivorous arthropods, and 4,500,003 grasses and forb plants. For a primer of the role of birds in controlling grasshoppers, go to the U.S. Dept. of Agriculture website, www.sidney.ars.usda.gov/grasshopper/Handbook/I/i_10.htm

   a. Outline a simple food chain.

   Producer grasses and forbs (4,500,003) → grasshoppers and other herbivorous insects (4,608) → mice (22) and meadow larks (6) → snakes (5) → hawk (1).

   b. Outline a food web.

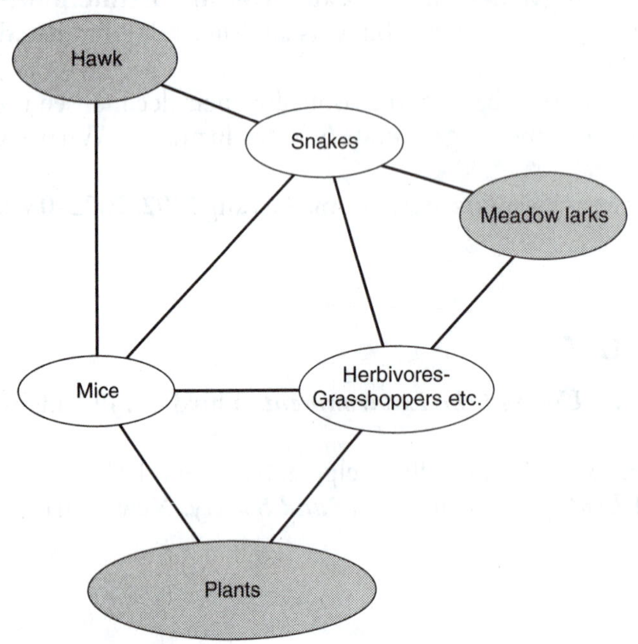

   c. Draw a pyramid of numbers. Make the size of each trophic level roughly proportional to the numbers represented.

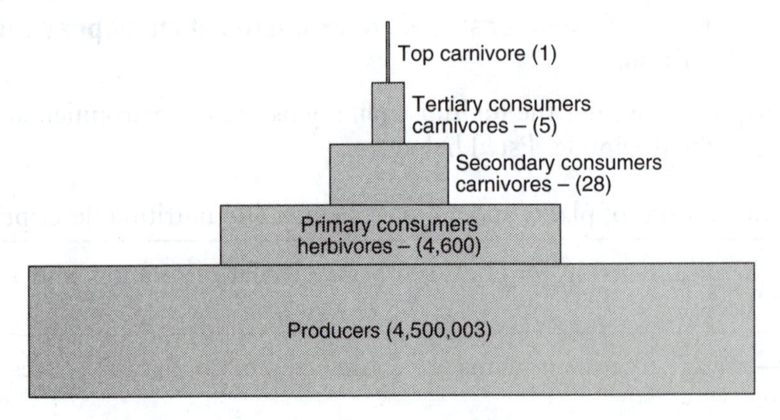

d. Why is the pyramid of numbers of an old-growth forest inverted? *Because the number of large trees is small.*

e. Applying the assumptions that in a given time period the plants in c. (above), 1) captured in photosynthesis 2% of 10 million calories of sunlight energy available and 2) an average transfer rate of 7% for each trophic (feeding) level; how many calories can the hawk receive? *10,000,000 calories (×) .02 captured by photosynthesis = 200,000 cal. (×) .07 transferred to herbivores = 14,000 cal. (×) .07 transferred to carnivores = 980 cal. transferred to tertiary consumer carnivores (×) .07 transferred to top carnivore = 68.6 cal.*

f. Why is a pyramid of biomass of a dense forest inverted? *Because some individual trees can weigh more than a ton.*

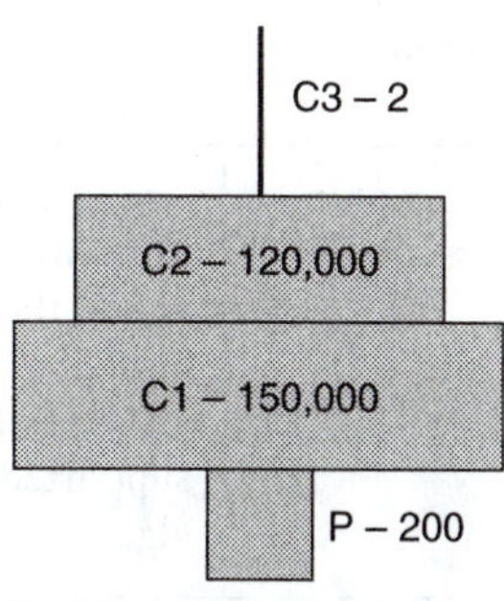

Deciduous Forest – Summer

4. Why is the pyramid of energy a more accurate representation of a simple food chain than a pyramid of biomass? *The various types of chemicals that store energy in the bodies of organisms vary in energy content, with the highest being fats and less in proteins and carbohydrates. Therefore energy content is a better measure of ecosystem structure than weight/biomass.*

5. The environment consists of the *biotic* living and the *abiotic* nonliving or physical.

6. Consider what would happen to life on Earth if satellite analysis detected a consistent worldwide trend of decreasing amounts of chlorophyll. Think of four major effects.

   1. Carbon dioxide levels and global surface temperatures would rise as photosynthesis decreased.
   2. A decrease in atmospheric oxygen would occur as photosynthesis increased.
   3. Unmanaged, natural ecosystems would collapse with concomitant extinction of species.
   4. Increased heating of air above the surface would take place due to a lack of energy absorption by the plants and algae with increased severity of storms resulting.

7. Speculate on the relative biological productivity of the following ecosystems by rating equal units of each: salt marsh, open (pelagic ocean, outside the continental shelf), a Saskatchewan wheat field, a farm in south Florida, and a tropical rain forest. Things to estimate are the annual primary productivity, biodiversity, standing biomass, nutrient inflow, and energy inflow.

Rank	Primary productivity: estimate High, Low, Moderate	Biodiversity: estimate High, Low, Moderate	Standing biomass: estimate High, Low, Moderate	Natural nutrient inflow: estimate High, Low, Moderate	Energy inflow: estimate High, Low, Moderate
1. Brackish marsh estuary	High. Cord grass starts a decom-position food chain Because of its high silica content, few herbivores eat it.	Low/Moderate. Cord grass is a monoculture algae and animals have greater diversity.	High	High. Fertilizer runoff and sewage can be naturally recycled into new plant growth.	High. Annual amounts of sunlight energy very high. Little or no shading occurs.
2. Open ocean	Low. Substrate for plants and animal habitats is rare except in rafts of kelp (sargassum) near the equator.	Low except in sargassum.	Low. Because nutrient influx is low, there is a small amount of supported algae.	Low	High. Most of the energy goes into heating up the water.
3. Saskatchewan wheat	Annually Moderate High in a short growing season.	Low. What is raised in a monoculture.	Low/ Moderate annually.	Low. Subsidized with animal manure, chemical fertilizers, and fossil fuel energy.	Annually Low/Moderate, seasonally high.
4. South Florida farm	High	Low. Although there is seasonal diversity of crops.	Moderate	Low. Farms are subsidized by chemical fertilizers and fossil fuel energy.	Moderate High. Being closer to the equator helps.
5. Tropical rain forest	Highest. Being on or near the equator.	Highest diversities of plants and animals.	Highest annual standing biomass.	High. Decomposition of materials is very efficient and very rapid.	Highest

Pop Quiz Answers

MATCHING

1. o
2. f
3. h
4. b
5. l
6. i
7. n
8. e

9. d
10. c
11. m
12. g
13. l
14. a
15. k

MULTIPLE CHOICE

1. a
2. b
3. c
4. e
5. c
6. b
7. c
8. d
9. e
10. c
11. b
12. e
13. a
14. e
15. d
16. e

5

Ecosystems and Living Organisms

LEARNING OBJECTIVES

After you have studied this chapter you should be able to:

1. Explain the four premises of evolution by natural selection as proposed by Charles Darwin.
2. Describe ecological succession and distinguish between primary and secondary succession.
3. Discuss an example of a keystone species.
4. Explain symbiosis and distinguish among mutualism, commensalism, and parasitism.
5. Define predation and describe the effects of natural selection on predator-prey relationships.
6. Define competition and distinguish between intraspecific and interspecific competition.
7. Describe the factors that contribute to an organism's ecological niche and distinguish between fundamental niche and realized niche.
8. Give several examples of limiting factors and discuss how they might affect an organism's ecological niche.
9. Relate the concepts of competitive exclusion and resource partitioning.
10. Summarize the main determinants of species richness in a community and describe factors associated with high species richness.
11. Give several examples of ecosystem services.

SEEING THE FOREST

1. The collapse of Lake Victoria, the world's largest tropical lake, as a food producer and the decline of the health of humans living nearby are due to many complex interactions, ecological and economical in nature. See various reports on its decline at allafrica.com.

Fill-in the table of seven factors that led to the decline of this naturally productive ecosystem and the resulting change brought about by each.

Change	Result
1.	Algal bloom
2. Rotting algae	
3.	Decline of plant eating cichlid fishes
4. Forest cutting for fuel and drying racks	
5. Increased turbidity of the lake water	
6.	Water hyacinths clogging waterways

SEEING THE TREES

VOCAB

The first law of ecology is that everything is related to everything else.
Barry Commoner

1. Evolution—literally, an "unrolling," as a scroll. The concept that life has changed in the history of the Earth from early simple forms to a succession of more stable and complex communities that adapt to open ecological niches.
2. Natural selection—Charles Darwin postulated that organisms with functional heritable variations would compete successfully with organisms that were not so favored, adapt to new eco-

logical niches, and change in anatomy and behavior accordingly. Darwin interpreted the domestication of plants and animals as evidence that natural selection could occur.

3. Overproduction—a concept that Darwin borrowed from the "dismal economist" Thomas Malthus, who concluded that although the human race had the potential to overpopulate the Earth, its increase in population was held somewhat in check by starvation, disease, and war. Similarly, populations of organisms have high reproduction potentials but do no overrun the Earth because of competition for food habitat, mates, and disease. Therefore, Darwin concluded that there was a "struggle for existence."
4. Heritable variation—Darwin assumed that heritable variations (what we today call genes and their expressions) enhance a species' survival and success of reproduction. Therefore, the survivors are able to pass those variations to offspring. Success of reproduction is measured by the relative number and viability of the offspring. The term *neo-Darwinism* is used to reflect the incorporation of modern genetics into the concept of evolution by natural selection.
5. Differential reproductive success—there is a "survival of the fittest" of the heritable variations and the organisms that possess them. If an organism possesses better heritable variations for gathering food or protecting young, then that organism will outcompete similar organisms without those variations, and it will survive.
6. Taxonomy—classification of living organisms that include, from general to specific, Kingdom, Phylum or Division, Class, Order, Family, Genus, and Species.
7. Ecological succession—a process of community development and changes that follows a major disturbance of the physical environment.
8. Sere—a stage in ecological succession that is usually named according to the dominant plant form—grasses, shrubs, pines, and deciduous hardwoods.
9. Climax community—a relatively stable community that represents the last stage of development in a given physical environment. As a general rule, it has the greatest biodiversity as compared to earlier stages.
10. Primary succession—a succession of communities that begins on bare rock with rock erosion and soil formation by lichens or the rooting of "sea oat" plants in sand dunes.
11. Secondary succession—"old field" succession that begins with land that has been stripped of plant life, with a nutrient-poor soil remaining.
12. Pioneer community—the first organisms to occupy or reoccupy a habitat.
13. Keystone species—species that are very important to the ecosystem, they may be dominant species like saguaro cactus in the Sonoran Desert or less obvious, such as ravens and wolves in Yellowstone National Park.
14. Symbiosis—an intimate living relationship between two species.
15. Coevolution—if two species are symbiotic, they will evolve together—the parasites of primates are quite distinct from the parasites of other zoological orders but correlated closely to each other.
16. Mutualism—both symbionts benefit from the relationship. The relationship may be obligate, only occur in symbiosis (some lichens); or nonobligate (protocooperation), where both organisms can survive independently (other lichens).

17. Commensalism—one symbiont is benefited and the other unaffected—certain colon amoebae live in the human alimentary tract, eating bacteria, without causing any benefit or harm. Spanish moss is a botanical relative of the pineapple that grows on tree limbs in the U.S. Atlantic coastal plain.
18. Parasitism—the smaller symbiont benefits by robbing calories from its host. Disease organisms use this lifestyle.
19. Nest or brood parasitism—the parasite essentially relies on the host bird to hatch and raise offspring that are typically larger than the host's offspring. The relationship may be obligate—the European cuckoo; or nonobligate where the parasite can raise its own offspring—North American cuckoos.
20. Predation—the typically larger symbiont benefits by devouring the smaller prey species. On a short-term basis, the host is negatively affected, but on the long term, the health of the prey population is enhanced by preventing its overpopulation and ensuring the "survival of the fittest" genetic makeup.
21. Amensalism—the antibiotic effects of chemicals produced by plants that poison attacking predators or by fungi and bacteria as they compete for resources.
22. Competition—members of the same species (intraspecific) or different species (interspecific) compete indirectly for resources or directly through aggression.
23. Pathogen—parasitic organism that causes disease of its host.
24. Ecological niche—the occupation of an organism or its role in an ecosystem that takes into account all the chemical and environmental requirements for its survival.
25. Fundamental niche—the idealized niche of all the activities that a species could possibly do including exploitation of various food resources and habitats.
26. Realized niche—the actual activities will be limited by competition—several species of birds compete for seeds, the birds then specialize as to the type seed they eat or the level of the forest floor and canopy where they nest.
27. Limiting factor—a resource that is scarce in amount that restricts an ecological niche.
28. Competitive exclusion—when species compete for a specific niche, one will be excluded by the competition. More exact, the competing species that survive will evolve specializations (resource partitioning) that subdivide the niche.
29. Species richness—the number of species in a community or biodiversity that is directly proportional to the niches available. Generally, community stability is enhanced by species richness.
30. Ecotone—where two communities or habitats meet—field and forest—where species richness is high (edge effect).

Questions to Ponder

1. Fill-in the table below listing and describing the six kingdoms of living organisms. Two different taxonomic terms, commonly used, are used below as indicated by /.

Kingdom	Characteristics	Examples
1.	Uses radiant energy for autotrophic nutrition, cells have nuclei and cell walls of cellulose and lignin tissues are present.	
2. Archaebacteria/Archaea		Thermophile bacteria in Yellowstone N.P. geyser pools, methanogens in intestines, halophiles (salt loving) in the Dead Sea.
3. Eubacteria/Bacteria (heterotrophic)		
4. Eubacteria/Bacteria: Cyanobacteria (blue-green bacteria)		
5.		Protozoa and algae
6.	Eukaryotic cells with cell walls, tissues and heterotrophic, saprotrophic nutrition. In ecosystems they serve as decomposers.	
7. Animalia		

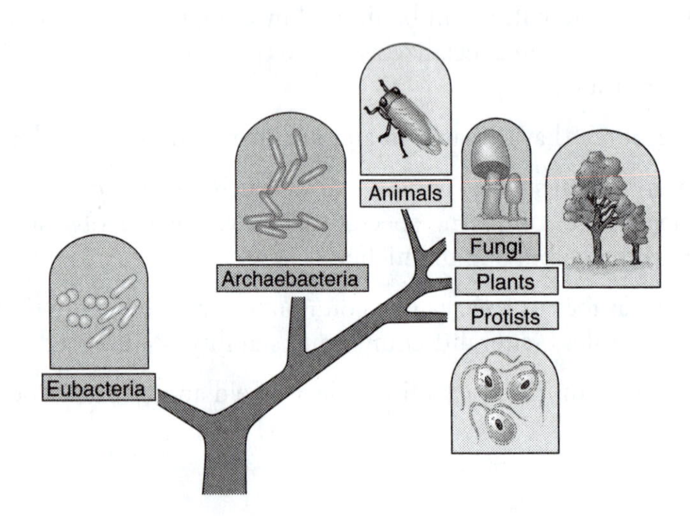

2. Fill-in the following table that illustrates the processes and stages (seres) of ecological succession.

Fashion entries in the right two columns according to your local flora and fauna. What follows below is relevant to the temperate deciduous forest.

Stage	Processes of transformation	Characteristic organisms	Organisms in the next successional stages (assuming later seres appear)	Stability of community (assuming later seres appear)
	Acids from lichens or roots erode rock, making soil. Humus is added by decomposing organisms.			Unstable, biodiversity is low.
Secondary succession—grasses and forbs (weeds)	As humus is added to the soil by decomposing plants and roots break up the soil, plants appear that are taller and have deeper root systems. Animals that find habitat or cover will migrate into the area.	Crab grass, fire weed, horse grass and other forbs. Herbivores, insectivores, snakes, and top carnivore hawks.		
Secondary succession—shrubs		Shrub or dwarf trees such as red cedar and wax leaf in eastern U.S., sparrows, mockingbirds, scrub jays.		
Secondary succession—pines		In eastern U.S., loblolly, slash, yellow and black pines. Pine warblers, kingbirds, crows.		
Secondary succession—climax hardwoods				Most stable, biodiversity is high.

3. Fill-in the table of symbiotic and competitive interactions between species. Species #1 is generally the smaller symbiont (parasite, commensal). Species #2 is generally the larger symbiont (prey or host) or other interacting species. Rate each as benefited positively, negatively, or not affected as to growth, survival, or other attribute as follows:

positively affected, = **+** ; not affected = **0** ; **negatively affected = –** . (After Odum, *Fundamentals of Ecology*, 1964).

Type of interaction	Species #1	Species #2	General nature of interaction: Short term and long term	Example
Neutralism	0	0	There is no direct effect apparent from interaction. Long-term effects are likely present as all creatures are "connected."	

Parasitism			Short-term effects may be limited to calorie robbing (e.g., intestinal tapeworms) or the death of the host (e.g., AIDS). The long-term effects of calorie robbing may make the host a better forager in the winter, when parasite populations tend to fall.	
Predation				
Commensalism			A long-term relationship may develop into mutualism or parasitism.	
Nonobligate mutualism/ proto-cooperation				In certain lichens, some algae and fungi can live independently or in symbiosis.
Obligate mutualism				
Competitive chemical inhibition (amensalism).			The "antibiotic effect." One species secures greater resources by limiting the growth of other species or competitors of the same species (creosote and sage brush). Long term: the antibiotics become more precisely targeted to competing species.	
Brood or nest parasitism			Long-term effects include coevolution of the brood parasite to include laying eggs similar to the host egg and having mouth patterns that mimic the patterns seen in the host species' hatchlings (African honey guide).	
Indirect resource competition				
Direct interference competition			Short-term aggression is usually directed to members of the same species competing for food, habitat, and mates. Birds are territorial.	

4. List different examples of the services of various ecosystems. See the text Table 5-1.

Ecosystem	Example
Forest	
Freshwater streams and lakes	
Freshwater marshlands	
Grassland	
Coasts	
Agricultural	
Brackish estuaries	
Continental ocean shelf	

5. List six organisms and different defenses that they use to prevent themselves from being eaten.

Organism	Defense
1.	
2.	
3.	
4.	
5.	
6.	

POP QUIZ

Matching

Match the following terms and definitions

____1. This invasive plant spreads in sewage, polluted tropical lakes and canals, making navigation difficult:	a. realized niche
____2. In thirty years, a southern U.S. grassy field becomes a pine and turkey oak forest, the process is called:	b. warning coloration
____3. These organisms, growing on the granite of Stone Mountain, Georgia, are turning rock into soil:	c. make ammonia fertilizer from atmospheric nitrogen
____4. As grasses invade old fields and die, they enrich the soil with water and mineral binding:	d. fundamental niche
____5. Naturally occurring intestinal bacteria make Vitamin K, necessary for proper blood clotting. This is an example of:	e. pathogen
____6. Mutualistic algae of corals:	f. humus
____7. Mutualistic root fungi of pines and other plants:	g. mutualism
____8. Mutualistic rhizobia bacteria of legume plants do this:	h. resource partitioning
____9. Much of the evidence supporting the concept of keystone species is unreliable in the scientific method because it is ______.	i. succession
____10. A parasite, like the plague bacillus, that causes disease is a(n) ______.	j. zooanthellae
____11. The scarlet king snake mimicing the appearance of the poisonous coral snake is an example of:	k. lichens
____12. The first seed-eating bird to reinhabit the volcanic island Krakatoa was occupying its ______.	l. Gause's rule
____13. In the U.S. gold finches, house finches, and cardinals compete for seeds. Each species is in its ______.	m. mycorrhizae
____14. When *Paramecium caudatum* and *P. aurelia* compete in a limited artificial habitat, *P. caudatum* dies out:	n. water hyacinths
____15. Competing warblers survive by varying the time of feeding and the height in the tree used for feeding and nesting. This is an example of:	o. anecdotal

Multiple Choice

1. The most important event that began the deterioration of Lake Victoria was:
   a. introducing Nile perch.
   b. cutting of surrounding forest creating erosion and water turbidity
   c. sewage pollution that caused algal blooms
   d. invasion of water hyacinths
   e. industrial pollution

2. An ecological implication of the theory of evolution is:
   a. Some plants and animals will always overpopulate their ecosystem and use up its resources.
   b. Direct interference competition or aggression will become the dominant behavior of animals.
   c. Heritable variations may become adaptations allowing for resource partitioning.
   d. Most animals will have big teeth and claws for defense.
   e. Parasitic disease organisms will usually kill their hosts to maintain host population control.

3. In the history of the Earth, new species have appeared and extinctions have occurred. Which of the following explanations are likely to be correct in light of the theory of evolution?
   a. New species appear and adapt rapidly to fill niches opened by a mass extinction event.
   b. If a species does not have heritable variation to allow for niche adaptation, that species will become extinct.
   c. Most species survive by adapting strategies to compete indirectly for resources.
   d. The fossils of *Triceratops* and duckbill dinosaurs will not be found in the same rock layer as buffalo and hippopotamus.
   e. All of the above

4. A soil organism that does not have a nucleus and membrane-bound organelles and serves as a decomposer is in this Kingdom:
   a. Archaebacteria
   b. Bacteria/eubacteria
   c. Protista
   d. Fungi
   e. all of the above

5. This Kingdom can produce enzymes that break down dead or decaying tissues.
   a. Bacteria/eubacteria
   b. Fungi
   c. Protista
   d. Plantae
   e. a and b only

6. Members of this Kingdom can trap radiant energy to use for making chemical energy stores (food).
   a. Eubacteria/cyanobacteria
   b. Protista/algae
   c. Plantae
   d. Fungi
   e. a, b, and c

7. Members of this Kingdom are found in high concentrations of salt such as the Dead Sea, in hot springs, in high concentrations of sulfur, and in rocks hundreds of meters below the surface.
   a. Archaebacteria (archaea)
   b. Eubacteria
   c. Protista/heterotrophic protozoa
   d. Fungi
   e. all of the above

8. These organisms are considered pioneers in primary succession:
   a. lichens
   b. mosses
   c. crab grass
   d. horse weed
   e. a and b

9. After the widespread fires in Yellowstone N.P., trout lily and some hardy weeds (forbs) appeared. The process is best called:
   a. primary succession
   b. secondary succession
   c. old field succession
   d. bare rock succession
   e. all of the above

10. As succession in a forest proceeds, which process occurs?
    a. Grasses and forbs break up the soil, making mineral nutrients available.
    b. When pioneer plants decompose, humus is added to the soil.
    c. Taller plants with deeper roots appear as the soil is prepared for them.
    d. Generally, species richness increases.
    e. All of the above are correct.

11. Should fig trees disappear in a tropical rain forest, this animal would also disappear:
    a. fruit bats
    b. monkeys
    c. humans
    d. jaguars
    e. a and b only

12. Which statement is an accurate interpretation of the graph below?
    a. *P. caudatum* has an early advantage of *P. aurelia* in population growth.
    b. Because *P. aurelia* is a predator, it ate up the *P. caudatum*.
    c. *P. aurelia* exploited the niche better than *P. caudatum*.
    d. a and b only.
    e. All of the above are correct.

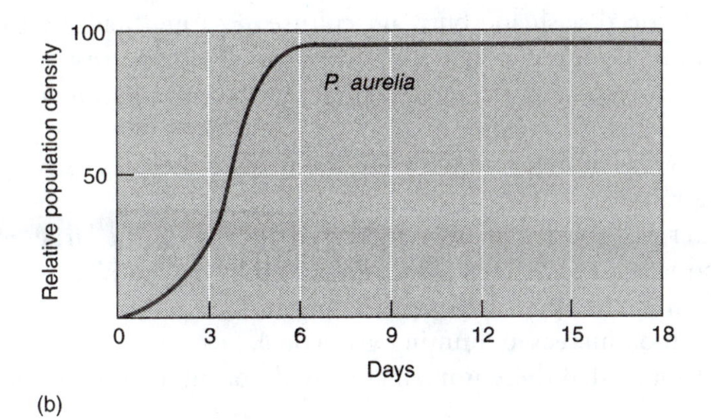

(b)

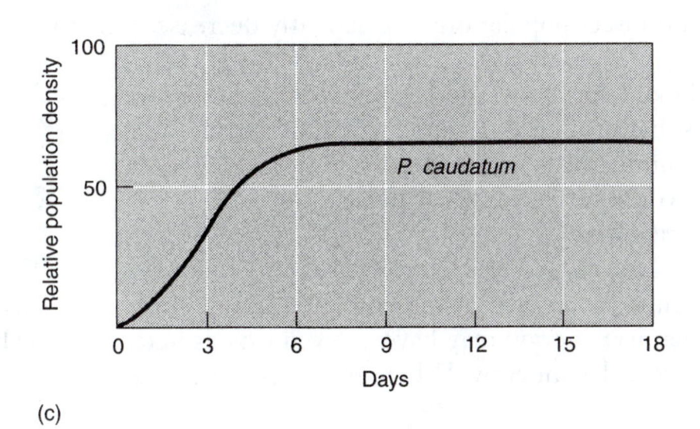

(c)

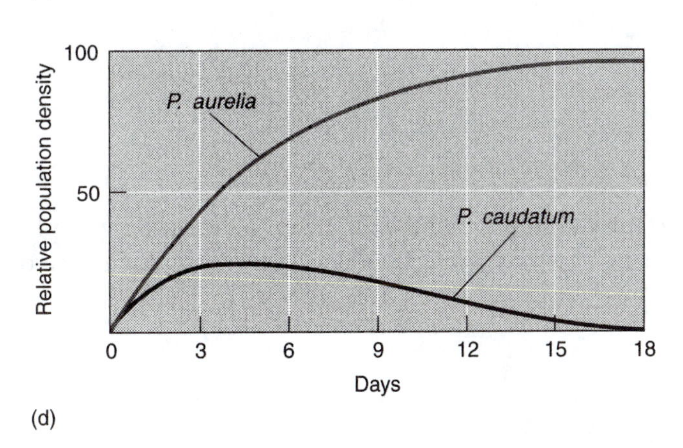

(d)

13. Which of the following ecosystems has the greatest species richness?
    a. tropical rain forest
    b. boreal coniferous forest
    c. deciduous forest
    d. brackish marsh estuary
    e. pelagic ocean

14. Native Americans practiced slash and burn agriculture, creating fields that were surrounded by forest. Which statement is correct concerning what happened as a result?
    a. Disturbance of the forest habitat created more contact between human and deer as the animals tried to find their former homes.
    b. The ecotone effect was more pronounced in the fields created as seen by a large increase in food-crop-eating rabbits.
    c. Hunting got better and species richness increased because of the larger area of contact between field and forest ecosystems. The crops produced were plentiful and maintained over a long period of time.
    d. Even as the plant food increased, hunting got worse.
    e. Deciduous forest succeeded there immediately without intervening stages.

15. If forest and grassland insect populations permanently decreased, the following effects would be seen:
    a. Wind-pollinated plant species would become dominant.
    b. Flowering plants that produced perfume would decrease.
    c. Nectar-eating hummingbirds would increase.
    d. Surviving plants would not have showy petals.
    e. All of the above are correct.

16. Biologist Lynn Margulis proposed that the mitochondria of eukaryotic cells (that metabolize glucose and ATP chemical energy) may have evolved first as heterotrophic bacteria and later lost their ability to live independently. Which term best describes that early symbiotic relationship?
    a. mutualism
    b. commensalism
    c. parasitism
    d. competition
    e. predation

17. Spanish moss and certain vines derive nutrients from the stem flow of plants that otherwise would go to the soil or be eaten by bacteria. This represents which type of symbiosis?
    a. amensalism
    b. commensalism
    c. parasitism
    d. competition
    e. predation

18. Mistletoe derives nutrients by penetrating tree bark as well as doing photosynthesis. This makes mistletoe a(n):
    a. amensal
    b. commensal
    c. predator
    d. parasite
    e. all of the above

19. The principal difference between parasitism and predation is:
    a. The parasite robs calories from the host and the predator does not.
    b. The parasite is much smaller than its host, the predator is larger by comparison.

c. The parasite never kills its host, and the predator always kills.
d. Only the predator actively seeks its prey.
e. Trick question alert! All of the above are true.

20. The reason why Lyme disease increases after a good growing year for oak trees. where their competitors are reduced in number by disease or insect attack, is:
    a. The Lyme bacterium grows in the tick parasites of oak trees.
    b. More acorn food available increases the mice population that serves as intermediate hosts of the Lyme bacterium-carrying ticks.
    c. Deer eat the acorns and acquire bacteria from the ticks that at one time parasitized the mice.
    d. b and c only
    e. a, b, and c

CYBER SURFIN'

1. For information on the cowbird's contribution (along with habitat loss) to the 40%+ decline in songbirds over the last 40 years in the eastern U.S., go to: **www.inhs.uiuc.edu/chf/pub/surveyreports/nov-dec97/cowbird.html** and **www.sciencenews.org/sn_arc98/5_30_98/bobl.htm**
2. From the study guide author's website is a review of symbiosis, see **www.gpc.edu/~jaliff/symbiorb.htm**
3. For those mathematically inclined, get familiar with the Shannon-Wiener index of diversity of an ecosystem. Do a survey of your backyard plants. Even if you do not know what their species name is, designate each one as tree #1, flower #1, grass #1, etc. Go to the Maryland Sea Grant institution at **www.mdsg.umd.edu/Education/biofilm/diverse.htm**. There are fill-in tables to enter and compute your data.
4. Simulate the effects of lion predation on the health of their prey. Read the following: **www.nap.edu/readingroom/books/wolves/** and do the predator/prey population simulation at **www.messiah.edu/hpages/facstaff/deroos/CSC171/PredPrey/PPIntro.htm**
5. To connect the concepts of population growth and evolution, go to **www.nap.edu/readingroom/books/evolution98/evol6-h.html** and do the exercises.

'LIKE' BOOKING IT

1. Wilson EO. ***Sociobiology: A New Synthesis, Abridged edition.*** Cambridge, Mass.: Harvard University Press, 1980.
2. Odum EP. ***Fundamentals of Ecology.*** Philadelphia: Saunders, 1971.
3. Odum EP. ***Ecology: A Bridge Between Science and Society.*** New Zealand: MacMillan, 2000.
4. Gould SJ. ***The Structure of Evolutionary Theory.*** Cambridge, Mass.: Harvard University Press, 2002.
5. Zimmer C, Gould SJ (introduction) and Hutton R. ***Evolution: The Triumph of an Idea.*** New York: Harper-Collins, 2001.

ANSWERS

Seeing the Forest

The collapse of Lake Victoria, the world's largest tropical lake, as a food producer and the decline of the health of humans living nearby are due to many complex interactions, ecological and economical in nature. See various reports on its decline at **http://allafrica.com.**

Fill-in the table of seven factors that led to the decline of this naturally productive ecosystem and the resulting change brought about by each. Answers are in italics.

Change	Result
1. *Sewage pollution and elimination of algae-eating fishes.*	*Algal bloom*
2. Rotting algae	*Oxygen debt and fish kills, odors.*
3. *Importation of nile perch.*	Decline of plant eating cichlid fishes and algae eating fishes.
4. Forest cutting for fuel and drying racks	*Increased soil erosion.*
5. Increased turbidity of the lake water	*Turbidity reduces photosynthesis and the production of food for aquatic food chains.*
6. *An invasive species entered Lake Victoria.*	Water hyacinths clogging waterways

Questions to Ponder (Many of these answers are examples of many possible responses, please don't memorize them.)

1. Fill-in the table below listing and describing the six Kingdoms of living organisms.

Kingdom	Characteristics	Examples
1. Plantae	Uses radiant energy for autotrophic nutrition, cells have nuclei and cell walls of cellulose and lignin. Tissues are present.	*Mosses, ferns, grasses, trees, flowering plants*
2. Archaebacteria/Archaea	*Prokaryotes that mostly live in extreme environments. Includes chemosynthetic (rock eaters) and photosynthetic forms. Genetically and biochemically different from eubacteria.*	*Thermophile archaea in Yellowstone N.P. geyser pools, methanogens in intestines, halophiles (salt loving) in the Dead Sea*
3. Eubacteria/Bacteria (heterotrophic)	Prokaryotic cells with simple carbohydrate cell walls and heterotrophic, saprotrophic nutrition. In ecosystems, they serve as decomposers.	*Staphylococcus skin bacteria*

4. Eubacteria/Bacteria: Cyanobacteria (blue-green bacteria)	*Prokaryotic single or colonial cells with cell walls and chlorophyll photosynthetic pigments.*	*Oscillatoria*—a filamentous form. See Chapter 2.
5. *Protista*	*Eukaryotic single or colonial cells that include heterotrophic protozoa and photosynthetic algae.*	*Protozoa and algae*
6. *Fungi*	*Eukaryotic cells with cell walls, tissues, and heterotrophic, saprotrophic nutrition. In ecosystems they serve as decomposers.*	*Amoeba, Paramecium, Volvox, Euglena*
7. Animalia	*Eukaryotic cells with tissues.*	*Us!*

2. Fill-in the following table that illustrates the processes and stages (series) of ecological succession.

   Fashion entries in the right two columns according to your local flora and fauna. What follows below is relevant to the temperate deciduous forest.

Stage	**Processes of transformation**	**Characteristic organisms**	**Organisms in the next successional stages (assuming later seres appear)**	**Stability of community (assuming later seres appear)**
Primary succession	Acids from lichens or roots erode rock, making soil. Humus is added by decomposing organisms.	*Lichens, sea oats in sand dunes*	*Mosses, ferns in cool climates. Mosses or grasses in warm climates.*	Unstable, biodiversity is low.
Secondary succession—grasses and forbs (weeds)	As humus is added to the soil by decomposing plants and roots break up the soil, plants appear that are taller and have deeper root systems. Animals that find habitat or cover will migrate into the area.	Crab grass, fire weed, horse grass and other forbs. Herbivores, insectivores, snakes, and top carnivore hawks.	*Heather, shrubs, and birds that nest in shrubs.*	*Unstable, biodiversity is low.*
Secondary succession—shrubs	*Decomposing plants and roots break up the soil, plants appear that are taller and have deeper root systems. Animals that find habitat or cover will migrate into the area.*	*Shrubs or dwarf trees such as wax leaf and red cedar in the eastern U.S., sparrows, mockingbirds, scrub jays*	*Pines, warblers, crows, ravens, elk, deer*	*More stable, biodiversity is increasing.*

Secondary succession—pines	*Pines shed needles that represent mulch for hardwood seedlings. Pine root penetrate into deeper soil layers.*	In eastern U.S., loblolly, slash, yellow and black pines. Pine warblers, kingbirds, crows. Pine beetles attack stressed trees.	*Maples, birch, oaks, poplars, etc. Deer, wolves. Pines are not generally tolerant of shade, so hardwoods succeed them.*	*More stable, biodiversity is increasing.*
Secondary succession—climax hardwoods	N/A	*The community is relatively stable with occasional disturbances like gypsy moths and Dutch elm fungus.*	*The community will be dominated by very large trees in a century or more.*	Most stable, biodiversity is high.

3. Fill-in the table of symbiotic and competitive interactions between species. Species #1 is generally the smaller symbiont (parasite, commensal). Species # 2 is generally the larger symbiont (prey or host) or other interacting species. Rate each as benefited positively, negatively, or not affected as to growth, survival, or other attribute as follows:

positively affected = + ; not affected = 0 ; **negatively affected** = – . (After Odum, *Fundamentals of Ecology*, 1964).

Type of interaction	Species #1	Species #2	General nature of interaction: Short term and long term	Examples
Neutralism	0	0	There is no direct effect apparent from interaction. Long-term effects are likely present as all creatures are "connected."	*Polar bears and penguins?*
Parasitism	+	–	Short-term effects may be limited to calorie robbing (e.g., intestinal tapeworms) or the death of the host (e.g., AIDS). The long-term effects of calorie robbing may make the host a better forager in the winter, when parasite populations tend to fall. *Generally, parasites kill hosts when the host population is overcrowded or weakened, the parasite has a short evolutionary relationship with the host, or if the death of the host is necessary for continuing its life cycle.*	*HIV, Pasturella pestis (plague), beef tapeworm, mosquitoes, lice, ticks. Parasitism may be basically continuous (tapeworm) or intermittent (mosquito).*
Predation	+	–	*Short-term effect on herbivorous prey is its death. Long-term effects improve host health by preventing overpopulation that destroys the plants in their food chain, and tends to weed out weaker heritable variations (genes and their expressions).*	*Foxes preying on rabbits, minnows eating Daphnia water fleas.*

Commensalism	+	0	A long-term relationship may develop into mutualism or parasitism.	*Many bacteria-eating amoebae in alimentary tracts. Hermit crabs "planting" stinging sea anemones on their borrowed shells*
Nonobligate mutualism/ protocooperation)	+	+	*The "grand balance of life" is reflected in a lichen. Because it is an aerobic heterotroph, the fungus supplies the alga with carbon dioxide and water, the alga responds by using photosynthesis to make food and oxygen. The wastes of one are the food of the other.*	In certain lichens, some algae and fungi can live independently or in symbiosis. Many times, fungal hyphae penetrate the algal cells, combining mutualism and parasitism.
Obligate mutualism	+	+	*Each organism requires association with the other.*	*The bacteria that produce the enzymes cattle need to digest the cellulose in straw, are found embedded in the flagellated protozoa of termites. The bacteria help feed the termites and their protozoa, and the termites supply lots of wood fiber.*
Competitive chemical inhibition (amensalism)	+ *or* –	– –	The "antibiotic effect." One species secures greater resources by limiting the growth of other species or competitors of the same species (creosote and sage brush). Long term: the antibiotics become more precisely targeted to competing species.	*Antibiotics are substances produced by an organism to combat another's growth. A good example is the normal flora concept. Bacteria make acids that inhibit the fungi or fungal-like bacteria that compete for resources. Fungi make antibiotic chemicals that inhibit the bacteria.*

Brood or nest parasitism	+	–	Long-term effects include coevolution of the brood parasite to include laying eggs similar to the host egg and having mouth patterns that mimic the patterns seen in the host species' hatchlings (African honey guide).	The African honey guide bird not only does protocooperation with humans and honey badgers, leading them to hives that the bird cannot get to, but it is a highly adapted brood parasite that kills the host's hatchlings and mimics the gaping mouth patterns of the host's natural offspring.
Indirect resource competition	–	–	*Both populations restrict the idealized niche of the other.*	*Various species of warblers and finches, respectively.*
Direct interference competition	– or +	– or +	*Short-term aggression is usually directed to members of the same species competing for food, habitat, and mates. Birds are territorial. The energy devoted to aggression is a negative effect. Some animals like hippos may die in territorial fighting.*	*Deer fighting with antlers, bighorn sheep butting heads.*

4. List different examples of the services of various ecosystems. See the text Table 5-1.

Ecosystem	**Example**
Forest	Clean air, lower carbon dioxide
Freshwater streams and lakes	Clean water
Freshwater marshlands	Changing ground water aquifers
Grassland	Grazing for herbivores
Coasts	Protection of inland areas from flooding from the ocean
Agricultural	Food

Brackish estuaries	Natural "sewage treatment plants" that cycle fertilizer runoff and other nutrients into cord grass and algae for marine food chains.
Continental ocean shelf	The ocean floor, particularly with coral reefs, serves as a substrate for kelp and various animals to feed and seek cover.

5. List six organisms and different defenses that they use to prevent themselves from being eaten.

Organism	Defense
Scarlet king snake	"Batesian" mimicry of the warning coloration of the poisonous coral snake.
White tailed deer	Flight and raising white underside of tail as a "warning flag" to others in the herd.
Coffee plant	Caffeine in the seeds (beans) that poisons attacking insects.
Tobacco plant	Nicotine in the leaves that poisons attacking insects.
Distraction display	Ducks and killdeer birds do a broken wing act to lead predators away from nests.
Death display	Some snakes and possums play dead. Some people attacked by bears do this also. Sometimes it works if the bear is not hungry and is protecting its cubs.

Pop Quiz Answers

Matching

1. n
2. i
3. k
4. f
5. h
6. j
7. m
8. c
9. o
10. e
11. b
12. d
13. a
14. l
15. g

Multiple Choice

1. a
2. c
3. e

4. b
5. e
6. e
7. a
8. e
9. b
10. e
11. b
12. c
13. a
14. c
15. e
16. b
17. d
18. b
19. d

6

Ecosystems and the Physical Environment

LEARNING OBJECTIVES

After you have studied this chapter you should be able to:

1. Diagram the carbon, nitrogen, sulfur, and hydrologic cycles.
2. Describe how humans have influenced the carbon, nitrogen, phosphorus, and sulfur cycles.
3. Summarize the effects of solar energy on the Earth's temperature, including the influence of albedos (reflectivity) of various surfaces.
4. Discuss the roles of solar energy and coriolis effect in the production of global air and water flow patterns.
5. Define El Niño-Southern Oscillation (ENSO) and La Niña and describe some of their effects.
6. Distinguish between weather and climate and give three causes of regional precipitation.
7. Contrast tornadoes and tropical cyclones.
8. Define plate tectonics and explain its relationship to earthquakes and volcanic eruptions.

SEEING THE FOREST

1. Biologists James Lovelock and Lynn Margulis proposed that because the Earth has homeostatic (self-correcting, balancing) systems, it can be considered a living organism with "geophysiology."

   Fill-in the table of six materials or environment conditions that appear to be homeostatically regulated by *Gaia*.

	Material or Condition
1.	
2.	
3.	
4.	
5.	
6.	

SEEING THE TREES

VOCAB

Nature is painting for us, day after day, pictures of infinite beauty if only we have the eyes to see them.
John Ruskin

1. *Gaia* hypothesis—the Earth behaving as a living organism, adjusting environmental parameters to maintain life.
2. Negative feedback loop—an increase or decrease of a material or environmental condition causes an integrating center (the Earth) to produce a response that will drive the input in the opposite direction.

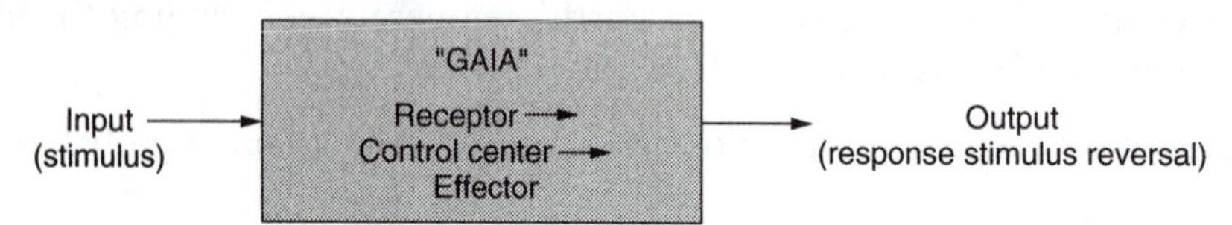

3. Biogeochemical cycles—the movement of materials from the biotic to the abiotic environment—interchanges among the biotic, geologic, atmospheric, and aquatic pools of materials.
4. Fossil fuels—geologic stores of carbon and carbon compounds that are used for energy production; the carbon atoms at one time were present in the biotic and atmospheric pools.
5. Combustion—the oxidation of carbon and carbon compounds to produce carbon dioxide and water as products:

$$C \text{ (carbon in coal)} + O_2 \rightarrow CO_2, \text{ or}$$

$$HxCx \text{ (hydrocarbons)} + O_2 \rightarrow CO_2 + H_2O$$

   Smoke and soot particles are unburned/incompletely combusted carbon or carbon compounds.

6. Nitrogen fixation—atmospheric nitrogen is converted to ammonia fertilizer by mutualistic rhizobia bacteria living on the roots of legumes and other plants. Nitrogen is moved from the atmospheric to the biotic pool.

7. Nitrification—the bacterial conversion of ammonia (NH_3) or ammonium (NH_4^+) ions to nitrites (NO_3^-) and on to nitrates (NO_3^-) that plants use to make proteins and other materials.
8. Ammonification—the conversion of proteins and other nitrogen-containing compounds to ammonia and ammonium ions by putrefying or ammonia bacteria.
9. Denitrification—the reduction of nitrate to gaseous nitrogen done by denitrifying bacteria.
10. Nitrogen oxides—N_2O (nitrous oxide), NO (nitric oxide) and NO_2 (nitrogen dioxide) formed in automobile engines and coal-burning power plants where nitrogen is combined at high temperatures and pressures with oxygen. Nitrogen oxides contribute to global warming, acid rain by forming HNO_3 nitric acid, and smog formation.
11. Transpiration—water loss from the leaves of plants. As water evaporates from the leaves, cohesive water molecules move up from the roots to the leaves.
12. Estuary—a place where the freshwater runoff from continents or islands mixes with ocean saltwater. The term *brackish* also refers to the mixing of fresh and saltwater.
13. Watershed—the area of land drained by the runoff of water.
14. Aerosols—particulate air pollution that includes fly ash, carbon, nitrates, sulfates, and mineral dusts. An increase in particulate air pollution reflects sunlight back into the upper atmosphere and space, resulting in an atmospheric cooling effect that, to some degree, counteracts global warming.
15. Albedo—the reflectivity of the Earth's surface.
16. Troposphere—the layer of the atmosphere that extends from the Earth's surface to an altitude of 10 km (6.2 miles).

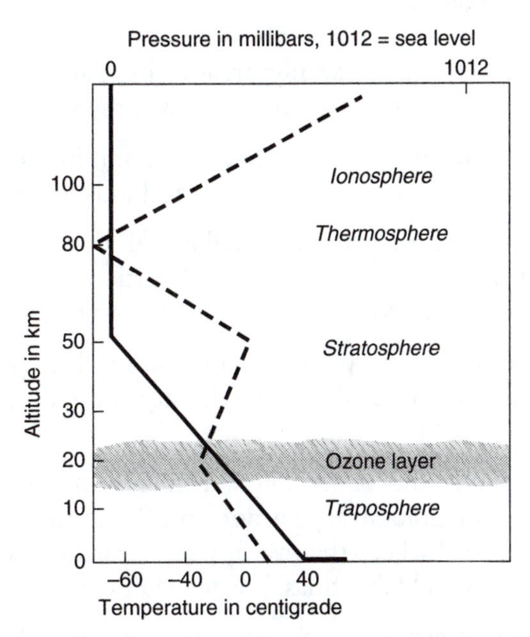

17. Temperature lapse rate—air temperature decreases about 6 °C/1 km increase in altitude (about 3 F°/1,000 ft. altitude if the tropospheric air is moist, and 5 °F/1,000 ft. if the air is dry).
18. Stratosphere—this layer extends from 10 km above the troposphere to an altitude of 45 km. The ozone (O_3) layer is found here.

19. Meosphere—the layer extends from 45 km of altitude to 80 km.

20. Thermosphere—from 80–500 km above the surface, temperatures steadily rise because of the X-ray and ultraviolet energy absorbed by gas molecules and reemitted as light (the auroras) or infrared (heat).

21. Exosphere—it extends from 500 km and merges with space.

22. Atmospheric circulation—air generally is heated at the equator, rises and cools, and falls back to the surface. At each increasing 30° level of latitude, a Hadley cell repeats that pattern.

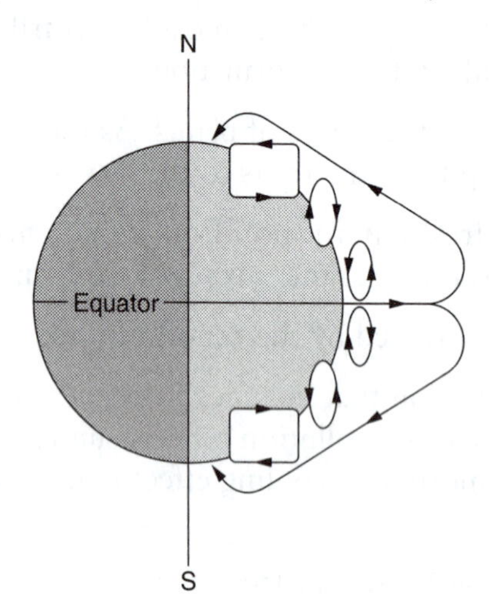

Wind circulation from the equator to the poles and in cells at 30 degrees of latitude from the equator.

23. Winds—because the Earth is heated unevenly, areas of high and low air pressure develop, with air moving horizontally from high pressure areas to low pressure areas.

24. Coriolis effect—the rotation of the Earth causes a deflection in the wind to the right in the Northern Hemisphere and to the left in the Southern Hemisphere.

25. Prevailing winds—the combined effects of the Hadley cell position and the coriolis effect produce winds that fairly consistently blow in a certain direction. Near the equator, the trade winds were used by the Spanish to communicate with the colonies in the Western Hemisphere.

26. Currents and gyres—patterns of atmospheric circulation produce parallel circulation patterns in the ocean's surface. A gyre is a circular flow. The density and temperature of seawater affects deep ocean currents.

27. El Niño/Southern Oscillation Event—a periodic warming of the surface waters of the tropical Pacific Ocean near the coast of Colombia, Ecuador, and Peru resulting from the weakening of the westerly trade winds that normally keep that warm water near Indonesia, the West Indies, and Australia. Because it occurs around Christmas, it is named in Spanish, after the boy Christ child.

28. ENSO effects—The warm surface water prevents the upwelling of nutrient-rich water that feed anchovies and other fishes that are economically important to the region. Air circulation storm and rainfall patterns are affected in the U.S., Canada, and globally because the position of the jet stream changes.

29. La Niña—surface water cools rapidly in the eastern Pacific Ocean, and the westerly trade winds

are strong. Because the conditions are opposite and represent a natural oscillation from El Niño conditions, the phenomenon is called La Niña, the little girl child.

30. Weather—atmospheric conditions at a specific place and time.

31. Climate—general meteorological conditions of a large geographical region over a large amount of time.

32. Rain shadow—when moist oceanic air moves over the Pacific Northwest coast in a prevailing easterly direction, it is forced to rise over the coastal mountain range. As the air cools, moisture condenses and produces large amounts of rain, like squeezing a sponge. On the eastern slopes, the air is much drier as a result. West Washington State is wet with dense coniferous forest vegetation, and east WA is drier semidesert. The dry area is the "rain shadow."

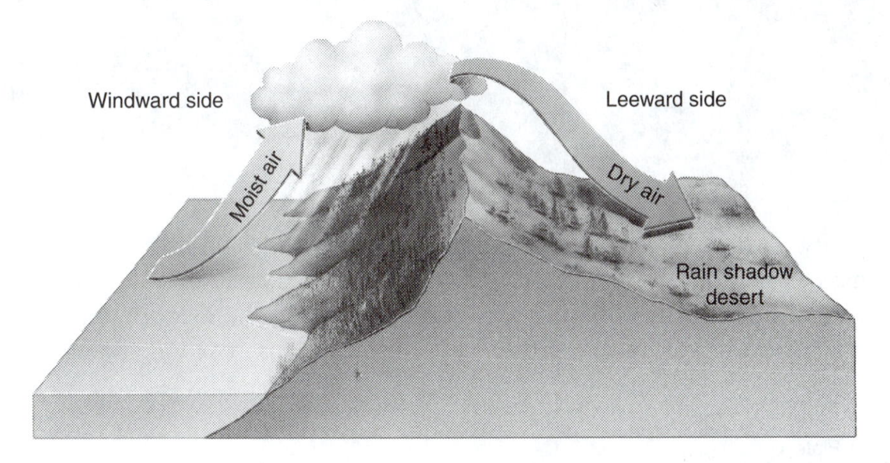

33. Tornado—a powerful rotating funnel of air at the trailing edge of a thunderstorm formed when a mass of cool air collides and displaces warm moist air.

34. Tropical cyclone—specifically in the Northern Hemisphere, a hurricane is a large area of low air pressure that circulates air counterclockwise towards its center or "eye." They form in the Atlantic Ocean off the coast of equatorial Africa and move west and north to near 30° latitude. High rainfall, tornadoes, and wind cause major damage to land masses.

35. Plate tectonics—the meteorologist Alfred Wegener proposed in 1912 that the continents of the Earth's crust drift on soft and molten rock (magma) of the mantle.

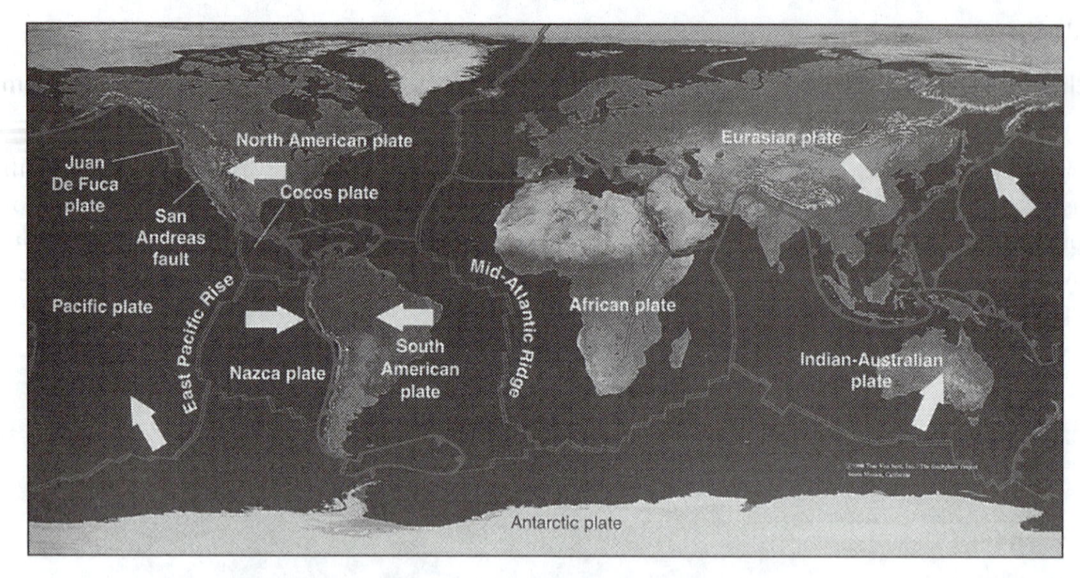

36. Volcanic activity—where plates collide and one subducts under another can produce volcanoes. The K-T mass extinction of organisms, including dinosaurs, may have happened as a result of the Indian subcontinent plate colliding with Asia, producing the Deccan volcanic shield. The solar-screening effect of volcanic dust then caused the collapse of food chains.

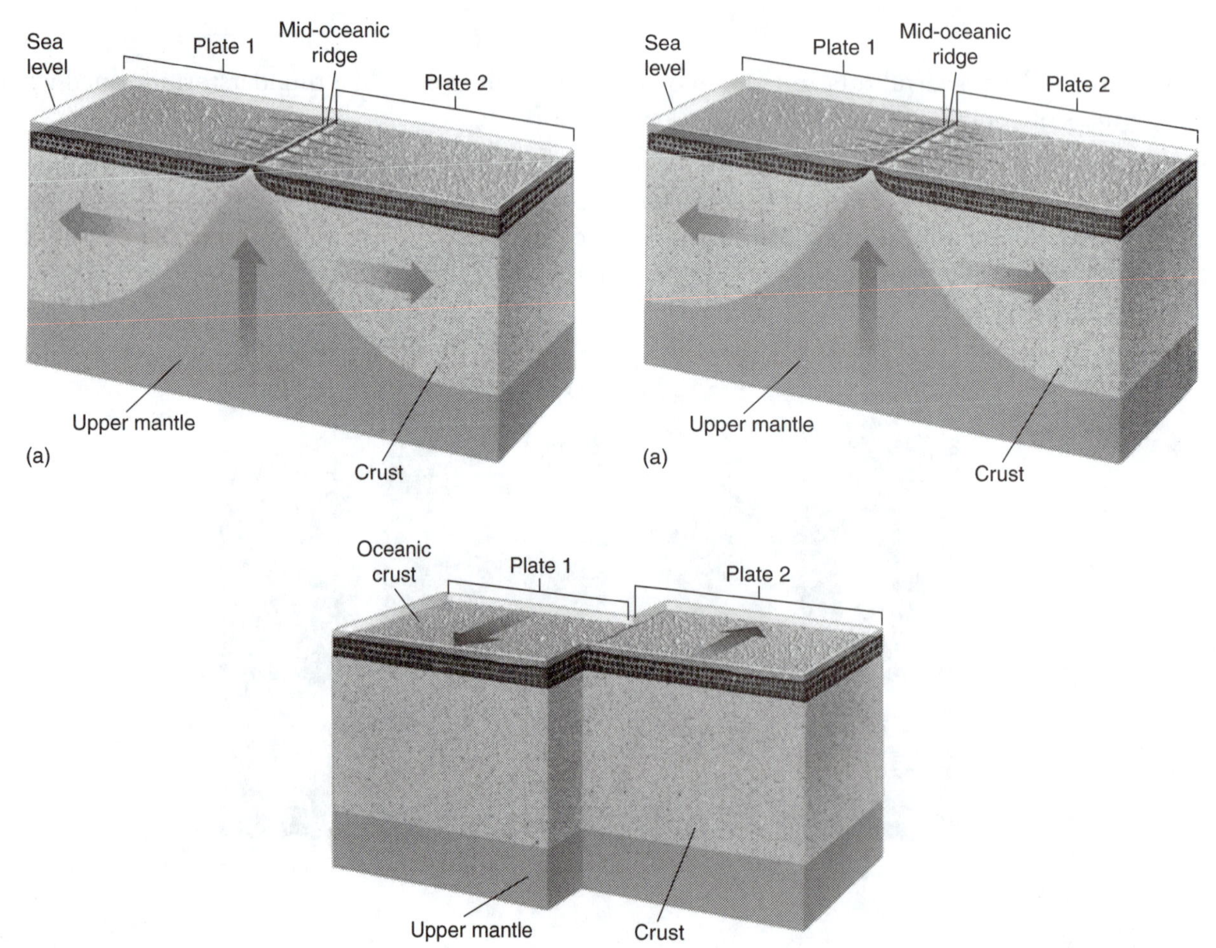

37. Earthquake—energy is stored in areas where there is subsurface contact between geological plates or in rock layers where the boundary rocks are moving in different directions. Fault lines are produced as a result—the San Andreas fault in California.

38. Richter scale—Charles Richter, in 1935, devised a scale of energy released by earthquakes. Each succeeding number represents 30 times the energy of the previous level.

Questions to Ponder

1. Fill-in the table below of acid rain pollutant gases, their acids, and sources.

Gas	Acid	Source of Polluting Gas
	Carbonic (H_2CO_3)	
		High-compression automobile engines: Notably high horsepower gasoline and diesel engines, high pressure coal furnaces
SO_2		

2. Indicate the effects of each process in the table below.

Process	Material(s) used up (reactants)	Materials(s) produced (products)
Photosynthesis	Carbon dioxide and water	
Chemosynthesis by sulfur bacteria		Sulfates (SO_4^-)
Decomposition of a log by heterotrophic bacteria		Carbon dioxide and water, _____ (minerals) released to the soil and humus.
Cellular respiration of animals		Carbon dioxide and water
Nitrogen fixation		
Putrefaction by ammonia bacteria		Ammonia fertilizer
Nitrification	Ammonia	
Denitrification		Nitrogen gas
Bacteria decomposition of sewage in a treatment plant		_____ + potassium, phosphorus, and nitrate released into "gray water."

3. Fill in the table below, representing the typical meteorological conditions at specific localities during certain meteorological events.

Location	El Niño or La Niña event?	Temperature—rate as Low, Moderate, High	Amount of precipitation—rate as Low, Moderate, High	Environmental effects
Indonesia	El Niño			Wildfires, drought
Indonesia	La Niña			
Coast of Ecuador, South America	El Niño			
Coast of Ecuador, South America	La Niña			
Northwest Pacific coast of North America	El Niño			
Northwest Pacific coast of North America	La Niña			
Pacific coast of southern California	El Niño			Mudslides
Pacific coast of southern California	La Niña			Wildfires and drought
North America and Europe	Explosion of the Indonesian island volcano of Krakatoa in 1883*			Sunamis and ______.

* See **www.drgeorgepc.com/Vocano1883Krakatoa.html**

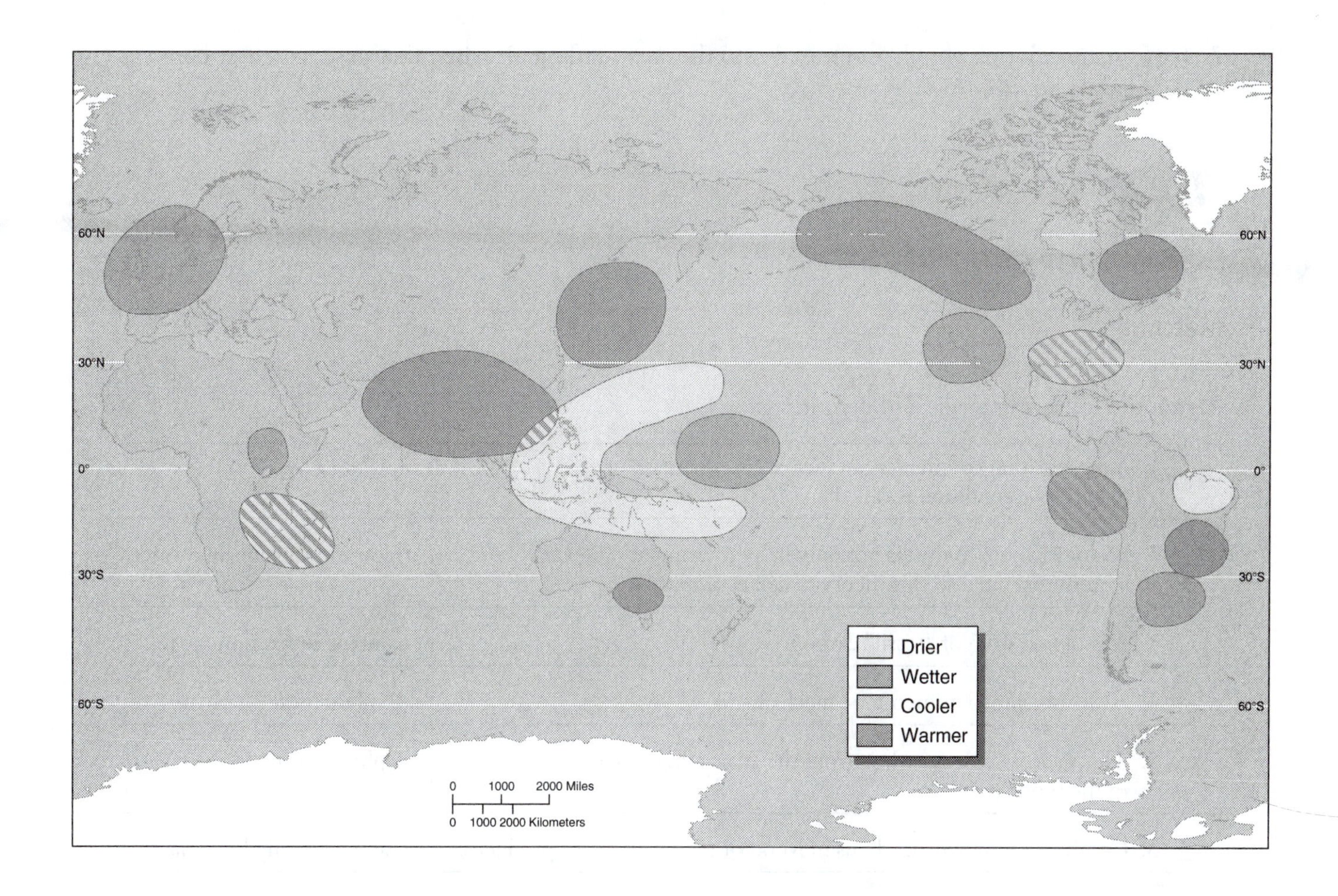

4. Compare the solar energy absorbed by a tropical rain forest as compared to a desert.
5. If 30 kg of fertilizer are required annually for each person in developed countries and 5 kg/capita in undeveloped countries, compare the requirements of the populations of Zambia and Georgia (U.S.).
6. How many calories of energy have to be added to water to transform 1 gram of liquid at 0°C to a gas at 100°C?

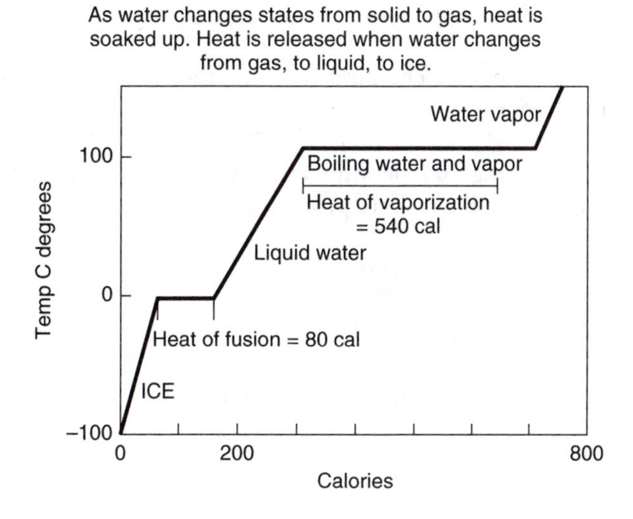

7. How many calories would be released into the surrounding air when that gaseous water condenses to a liquid?

POP QUIZ

Matching

Match the following terms and definitions

____1. A process that counteracts an increase or decrease of a material or condition is called:	a. ENSO
____2. Biologist Lynn Margulis not only helped formulate the *Gaia* hypothesis, but also this theory of cell evolution:	b. carbon dioxide and water
____3. A geologic store of "fossil" carbon:	c. nitrogen fixation
____4. *Gaia's* presumed homeostatic mechanisms:	d. subduction
____5. This process moves carbon from the atmospheric pool to the biotic pool:	e. phosphorus
____6. Combustion of hydrocarbons in wood or gasoline always produces:	f. ammonifying bacteria
____7. Nitrogen is needed by plants to make ________?	g. negative feedback homeostasis
____8. Process that splits nitrogen (N_2) molecules into separate atoms that are combined with hydrogen to make ammonia:	h. geophysiology
____9. Soybean plants make seeds high in protein because of this growing in their roots:	i. fault
____10. A mouse carcass that stinks is being attacked by ________?	j. coal
____11. Plants need this to make the energy chemical ATP and the chemical of inheritance DNA:	k. proteins
____12. When combustion occurs at high temperature and pressure, these noncarbon-containing pollutants are produced:	l. rhizobia bacteria
____13. Where two tectonic plates make contact on the land:	m. photosynthesis
____14. When one tectonic plate flows under another:	n. nitrogen oxides
____15. Periodic cooling of waters in the Indonesian Pacific Ocean:	o. endosymbiotic

Multiple Choice

1. In the theoretical scenario that the K-T mass extinction of dinosaurs was caused by a large meteor impact, which statement would be correct?
   a. Fires would have greatly increased carbon dioxide levels in the atmosphere.
   b. As the crust and mantle were displaced, volcanic action would have increased releasing more carbon dioxide into the atmosphere.
   c. Eventually, excess carbon dioxide would have favored photosynthetic organisms to do a negative feedback loop.
   d. Fires and volcanism would have increased solar radiation blocking dust and decreased surface temperatures for a few years.
   e. All of the above could occur.

2. Which process is correctly matched with its reactants or products?
   a. combustion/oxygen product
   b. photosynthesis/carbon dioxide reactant
   c. nitrogen fixation/nitrogen gas product
   d. denitrification/nitrogen gas reactant
   e. ammonification/ammonia reactant

3. Which of the processes below does not add to an increase of carbon dioxide to the atmosphere? See text Figure 6-2.
   a. cellular respiration
   b. decomposition
   c. photosynthesis
   d. the metabolism of heterotrophic bacteria
   e. combustion of fossil fuels

4. The industrial revolution and its aftermath led to:
   a. increased mining and drilling for fossil fuels
   b. increased carbon dioxide levels in the atmosphere
   c. increased sulfur levels in the soil globally
   d. a and b
   e. a, b, and c

5. Which of the following processes lead to an increase in atmospheric carbon?
   a. combustion of fossil fuels
   b. cutting of forests
   c. photosynthesis
   d. a and b only
   e. a, b, and c

6. Too much nitrogen in an aquatic environment leads to:
   a. algal bloom
   b. rotting algae
   c. increase in decomposing bacteria
   d. decreasing oxygen and fish kill
   e. all of the above

7. Which bacteria *complete* the process of making the nitrogen that is absorbed by most non-legume plant roots?
   a. ammonifying
   b. nitrite bacteria
   c. nitrate bacteria
   d. rhizobia
   e. denitrifying

8. As a result of farming and combustion, which solar-radiation-blocking aerosol pollutant has increased in the air?
   a. carbon in smoke
   b. sulfates
   c. mineral dust
   d. nitrates
   e. all of the above

9. Which factor below contributes to the low temperatures at the North Pole as compared to the equator?
   a. The North Pole is always further from the sun than the equator.
   b. Sunlight is reflected back into space by polar ice.
   c. The sun does not rise in December because of the Earth's annual 23.5° wobble.
   d. A given unit of solar radiation striking the atmosphere is spread out over a larger area than at the equator.
   e. All of the above are correct.

10. If global warming continues to melt the polar ice, this could result:
    a. A loss of albedo would create a positive feedback loop causing further increases in global temperature.
    b. Water would rise to cover the entire Earth.
    c. Surface temperatures would actually increase due to increased cloud formation.
    d. a and b only
    e. a, b, and c

11. For five years after Krakatoa blew up, Europe had lower crop yields. The reason was:
    a. rapidly increasing carbon dioxide levels
    b. solar radiation blocking dust
    c. acid rain
    d. an increase in infrared radiation
    e. all of the above

12. Which list of the layers of the atmosphere are in correct ascending or descending order?
    a. troposphere-mesosphere-exosphere-stratosphere-thermosphere
    b. ozone layer-troposphere-thermosphere-stratosphere-exosphere
    c. troposphere-stratosphere-mesosphere-thermosphere-exosphere
    d. exosphere-mesosphere-stratosphere- thermosphere-troposphere
    e. Trick question alert! None of these are correct.

13. The ultraviolet radiation screen, the ozone layer, is found here:
    a. troposphere

b. stratosphere
c. mesosphere
d. thermosphere
e. exosphere

14. Which statement is an *incorrect* interpretation of the figure below?
a. Sargassum seaweed communities are likely to flourish in the warm current of the southern gulf stream.
b. The warmer, less dense waters are more salty that the colder, denser waters.
c. Colder, denser water sinks.
d. Warmer, less dense water rises.
e. Cold polar ice and water from the Arctic Ocean near Greenland make the gulf stream water sink.

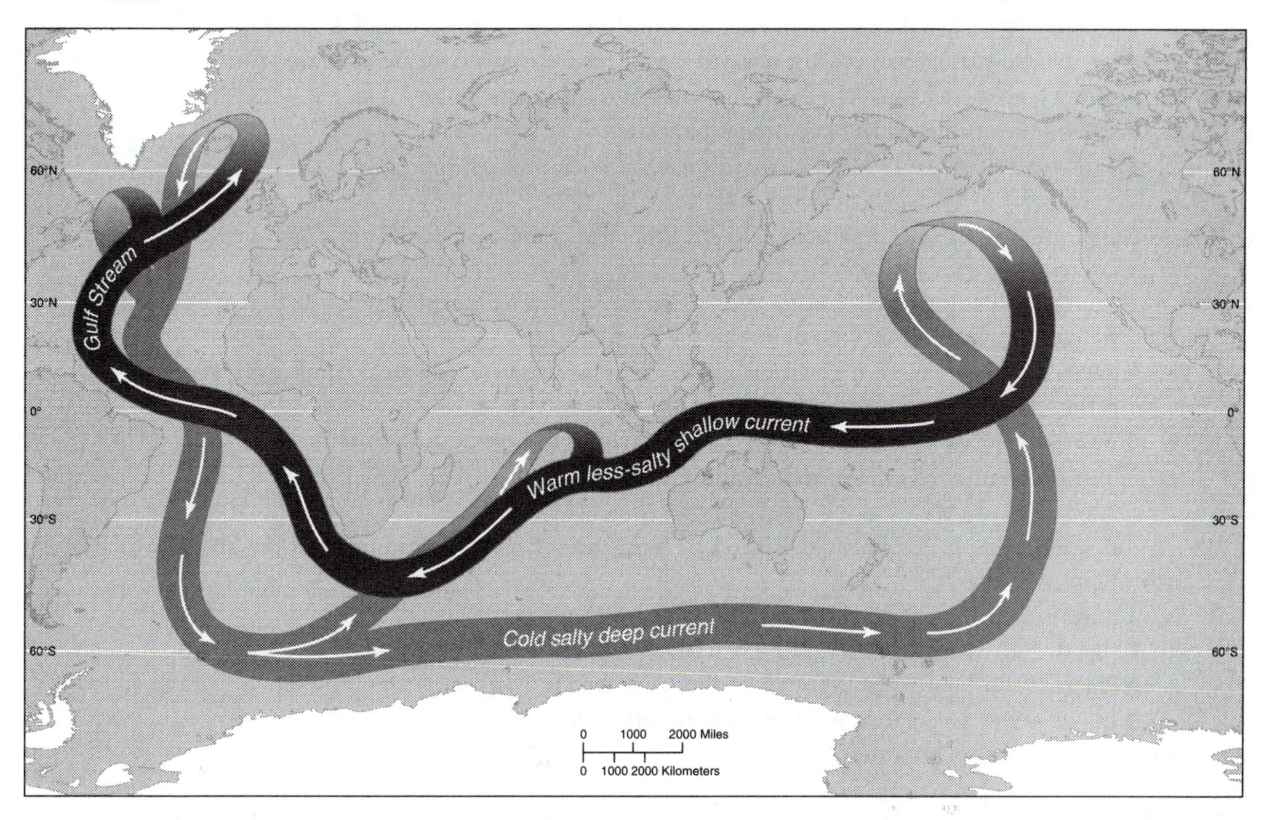

15. Increasingly, phosphate pollution in streams that contributes to algal blooms, comes from:
a. pig and chicken factory farming
b. coal-burning power plants
c. steel manufacturing
d. nuclear wastes
e. all of the above

16. Which gas(es) contributes to acid rain, respiratory diseases, global warming, and smog formation?
a. sulfur dioxide
b. carbon dioxide
c. nitrogen oxides

d. methane
e. ozone

17. The wind typically blows, at 2,500 feet above the surface, 45° to the right of the surface wind in the U.S. The reason why is:
a. jet stream effects
b. coriolis effect of the Earth's rotation
c. This is a trade wind deflecting toward the equator.
d. Hadley cell flow
e. all of the above

18. The 1906 earthquake at the San Andreas fault in San Francisco measured 8.3 on the Richter scale. Which of the following responses are correct about that event?
a. The energy released would have been 27,000 times more than a 4.3 level quake.
b. The Pacific plate was moving northwest and the North American plate southeast.
c. Energy was stored in at the boundary of the countermoving rocks.
d. Earthquakes occur as the boundary breaks.
e. All of the above are correct.

19. As moist air flows over the Rocky Mountains from west to east, climate and vegetation are affected in the following way:
a. The western slope of the mountain is drier than the eastern slope.
b. The eastern slope is wetter than the western slope.
c. As water changes from a gas to a liquid, heat is incorporated into the liquid, further decreasing the cooling temperature of the water.
d. The eastern slope is drier leading to semiarid grassland plains further east.
e. The rain shadow is on the western slope.

20. In the *Wizard of Oz*, Dorothy dreamed of being in a house rotating in a tornado. The reason why Kansas is the setting is:
a. There is lots of moist, warm air there, more than farther east.
b. Cold dry air from Canada collides with moist Gulf of Mexico air.
c. Isolated thunderstorms occur there because of the convective heating of the wheat fields.
d. The air there is warmer with increasing altitude.
e. All of the above are correct.

21. Pilots report that when they are flying low over fields they have updrafts and over forests downdrafts. Which statement accurately explains either phenomenon?
a. Radiation is being absorbed by photosynthesis.
b. The albedo of forests is higher.
c. The albedo of fields is lower.
d. Heat is reflected by the forest.
e. Transpiration in the field is high.

CYBER SURFIN'

1. For an informative overview of nitrogen fixation, go to the Biology Teaching Organization site at the University of Edinburgh site in the U.K.: **helios.bto.ed.ac.uk/bto/microbes/nitrogen.htm**
2. Tag along on a field trip to the San Andreas fault, courtesy of University of California Santa Cruz, Earth Sciences: **emerald.ucsc.edu/~es10/fieldtripSAF/SAF.html**
3. From the U.S. Geological Service, good information on earthquakes: **geology.er.usgs.gov/eastern/earthquakes**
4. For the exploration of "cyber scientists," the National Geographic offers *Riddles of a Changing Climate:* **tectonic.nationalgeographic.com/2000/physical/climate/main.html**
5. Go to the author's website for more on meteorology: **www.gpc.edu/~jaliff/airrb.htm**

'LIKE' BOOKING IT

1. Lutgens FK, Trabuck EJ, and Tasa D. ***The Atmosphere: An Introduction to Meteorology.*** 8th ed. Upper Saddle River, NJ: Prentice Hall, 2000.
2. Bonan GB. ***Ecological Climatology.*** Cambridge, U.K.: Cambridge University Press, 2002.
3. Walker BH, and Steffen W (Eds). ***Global Change and Terrestrial Ecosystems.*** Cambridge, U.K.: Cambridge University Press, 2002.
4. Doering OC, Randolph JC, Southworth J, Pfeiffer RA, and Kress MC (Eds). ***Effects of Climate Change and Variability on Agricultural Production Systems.*** Dordrecht, The Netherlands: Kluwer Academic Publishers, 2002.
5. Glantz MH. ***Currents of Change: Impacts of El Niño and La Niña on Climate and Society.*** Cambridge, U.K.: Cambridge University Press, 2001.

ANSWERS

Seeing the Forest

1. Fill-in the table of six materials or environment conditions that appear to be homeostatically regulated by *Gaia.*

	Material or Condition
1.	Temperature of the atmosphere and soil
2.	Rainfall
3.	Atmospheric oxygen levels
4.	Atmospheric carbon dioxide
5.	Populations of organisms
6.	Atmospheric and biotic nitrogen levels

Questions to Ponder (Many of these answers are examples of many possible responses, please don't memorize them.)

1. Fill-in the table below of acid rain pollutant gases, their acids, and sources.

Gas	Acid	Source of polluting gas
CO_2	Carbonic (H_2CO_3)	Combustion of fossil fuels, gasoline, kerosene, natural gas, wood
N_2O, NO, NO_2 (Nitrogen oxides)	*Nitric (HNO_3)*	High-compression automobile engines: notably high horsepower gasoline and diesel engines, high pressure coal furnaces
SO_2	*Sulfuric (H_2SO_4)*	*Sulfur-contaminated gasoline and coal*

2. Indicate the effects of each process in the table below.

Process	Material(s) used up (reactants)	Materials(s) produced (products)
Photosynthesis	Carbon dioxide and water	Glucose (basic food for cells) and oxygen
Chemosynthesis by sulfur bacteria	*Electrons from H_2S (hydrogen sulfide) and oxygen*	Sulfuric acid sulfates (SO_4^-)
Decomposition of a log by heterotrophic bacteria	*Carbohydrates (mostly)*	*Carbon dioxide and water, potassium, phosphorus, and nitrate (minerals) released to the soil and humus.*
Cellular respiration of animals	*Glucose (basic food for cells) and oxygen*	Carbon dioxide and water
Nitrogen fixation	*Atmospheric N_2*	*Ammonia (NH_3)*
Putrefaction by ammonia bacteria	Proteins	Ammonia fertilizer
Nitrification	Ammonia	*Nitrate fertilizer*
Denitrification	*Nitrates and ammonia*	Nitrogen gas
Bacteria decomposition of sewage in a treatment plant	*Carbohydrates, fats, and proteins*	*carbon dioxide* + potassium, phosphorus, and nitrate released into "gray water."

3. Fill in the table below, representing the typical meterorological conditions at specific localities during certain meteorological events.

Location	El Niño or La Niña event?	Temperature—rate as Low, Moderate, High	Amount of precipitation—rate as Low, Moderate, High	Environmental effects
Indonesia	El Niño	*High, +3 to –4 degrees above normal*	*Low*	Wildfires, drought
Indonesia	La Niña	*–3 to –4 degrees lower*	*High*	*Floods, mud slides in deforested areas*
Coast of Ecuador, South America	El Niño	*Higher*	*Low*	*Weak trade winds prevent upwelling of nutrient-rich water—commercial anchovy fishing devastated.*
Coast of Ecuador, South America	La Niña	*Lower*	*Higher*	*Floods, mud slides in deforested areas*
Northwest Pacific coast of North America	El Niño	*Higher*	*Lower*	*No appreciable effects*
Northwest Pacific coast of North America	La Niña	*Lower*	*Higher*	*Floods, mud slides in deforested areas*
Pacific coast of southern California	El Niño	*Lower*	*Higher*	*Increased snowfall in the central and southern Sierra and Rocky Mountains. Spring thaw floods, mud slides in deforested areas*
Pacific coast of southern California	La Niña	*Higher*	*Lower*	Wildfires and drought
North America and Europe	Explosion of the Indonesian island volcano of Krakatoa in 1883*	*Lower*	*Higher*	*Sunamis and increased snowfall in winter. Lower temperatures and shorter growing seasons—starvation.*

* See **www.drgeorgepc.com/Vocano1883Krakatoa.html**

4. Compare the solar energy absorbed by a tropical rain forest as compared to a desert. *The amount of radiation absorbed by plants to be used in the process of photosynthesis is significant, about 1–4.5% of the energy available to the photosynthetic organism. Because the plants are closely crowded together in the tropical rain forest and their equatorial position allows for high levels of radiation, the tropical rain forest absorbs a great deal of solar radiation that is translated into high primary production. The albedo (reflectivity) of the desert is about 50% higher than the tropical rain forest and plant totals are low, there is less energy trapped at or near the surface.*
5. If 30 kg of fertilizer are required annually for each person in developed countries and 5 kg/capita in undeveloped countries, compare the requirements of the populations of Zambia and Georgia (U.S.). *Zambia, in 2000, had a population of approximately 10 million and Georgia 8 million. Zambians would use 10 million × 5 lb = 50 million pounds of chemical fertilizer, whereas Georgians would use 8 million × 30 lb = 240 million pounds.*
6. How many calories of energy have to be added to water to transform 1 gram of liquid at 0° C to a gas at 100° C? *640 calories—100 calories to raise the temperature from 0° to 100° C, plus 540 calories heat of vaporization to break the hydrogen bonds between the water molecules.*
7. How many calories would be released into the surrounding air when that gaseous water condenses to a liquid? *540 calories of heat would be released to the surrounding air.*

Pop Quiz

Matching

1. g
2. o
3. j
4. h
5. m
6. b
7. k
8. c
9. l
10. f
11. e
12. n
13. i
14. d
15. a

Multiple Choice

1. e
2. b
3. c
4. e
5. d
6. e
7. c
8. e

9. e
10. a
11. b
12. c
13. b
14. b
15. a
16. c
17. b
18. e
19. d
20. b
21. a

7

Major Ecosystems of the World

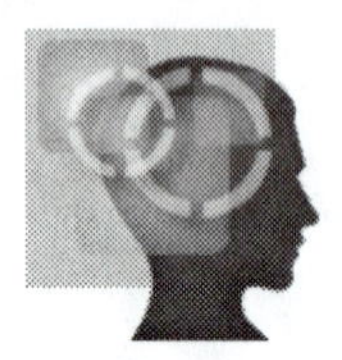

LEARNING OBJECTIVES

After you have studied this chapter you should be able to:

1. Define a biome and discuss how biomes are related to climate.
2. Explain the similarities and the changes in vegetation observed with increasing elevation and increasing latitude.
3. Briefly describe the nine major terrestrial biomes, giving attention to the climate, soil, and characteristic plants and animals.
4. Relate at least one human effect on each of the biomes discussed.
5. Summarize the important environmental factors that affect aquatic ecosystems.
6. Briefly describe the various freshwater, estuarine, and marine ecosystems giving attention to the environmental characteristics and representative organisms of each.
7. Relate at least one human effect in each of the aquatic ecosystems discussed.
8. Outline the environmental history of the Florida Everglades.

SEEING THE FOREST

Lightning strike fires can maintain ecosystems, such as the chapparal and dwarf forests of the U.S. west and the "piney woods" of the eastern U.S. coastal plain. Managed fires and wildfires started by humans also change the environment.

a. Fill-in the table of four effects of fires on the environment and indicate whether each is positive or negative.

Fire effect on the environment	Positive or negative?
1.	
2.	
3.	
4.	

b. Explain how the piney woods/pine barrens of the Atlantic coastal plain of the U.S. (New Jersey and southward) are maintained by periodic fires.

SEEING THE TREES

VOCAB

When one tugs at a single thing in nature, he finds it attached to the rest of the world.
John Muir

1. Natural fire—the combustion of plant biomass as a result of lightning strikes. Biomes with dry seasons are typically fire adapted or maintained. Prescribed or managed fires are ideally conducted under low temperature and higher moisture conditions that releases minerals in the soil—potassium in potash. Many wildfires are high-temperature "crown fires" where the tops of the trees catch fire and minerals may be vaporized.
2. Biome—a large geographical region that has similar climate, soil, and communities of plants and animals.
3. Vertical and latitudinal zonation—the changes of biomes that occur with increasing altitude are similar to those that occur as one advances toward the poles—tundra is found at the tops of the highest Rocky Mountains as it is found in northern Alaska and Canada.

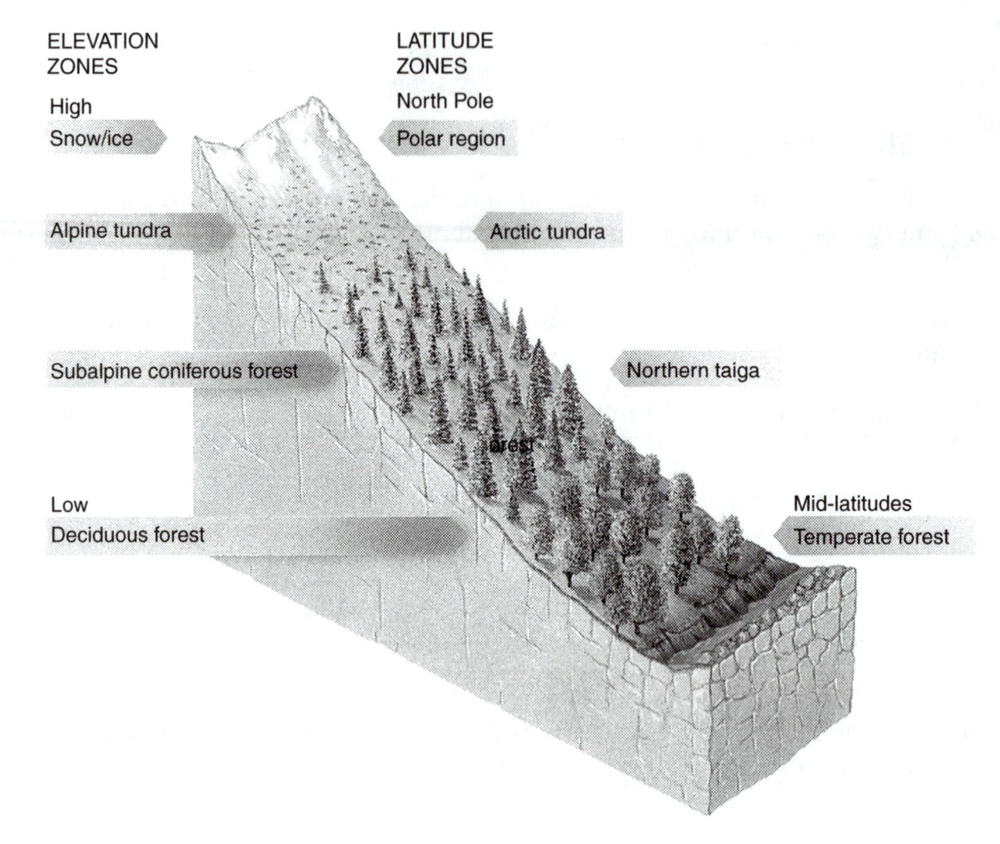

4. Tundra—a biome that occurs at the tops of the northern Appalachian and highest Rocky Mountains. The Arctic tundra is located near the Arctic Circle worldwide. It consists of lichens, mosses, grasses, sedges, and in sheltered areas, dwarf willows. In the brief summer, the surface is waterlogged. The rest of the year, the cold dry air produces precipitation amounts similar to a semiarid grassland. "Boom and bust" cycles of population numbers are seen.
5. Permafrost—a layer of permanently frozen soil beneath a surface that thaws periodically. The frozen soil cannot be penetrated by roots, therefore plants that grow there are adapted for shallow rooting, grow close to the surface; none are taller than 30 cm.
6. Taiga or boreal forest—covers most of Northern Canada, Asia, Europe, and some higher altitude regions of the U.S. This biome is characterized by precipitation of around 80 cm per year, high snowfall, very cold and long winters, short warm summers, and dark needle-matted forest floors. Coniferous evergreen plants have evergreen needles instead of deciduous leaves; the needles do photosynthesis through the winter when enough light is present. Needles have a natural antifreeze and small stomates (air openings) to counteract the freezing. Thick mats of spruce needles decompose slowly to produce acidic soils in the northern areas of the taiga. Especially hard or long winters may occur and generally long periods of time are required for the decomposition and recycling of detritus. Therefore, "boom-and-bust" cycles of the population levels of predators (such as lynxes, their snowshoe hare prey, or snowy owls and lemmings in the tundra) are well documented.
7. Temperate rain forest—This biome is found on the west coast of southern Alaska, British Columbia and the Olympic Peninsula of Washington State. Rainfall amounts to over 200 cm/year. The forests are dominated by giant conifers such as sitka spruce, giant cedars, hemlock, and grand firs. The forest is very thick, and the floor is quite dark in most places. There are mosses covering tree branches and ferns on the forest floor.

8. Temperate deciduous forest—Although there is considerable variation in this biome, the typical eastern deciduous forest is in Appalachia. It features a high rainfall (around 120 cm per year), warm summers, cool winters, and is dominated by maples, hickories, oaks, poplars, and at one time, elms (until Dutch elm fungus got most of them). In the fall, brilliant displays of colors occur as the leaves die and the trees enter winter dormancy. Sugar maple leaves turn red as their chlorophyll is enzymatically destroyed, displaying the red and yellow accessory pigments. Animals include squirrels, white-tailed deer, black bear, blue jays, and crow. Soils of the deciduous forest are typically deep and dark and useful in agriculture.

9. Temperate grassland—the "breadbasket" of the U.S. and Canada is characterized by approximately 30 cm of rain per year and cold winters. As one proceeds west from the Mississippi River, the grasslands have progressively shorter grasses (tall grasses changing to short bunch grasses) because the rainfall decreases westward toward the Rocky Mountains. Virgin grassland is rare. This biome is naturally maintained by lightning-started fires. Typical animals include large grazers such as buffalo and antelopes, and birds like grouse, prairie chickens, and meadow larks. The sparse rainfall there causes minerals and nutrients to leach downward only a short distance (a foot or more) where the grass root systems can pick them up and recycle them. It is typical to see a white layer of calcium and other minerals lying in a band at root level (A–B horizons). The cold winters preserve the organic nutrients in the frozen humus. Cultivated grasses such as corn, wheat, and barley do best as crops. The soil is generally deep and dark with a large amount of humus.

10. Chaparral/dwarf forest—chaparral or pinyon/juniper forest is characterized by short to medium-sized evergreen shrubs (the manzanita in California). In many localities around the world, there is considerable variation of the species of dominant trees or shrubs, but their 6–12 foot height is fairly uniform, along with an annual precipitation of about 45 cm per year. Sometimes this biome is called "dwarf or pigmy forest." Many plants of the chaparral are resinous, and their debris/litter builds up over time and burns readily. Some trees and shrubs depend on periodic fires (approximately every 15 years) to pop seeds out of cones or pods so that the dwarf forest can rejuvenate. In Utah, the pinyon pine produces three large, teardrop-shaped seeds/nuts that are still roasted for food. Native Americans enrich their breads by roasting the pine nuts to drive out the resins and turpentines; grinding to make a high-protein, vitamin and fat meal; and using that meal to enrich the low-protein, low-vitamin commercial flour available in stores. 6-foot tall dwarf gambel oaks are also seen. Typical animals include the mule deer, wood rat, jackrabbits, and many birds including towhees and jays.

11. Deserts—Deserts are ecosystems in which annual rainfall averages under 20 cm. There is usually a large difference in high day and low night temperatures (50°F or more). This difference will be less on cloudy days because the clouds will trap infrared radiation (heat) typically radiated back into space on cloudless nights. True deserts, like the famous Sahara Desert of Africa, have under 2 cm of rainfall/year and are usually dominated by sand or poorly weathered rock. Fewer plants or animals are found here because the sparse soil is high in minerals, but low in humus. Arizona is covered by desert and semidesert. You notice that in the lower altitudes south of Phoenix (about 1,000 feet above sea level), vegetation is sparse because of higher temperatures and less rainfall. As one travels northward and the altitude increases, more scrubs and palo verde trees appear. Deserts are also classified as to temperature and vegetation. A cool desert would include the sagebrush deserts of western Washington State (near Hanford), Idaho, Utah, Wyoming, and Colorado. Rainfall is low, but temperatures are moderate. Cool temperatures below freezing frequently occur in the winters, and there are few days with above 100°F temperatures in the summer. Further south in Arizona, warm deserts occur. Here the winter

temperatures rarely fall below freezing, and many summer days have temperatures above 100°F. In Phoenix in 1989, there were over 140 days with 100°F+ temperatures.

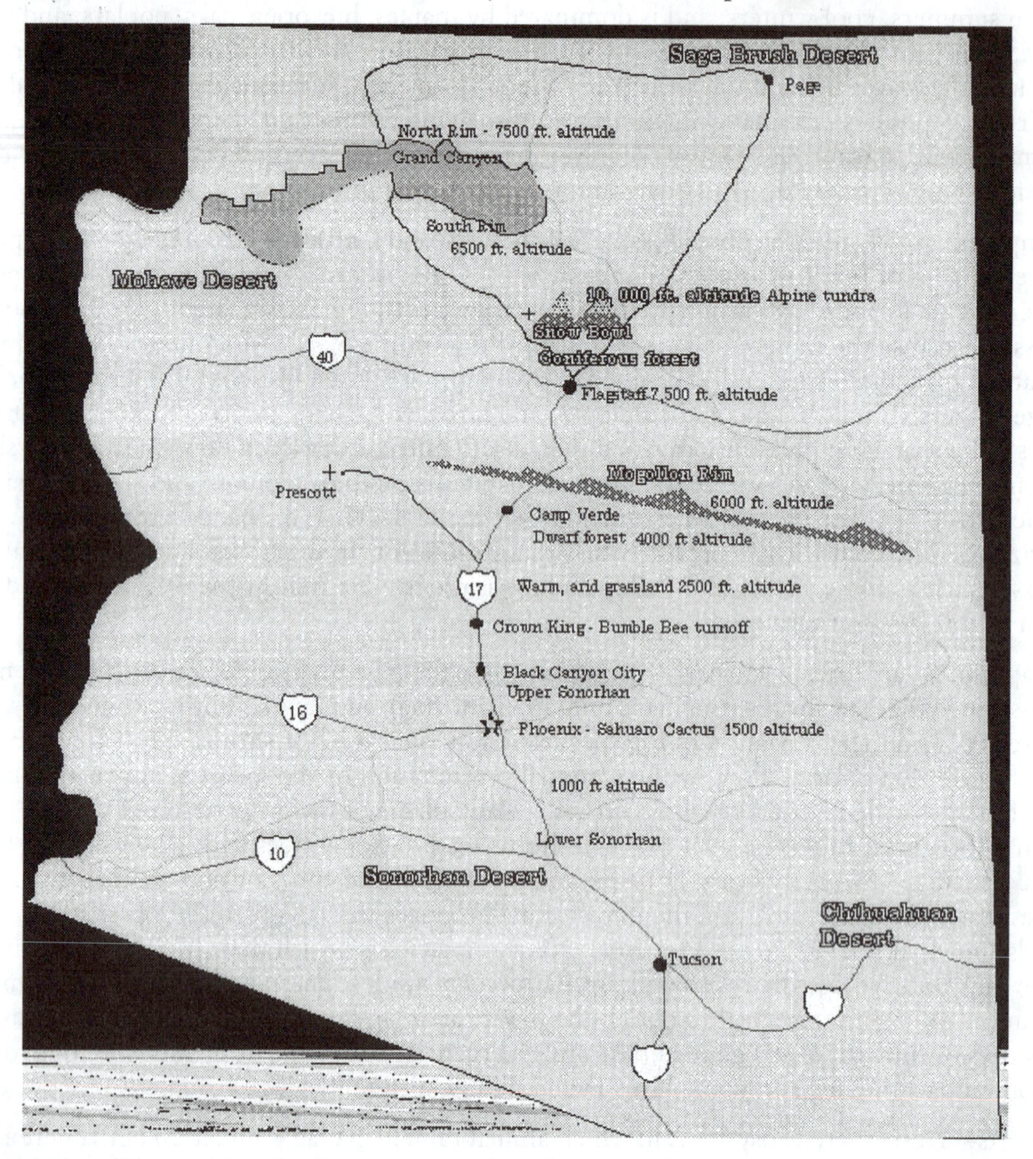

12. Tropical deciduous forest—This is found in parts of Central America and the Indian subcontinent where the temperatures are warm, but there is at least one dry season. Leaves are shed then. The wet season is called a monsoon. Rainfall varies between 85–150 cm annually. There are several subtypes of dry broadleaf forests including, from wetter to drier, semievergreen broadleaf, tropical deciduous broadleaf, and thornscrub.
13. Tropical savannah or grassland—This drier subtropical biome has about 90 cm of rainfall per year and pronounced dry seasons. There are small areas of tropical savannah in Central and South America. The most famous tropical savannah is the Serengeti Plain in Africa. Millions of antelope, giraffes, and wildebeests are preyed upon by lions, leopards, wild dogs, and hyenas. The scarce trees include acacias that are favored by baboons, and baobabs, that elephants tear up for stored water during droughts.
14. Tropical rain forest—The tropical rain forest is found in equatorial Central America and South America, Africa, Asia, Indonesia, New Guinea, and other South Pacific islands. Rainfall exceeds

300 cm per year. Tropical rain forests have thin soil but the tallest trees overall of any biome. Most of the nutrients of the ecosystem are tied up in growth above the ground. Hundreds of plants/trees compete for sunlight (strong year round at the equator), and the forest floor is quite dark. Species diversity—the greatest of any biome—occurs more in the tropical rain forest than in any biome so it is hard to list typical plants, but in South America trees include mahogany, rosewood, purple heart, Spanish cedar, and rubber. Large strangler fig (a woody vine) may grow completely around the lower trunk of a larger tree. Animals include crocodiles, jaguars, vine snakes, frogs, insects, monkeys, and brightly colored birds, such as parrots and toucans. Natives traditionally use "slash-and-burn" agriculture but crops will only be productive for three years or so and then the heavy rains will have washed away the thin soil. A thick, hard crust of a toxic aluminum alkaline clay called *laterite* underlies much of the thin rain forest soil.

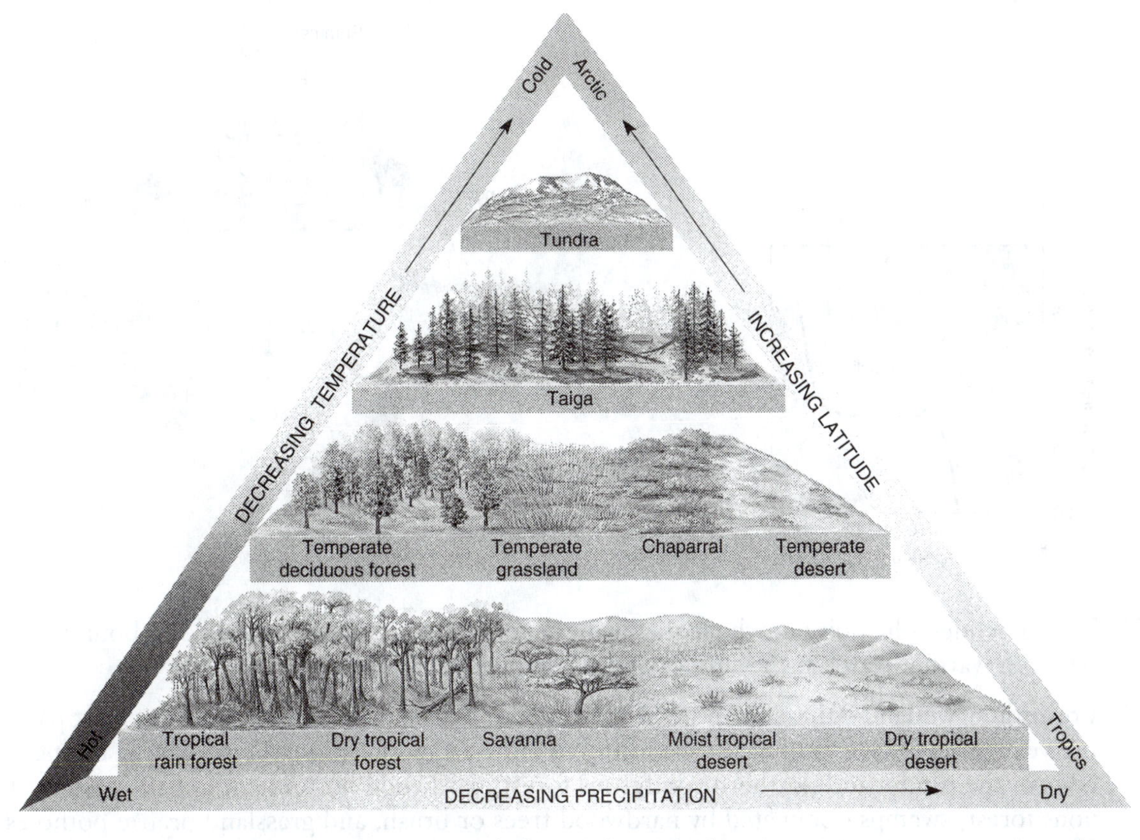

15. Plankton—microscopic cyanobacteria and algae (phytoplankton), protists, microscopic arthropods like water fleas (copepods and amphipods) and shrimps, arthropod larvae, and echinoderm larvae (collectively zooplankton) that drift in currents.
16. Nekton—strongly swimming organisms, such as bony fishes, squids, octopus, rays, sharks, turtles, and whales.
17. Benthos—bottom dwellers that are sessile (attached to the substrate) are sea anemones, corals, sponges, clams, and barnacles; those that burrow are tube worms, sand shrimp, and sea cucumbers; and those that walk on the sea floor are sea urchins, starfish, crawfish and lobsters.
18. Flowing water freshwater (riparian) ecosystem—a stream or river that varies greatly in the organisms present according to the speed of water flow, turbulence, nutrients present, and oxygen content.

19. Standing freshwater ecosystem—a pond, lake, or impoundment that consists of a littoral zone or shore where nutrients flow into the ecosystem, a limnetic zone of open water where plankton occur and do photosynthesis, and a profundal zone where light does not penetrate—bacterial decomposition occurs there.

20. Thermal stratification—because water is most dense at 4°C and therefore sinks to the bottom, there are fall and spring turnovers in lakes of the northern temperate and subarctic zones. The fall turnover occurs as surface water is cooled and sinks. The spring turnover occurs as ice (which is less dense than liquid water) becomes more dense as it melts and sinks. The sinking of water pushes less dense and warmer water up from the bottom. In this way, the bottom receives oxygenated water that the bacteria need, and the surface water receives minerals made available in the process of decomposition.

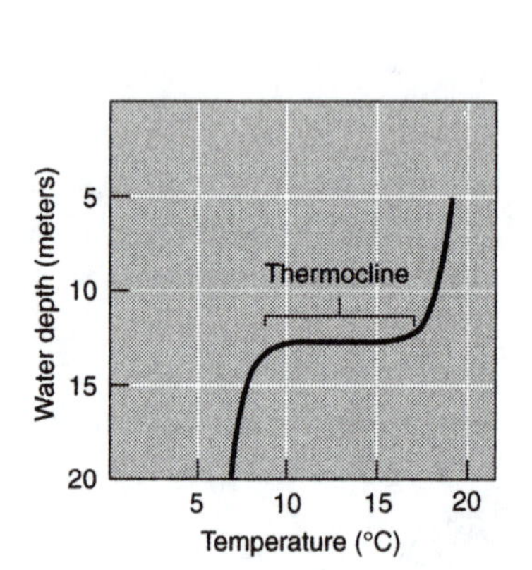

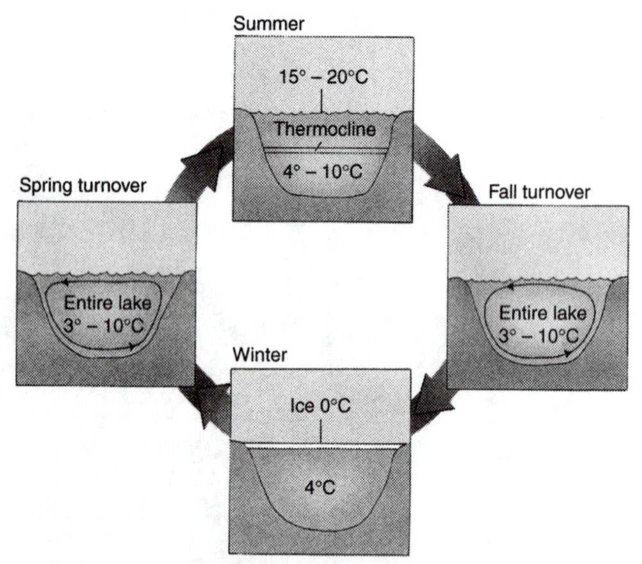

21. Thermocline—the point at which an abrupt change of temperature occurs or a boundary between warmer water at the surface and colder, deeper water.

22. Freshwater wetland—these ecosystems are covered by shallow standing water at least part of the year. They vary considerably according to latitude and location. They include the peat moss bogs of the tundra and northern coniferous forest, hardwood bottomlands of the eastern deciduous forest, swamps dominated by hardwood trees or brush, and grassland prairie potholes. Marshlands—both freshwater and saltwater—catch organic wastes and debris and turn them into rich soil and plant growth. Freshwater marshlands also charge groundwater aquifers. These benefits are called *natural ecosystem services.*

23. Estuary—an area where ocean water mixes with freshwater (brackish) that includes shores, mud flats, and tidal estuaries that include temperate salt marshes and tropical mangrove forests. They are highly productive, nurturing many commercially valuable organisms such as larval shrimp, oysters, clams, crabs, and flounder. They provide the same ecosystem services as freshwater marshes. In those areas where they have been filled in for real estate developments, the fishing industry has suffered or disappeared. The deed given to the colony of Georgia (U.S.) by the King of England gave "all the land between the tides" to the state, thereby allowing for salt marsh preservation.

24. Decomposition food chain (review)—detritus (decomposed debris) and recycled nutrients liberated by bacteria and fungi form a decomposition food chain. Grazing or filter-feeding organisms, such as shrimp larvae and oysters, feed on grass detritus or on the algae that grow from

the recycled nitrates, phosphates, and other minerals made available by decomposer bacteria and fungi.

25. Intertidal zone—the area of shore between high tide and low tide, consisting of sandy and rocky shores. Burrowing creatures (such as mole crabs, hermit crabs, snails, coquina clams, tube and sand worms) occupy sandy shores. Rocky shores offer places for brown algae (rockweed), mussels, and barnacles to anchor. The rockweed are not easily broken by wave action, but when fragmented, the broken pieces can regenerate asexually.
26. The oceanic benthic ecosystem—the ocean floor ecosystems are most diverse and productive in shallow waters where there are lots of surfaces or substrates for attachment. Below the area that light can penetrate, the abyssal and hadal benthic ecosystems depend on decomposition of detritus or marine snow falling from above.
27. The neritic province—the coastal or continental shelf zone extends roughly 300 miles from shore and is 200 m deep. It includes sandy bottom, sea grass beds, kelp forests, and coral reefs. The sea grass and kelp form the foundation of a decomposition food chain. Because of the light available and the bottom substrate on which to attach, these areas are much more productive than the open sea (pelagic zone) further out, the abyssal benthic zone (4,000–6,000 m depth) or the hadal benthic zone deeper. The coastal shelf lies within the depth of water penetrated by light, the euphotic zone.
28. Coral reefs—reefs are built up by the calcium carbonate skeletons of coelenterates (polyps) that are filter feeders of plankton. They include fringing reefs that are directly attached to a coast or island, atolls that circle an island and form a central lagoon of standing water, and barrier reefs that form lagoons, paralleling the coastlines of continents. Corals also harbor mutualistic algae (zooanthellae) that do photosynthesis. Only coral reefs with zooanthellae will produce skeletons. Many fishes find hiding places among the crevasses among the coral skeletons. The fish populations are normally high around reefs. The state of Florida sinks old ships and cars to begin reef formation. In Australia's Great Barrier Reef, the crown of thorns starfish has devoured large areas of the highly populated reef at times.
29. The abyssal and hadal benthic ecosystems—many creatures of the abyssal and hadal zones can make their own light from ATP (bioluminescence). Many bottom dwellers here feed on marine snow (falling detritus from above), and a decomposition food chain supports a small community of worms, squids, sharks, and fishes. Chemosynthetic and heterotrophic bacteria are plentiful in the sediments. You may have seen the "ironsicles"—the downward streaming rust columns of the sunken *Titanic*—that contain chemosynthetic iron bacteria.

Questions to Ponder

1. In the southeastern U.S., it is said that driving 350 miles north (at a similar altitude) is roughly equivalent to increasing altitude by 1,000 feet ascending the Appalachian Mountains. At an elevation of 3,500 feet and higher in North Carolina, one encounters increasing numbers of spruce trees similar to that encountered in northern Michigan or central Ontario, as well as similar temperatures and rainfall. Explain this phenomenon.
2. For the biomes and ecosystems below, fill in the table with information on the biomes' geographical location, climate (temperature and moisture), and dominant plants and animals. Rate temperature as very cold, cold, temperate, or warm (tropical), or hot. Rate moisture as annually very dry, dry, moderate, wet, or very wet; or periodically very dry and very wet.

Biome/ ecosystem	Location	Climate temperature	Climate moisture	Dominant plants	Dominant animals
Tropical rain forest					
Tropical deciduous forest					
Temperate rain forest (coniferous)					
Deciduous forest					
Taiga					
Chaparral					
Temperate grassland					
Tundra					
Tropical savannah					
Desert (Sahara)					
Semidesert (Phoenix, Arizona)					

3. Describe one major human effect on the biomes and water ecosystems listed below.

Biome	Cause	Effect
Tropical rain forest		
Freshwater flowing ecosystems		
Standing freshwater ecosystems (wetlands)		
Brackish estuaries		
Arctic tundra		
Taiga		
Chaparral		
Desert		
Eastern deciduous forest		

4. Offer three reasons why plants taller than 30 cm cannot grow in the tundra.

   1. ______________________________
   2. ______________________________
   3. ______________________________

5. Do some research and come up with five conditions that cause coral reef destruction.

   1. ______________________________
   2. ______________________________
   3. ______________________________
   4. ______________________________
   5. ______________________________

6. Explain why algal blooms may occur in spring and fall turnovers. What specifically is made available to the algae, and where does it come from?
7. Explain how are salt marshes and mangrove swamps important economic resources for the temperate and tropical Atlantic coast of North America.
8. Offer three environmental stresses threatening the Okefenokee Swamp of Georgia and the Florida Everglades.

   1. ______________________________
   2. ______________________________
   3. ______________________________

POP QUIZ

Matching

Match the following terms and definitions

1. A large geographical area of similar climate, vegetation and animal communities:	a. substrate present
2. Periodic fires maintain these biomes:	b. northern coniferous forest
3. A layer of permanently frozen soil:	c. atoll
4. The *spruce-moose* biome:	d. decomposition of grasses

5. These biomes have pronounced wet and dry seasons:	e. biome
6. When large amounts of moisture are available annually, these biomes result:	f. recycling minerals and organic wastes into plant growth
7. In the "bread basket" of North America, corn and wheat grow well because:	g. consistently high precipitation and sunlight
8. The diversity of the tropical rain forest is the greatest of biomes for this reason:	h. water is most dense at 4°C
9. The neritic continental shelf is more productive than the open pelagic ocean due to:	i. grasslands, chaparral
10. Abyssal and hadal benthic zones food chains are supported by:	j. nitrate, phosphate, potassium
11. Freshwater marshes and brackish water estuaries have food webs based on:	k. permafrost
12. A reef that encircles an island with a lagoon enclosed:	l. tropical savannah and tropical deciduous forest
13. Benthic bacteria decompose detritus from above, releasing this into the water:	m. sparse rainfall washes minerals to root level
14. The reason why lake turnovers occur:	n. marine snow
15. An ecosystem service of swamps, bogs and marshes:	o. temperate and tropical rain forests

Multiple Choice

1. Which statement below is correct in describing the role of fire in nature?
   a. Low temperature fires are needed for some seeds to generate in the chaparral.
   b. Low temperature fires return minerals to the soil.
   c. Lightning-strike fires prevent the "piney woods" central and south Atlantic coastal plain from becoming dominated by hardwoods.
   d. All natural and human-started fires are very damaging to trees.
   e. a, b, and c

2. The most important factor in determining the numbers and diversity of the populations of the biota of a temperate or tropical biome is:
   a. temperature
   b. moisture
   c. longitude
   d. permafrost
   e. All of these have equal importance.

3. Which of the following biomes have permanent permafrost throughout their range?
   a. alpine tundra
   b. taiga
   c. Arctic tundra
   d. any place with an altitude of greater that 600 ft. above sea level
   e. the North Pole

4. Which biome generally has a deep layer of nutrient-rich soil, plentiful humus with clay subsoil?
   a. tropical rain forest
   b. temperate grassland
   c. northern coniferous forest
   d. temperate deciduous forest
   e. tundra

5. Which biome is best for growing wheat?
   a. tropical rain forest
   b. temperate grassland
   c. northern coniferous forest
   d. temperate deciduous forest
   e. chaparral

6. Which biome is called a "dwarf forest" of trees and resinous shrubs renewed by periodic fires?
   a. tropical rain forest
   b. temperate grassland
   c. tropical savannah
   d. temperate deciduous forest
   e. chaparral

7. Long-term irrigation to raise crops in the desert is difficult for what reason?
   a. This top soil is washed away.
   b. Evaporation of surface water leaves salts behind.
   c. Desert soils are mineral poor.
   d. Few plants can tolerate the high temperatures.
   e. All of the above are correct.

8. Which of the following characteristics would enhance the survival of plants in the warm desert?
   a. hydrophilic proteins (aloe vera) that bind water
   b. antibiotics that inhibit competition for the sparse humus in soil and water
   c. very small leaves
   d. large body volume with a small surface for transpiration (cacti)
   e. all of the above

9. The famous Serengeti Plain in Africa that is populated by wildebeest, zebras, and lions is in which biome?
   a. tropical grassland with pronounced wet and dry seasons
   b. tropical rain forest with no wet or dry season
   c. tropical deciduous forest with no wet or dry season
   d. temperate grassland
   e. desert

10. This biome has the most diversity but thinnest soil:
    a. tropical savannah
    b. tropical rain forest
    c. tropical deciduous forest
    d. temperate grassland
    e. deciduous forest

11. The tropical rain forest is dominated by a thick growth of trees hundreds of feet above the soil. The best explanation for this is:
    a. competition for sunlight
    b. competition for rain
    c. it is cooler at the top of the canopy
    d. there is more water at the top of the canopy for the roots
    e. all of the above

12. Which ranking below best represented the annually driest to the wettest biomes?
    a. desert-deciduous forest-tundra-taiga-temperate rain forest
    b. desert-tundra-temperate deciduous forest-tropical deciduous forest
    c. tropical deciduous forest-temperate grassland-deciduous forest-tundra
    d. tropical deciduous forest-tropical savannah-temperate grassland-tropical rain forest
    e. Trick question alert! None of these are accurate.

13. The tropical rain forest offers this important ecosystem service:
    a. great conditions for sustained agriculture
    b. great plant diversity for therapeutic drug prospecting
    c. homeostatically balancing carbon dioxide levels
    d. moderating climate
    e. all excepting a

14. Which statement is an accurate description of the ecosystem services of a wetland?
    a. Groundwater aquifers are charged in freshwater habitats.
    b. Bacterial and fungal decomposition of grasses nourishes algae that form the foundation of an aquatic food chain.
    c. Nutrients from sewage and agricultural runoff are recycled into new plant growth.
    d. Habitat for waterfowl is provided for the aesthetic enjoyment by ecotourists and poets.
    e. All of the above are correct.

15. Headwaters of riparian flowing-water ecosystems are characterized by:
    a. high levels of oxygen
    b. slow water flow
    c. high turbidity
    d. warm water
    e. all of the above

16. The profundal zone of lakes and the abyssal and hadal zones of the ocean have this in common:
    a. marine snow
    b. high levels of oxygen

c. some light at the highest levels
d. bacterial decomposition and high BODs
e. all of the above

17. Which of the descriptions below are accurate regarding lake turnovers?
a. In the fall, cold water from below brings nutrients to the surface where the water is warmer.
b. Algal blooms result when the warmer water from below is displaced by colder surface water.
c. In the spring, the warmer water at the surface displaces the colder water from below.
d. When ice forms at the surface, the solid material sinks into the dense liquid.
e. All are correct.

18. Brackish water estuaries offer this ecosystem service:
a. recycling excess nitrate and phosphate from fertilizer runoff into new plant growth
b. providing decomposition food chains that support commercial fishing
c. providing a "natural sewage treatment plant"
d. serving as a stormwater surge barrier for inland ecosystems
e. all of the above

19. Ancient naturalists thought that corals were plants because they were brightly colored. Which statement below is an accurate description of the role of corals in a coral reef community?
a. Corals have photosynthetic algae living in their bodies.
b. Corals are animals that feed on plankton and larger fish larvae.
c. Corals grow when calcium carbonate is deposited on the substrate.
d. Fish populations will decline when the coral animals are eaten by a plague of crown of thorns starfish.
e. All of the above are correct.

20. The principal environmental stress threatening the Okefenokee Swamp and the Everglades is:
a. sewage pollution.
b. strip mining for phosphates.
c. diversion of the water to cities and the trans-Florida barge canal.
d. lightning-strike fires.
e. recreation in Lake Okeechobee

21. Which statement is correct regarding the graphs below?
a. Graph (a) can represent a tropical rain forest.
b. Graph (b) can represent tundra.
c. Graph (b) can represent a temperate grassland in North America.
d. Graph (b) represents a deciduous forest in Australia.
e. Graph (c) is a tropical deciduous forest in subequatorial South America.

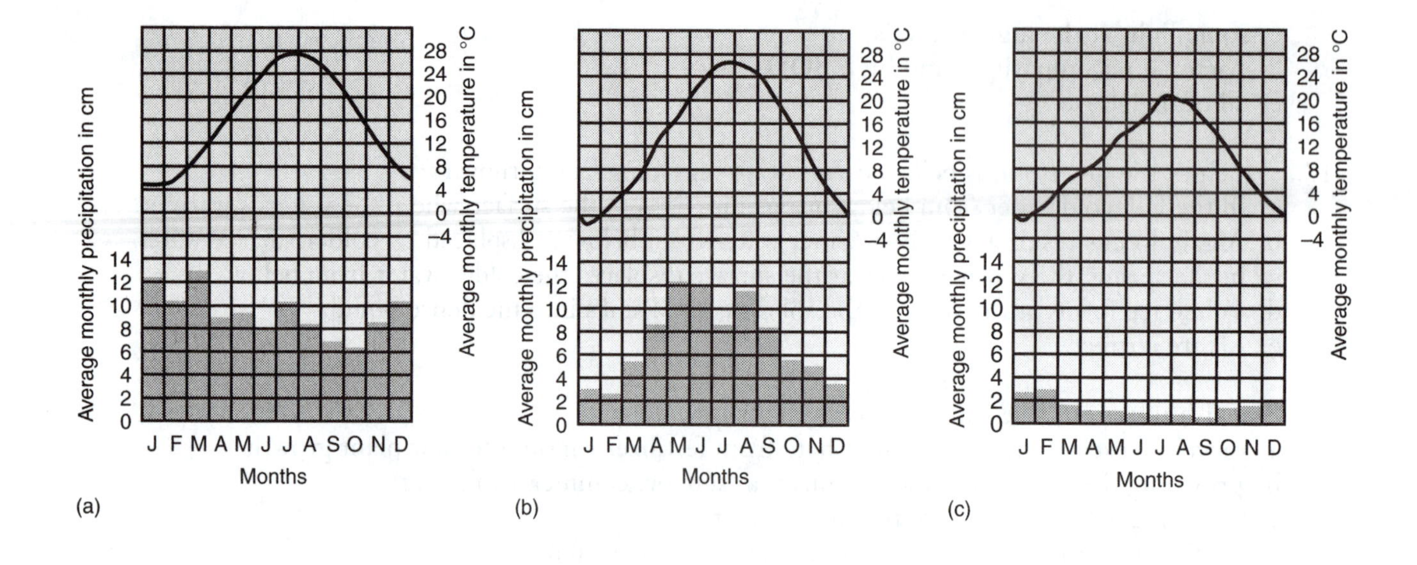

CYBER SURFIN'

1. For an essay on the use of fire by Native Americans, see: **wings.buffalo.edu/academic/department/anthropology/Documents/firebib**
2. From the Canadian Forest Service, a primer on the effects of fire on the age structure of the boreal forest and control of pine beetles, go to: **www.pfc.cfs.nrcan.gc.ca/entomology/mpb/management/index_e.html**
3. Go to the author's website for more information on biomes: **www.gpc.edu/~jaliff/biome.htm.** Also see Geography World: **members.aol.com/bowermanb/ecosystems.html**
4. This site reviews the role of fire to maintain the Florida Everglades: **www.nps.gov/ever/fire/firemgmt.htm**
5. The National Wildlife Federation looks at the Okefenokee Swamp of Georgia and Florida: **www.nwf.org/okefenokee/**
6. For more information on wetlands, from Purdue University, the Wetlands Education Center, go to: **agen521.www.ecn.purdue.edu/AGEN521/epadir/wetlands/menu.html**
7. From the Coral Reef Alliance, Working to Keep Coral Reefs Alive: **www.coralreefalliance.org/**

'LIKE' BOOKING IT

1. Douglas M. ***The Everglades: River of Grass*** (rev. 1988). Sarasota, FL: Pineapple Press, 1947.
2. Kingsolver B, and Belt AG. ***Last Stand: America's Virgin Land.*** Washington, DC: National Geographic Society, 2002.
3. Ward K, and Redford R. ***The Last Wilderness: Arctic National Wildlife Refuge.*** Cambridge, U.K.: WildLight Press, 2001.
4. Spalding MD, Ravilious C, Green EP. United Nations Environment Programme, and World Conservation Monitoring Centre. ***World Atlas of Coral Reefs.*** Berkeley, CA: University of California Press, 2001.

ANSWERS

Seeing the Forest

a. Fill-in the table of four effects of fires on the environment and indicate whether each is positive or negative.

Fire effect on the environment	Positive or negative?
1. Release of minerals into the soil	Positive
2. Opening seeds that have to be heated	Positive
3. Maintaining fire-maintained climax or subclimax communities	Positive
4. Vaporizing nutrients in high-temperature crown fires	Negative
5. Destruction of plants that hold soil, subsequent mudslides	Negative

b. Explain how the piney woods/pine barrens of the Atlantic coastal plain of the U.S. (New Jersey and southward) are maintained by periodic fires. When fires are set at ground level, the burning leaf litter and pine straw sets the bark of the young hardwoods on fire, killing them. Most pines trees can have their bark burn without killing the trees. Therefore, most hardwoods are prevented from succeeding the pines, and the pines continue to dominate the habitat in the "piney woods."

Questions to Ponder (Many of these answers are examples of many possible responses, please don't memorize them.)

1. In the southeastern U.S., it is said that driving 350 miles north (at a similar altitude) is roughly equivalent to increasing altitude by 1,000 feet ascending the Appalachian Mountains. At an elevation of 3,500 feet and higher in North Carolina, one encounters increasing numbers of spruce trees similar to that encountered in northern Michigan or central Ontario, as well as similar temperatures and rainfall. Explain this phenomenon. *The zonation of communities encountered with an increase of altitude, parallels that of increasing latitude (driving toward the North or South Pole). Therefore, tundra is found at 10,000+ feet elevations in the Colorado as well as the 6,000+ feet elevations of the northern Appalachian Mountains. When driving northward to the Cumberland Gap in Tennessee (4,000 feet elevation), one encounters a transition between deciduous forest at a lower altitude and coniferous forest dominated by spruce trees at higher altitudes. The decrease of temperature is responsible. The dominance of certain conifers, like spruce and hemlock, is due to a natural antifreeze in the needles (leaves).*

2. For the biomes and ecosystems below, fill in the table with information on the biomes' geographical location, climate (temperature and moisture), and dominant plants and animals. Rate temperature as very cold, cold, temperate, or warm (tropical) or hot. Rate moisture as annually very dry, dry, moderate, wet, or very wet, or periodically very dry and very wet.

Biome/ ecosystem	Location	Climate temperature	Climate moisture	Dominant plants	Dominant animals
Tropical rain *forest*	*Equatorial*	*Warm*	*Wet*	*Very tall rosewood and mahogany, strangler fig*	*Tigers, jaguars, monkeys, apes*
Tropical deciduous forest	*Equatorial*	*Hot*	*Periodically dry and wet*	*Teak, acacia*	*Deer, antelopes, leopards, lions, baboons*
Temperate rain forest (coniferous)	*West coast of Alaska, British Columbia, North latitude 50–60°*	*Cool*	*Wet*	*Sitka spruce*	*Black and brown bears*
Deciduous forest	*Eastern U.S. and Canada, 30–45°*	*Moderate*	*Moderate*	*Maples in the north, oaks and hickories in the south*	*Black bear, white-tailed deer, bobcats*
Taiga	*50–70°, north latitude in Alaska, Canada, Scandinavia, and Russia*	*Cold*	*Wet—because low temperatures inhibit evaporation.*	*Spruces, hemlocks, firs, larch*	*Wolves, reindeer in the winter, moose, elk, snowshoe hares, ptarmigans, squirrels, brown and black bears*
Chaparral	*20–30° north latitude, "Mediterranean," southern California, central Arizona*	*Warm*	*Dry, with a short wet season of two-three months.*	*Shrubs and small trees, dwarf oaks, Utah juniper, manzanita*	*Mule deer, scrub jays, jackrabbits, cougars, desert bighorn sheep*
Temperate grassland	*East of the Rocky Mountains from 30–50° north latitude.*	*Moderate*	*Dry-moderate*	*Tall and short prairie grasses in wetter and drier conditions, respectively*	*Wolves, antelope, buffalo, wild horses*
Tundra	*50° north latitude to the Arctic Circle.*	*Very cold most of the year*	*Arctic—Dry most of the year in cold air, some moderate summer moisture*	*A cold grassland of perennial forbs, willows, sedges, mosses, and lichens*	*Arctic fox, polar bear, reindeer in the summer, wolves*
Tropical savannah	*Equatorial*	*Warm-Hot*	*Dry, with a short wet season of one-two months.*	*A warm-hot grassland with sparse trees or thornscrub*	*Wildebeest, zebra, wild dogs, cape buffalo, crocodiles in rivers, elephants, leopards in sparse trees*

Desert (Sahara)	*Equatorial*	*Hot*	*Very dry, less than 2 cm per year precipitation*	*Sand dunes, rare plants except at springs (oasis)*	*Sidewinder-type snakes, some lizards, some foxes, camels*
Semidesert (Phoenix, Arizona)	*20–30° north latitude at elevations above 1,000 ft.*	*Warm-Hot*	*About 10–20 cm annually.*	*Saguaro, compass, and barrel cactuses, cholla, palo verde trees, mesquite, creosote bush*	*Rattlesnakes, kangaroo rats, scorpions, tarantulas, mule deer, jackrabbits*

3. Describe one major human effect on the biomes and water ecosystems listed below.

Biome	**Cause**	**Effect**
Tropical rain forest	*Conversion of tropical rain forest to grazing or farmlands*	*Soil erosion, loss of diversity of plants and animals, areas cleared are only productive for a short time*
Freshwater flowing ecosystems	*Sewage pollution from overfilled storm sewers, septic tank leakage, and pig farms*	*Foul-smelling water, algal blooms, increased transmission of E. coli H7:0157, hepatitis A virus*
Standing freshwater ecosystems (wetlands)	*Filling-in for real estate development, toxic pollution*	*Death of grass and other plants, reduction of water cleaning ecosystem services, decrease of waterfowl*
Brackish estuaries	*Filling-in for real estate development, toxic pollution*	*Death of grass and other plants, reduction of water cleaning ecosystem services, reduction of commercial fishing for oysters, clams, blue crabs, shrimps*
Arctic tundra	*Oil drilling, road building, global warming, pipe line construction*	*Reduction of reindeer and musk ox populations, followed by wolves declining*
Taiga	*Logging*	*Long times required to regenerate the forest, soil erosion, loss of diversity because many species replaced by monoculture*
Chaparral	*Wildfires*	*Extensive fires when resinous plants are crowded, mudslides*
Desert	*Recreational off-road vehicle traffic, real estate development*	*Loss of diversity of plants and animals*
Eastern deciduous forest	*Real estate development, logging*	*Loss of habitat for songbirds, increasing parasitism of crowded territorial nests by brood parasites like cowbirds. Soil erosion, siltation of streams*

4. Offer three reasons why plants taller than 30 cm cannot grow in the tundra.

    1. Cold winds kill buds.
    2. Deep roots cannot form due to permafrost.
    3. Water-logged surface soil liquefies in the short summer.

5. Do some research and come up with five conditions that cause coral reef destruction.

    1. Sewage pollution from the mouths of rivers
    2. Excessive UV radiation because of ozone depletion
    3. Global warming—increased water temperatures
    4. Toxic pollution
    5. Large populations of crown of thorns starfish, coral-eating parrot fish.

6. Explain why algal blooms may occur in spring and fall turnovers. What specifically is made available to the algae and where does it come from? *The lower profundal zone of the lake has decomposing biotic materials. In the fall turnover, those materials are forced to the surface when water cools to 4°C, its greatest density, and falls to the bottom of the lake. In the spring turnover, ice melts at the surface and as it warms to 4°C, it falls to the bottom, again forcing the warmer water from below to the surface. Bacterial decomposition at the bottom releases nutrient minerals (potassium, nitrate, phosphate) to the surface where algae utilize them as "fertilizers" for growth.*
7. Explain how are salt marshes and mangrove swamps important economic resources for the temperate and tropical Atlantic coast of North America. *The salt marsh or brackish estuary is characterized by a lush growth of grasses, the decomposition of which forms the basis of a food chain that includes many organisms, especially commercially valuable shellfish crabs and flounder.*
8. Offer three environmental stresses threatening the Okefenokee Swamp of Georgia and the Florida Everglades.

    1. Diversion of water for barge canals, irrigation, and the use of cities
    2. Phosphate mining and real estate development
    3. Toxic pollution, invasion of exotic grasses

Pop Quiz

Matching

1. e
2. i
3. k
4. b
5. l
6. o

7. m
8. g
9. a
10. n
11. d
12. c
13. j
14. h
15. f

Multiple Choice

1. e
2. b
3. c
4. d
5. b
6. e
7. b
8. e
9. a
10. b
11. a
12. b
13. e
14. e
15. a
16. d
17. b
18. e
19. e
20. c
21. c

8

Understanding Population Change

LEARNING OBJECTIVES

After you have studied this chapter you should be able to:

1. Describe the extent of the HIV/AIDS epidemic in sub-Saharan Africa.
2. Explain the four factors that produce changes in population size and solve simple problems involving these changes.
3. Define biotic potential (intrinsic rate of growth) and carrying capacity and explain the differences between J-shaped and S-shaped growth curves.
4. Distinguish between density-dependent and density-independent factors that affect population size and give examples of each.
5. Describe some of the density-dependent factors that may affect boom-or-bust population cycles.
6. Describe Type I, Type II and Type III survivorship curves.
7. Summarize the history of human population growth.
8. Identify Thomas Malthus, relate his ideas on human population growth, and explain why he may or may not be wrong.
9. Explain why it is impossible to answer precisely the question of how many people the Earth can support—Earth's carrying capacity for humans.
10. Explain how highly developed and developing countries differ in population characteristics, such as infant mortality rate, total fertility rate, replacement-level fertility, and age structure.
11. Briefly relate the history and controversies of U.S. immigration.

SEEING THE FOREST

1. Compare the HIV/AIDS infection rates in Nigeria, the Philippines, Canada, Argentina France, and Australia. Go to the World Factbook: **www.odci.gov/cia/publications/factbook/.**

Nation/Continent	HIV/AIDS infection rate
Nigeria/Africa	
Singapore/Southeast Asia	
Canada	
Argentina/South America	
France/Europe	
Australia	

SEEING THE TREES

VOCAB

Anyone who believes exponential growth can go on forever in a finite world is either a madman or an economist.

Kenneth Boulding

1. Population ecology—the study of the relationship of environmental conditions to population levels as follows: average age of individual; competition for food; habitat or reproduction; disease, predation, and antibiotic effects; and abiotic resource competition and limiting factors.
2. Population density—the number of individuals of a certain species per unit area or volume.
3. Population growth rate—the simple population growth rate is the annual *crude birth rate* (expressed in births/1000 individuals living) = decimal rate, minus annual *death rate* (expressed as deaths/1000)—birth rate of 30/1000 (0.030) – death rate of 15/1000 (0.015) = 0.015 or 1.5% per year. $r = b - d$.
4. Dispersal or migration—passive (seeds) or active movement from one area to another—respectively, fire ants came to Alabama in the 1950s in a shipment of logs from South America, cattle egrets were apparently blown off their migration course from Africa to South America to establish new populations in the Western Hemisphere.
5. Immigration—the positive effect on population size resulting from the gain of individuals dispersing to a given region.
6. Emigration—the negative effect on population size resulting from the loss of individuals migrating from a certain region.

7. True population growth rate—birth rate minus death rate plus net dispersal: *b – d + (immigration – emigration).*
8. Biotic potential—the intrinsic rate of growth is the maximum rate that a population could grow under ideal conditions, assuming that no death occurs.
9. Environmental resistance to population growth—population level is related to the average age of individuals; competition for food, habitat, or reproduction; disease, predation, and antibiotic effects; and abiotic resource competition and limiting factors.
10. Exponential population growth—when numbers of individuals are graphed against time, a J-shaped curve results as the time between doubling of the population decreases and the upward slope of the line increases.
11. Carrying capacity *(K)*—the largest population that can be sustained by the unsubsidized environment, represented by the plateau of a graph or peak of an S-shaped curve.
12. *r* strategy or *r* selected species—*r* strategists emphasize the reproduction (birth) of high numbers of offspring because of the high rates of death where the predominant number of eggs, larvae, or young die from predation or other adverse environmental conditions. Very generally, *r* strategists have these adaptations: small body size, they release large numbers of eggs or reproduce by fission, they produce live offspring that mature quickly, they are good opportunists—invading disturbed environments like old farm fields (previous chapter).
13. *K* strategy—*K* means constant or sustaining carrying capacity. *K* strategy adaptations generally include: they reproduce small numbers of live offspring that they protect, they have large body size that grows and matures slowly, and they live in stable environments such as subclimax or climax communities.
14. Survivorship—the proportion of individuals of a population that will survive to a specified age. *Type I* survivorship is characterized by a high proportion of surviving young (prereproductive) and of the age of reproduction, the proportion of survivors rapidly decreasing with age—humans. *Type II* survivorship has a more even survivorship at all age levels—poisonous animals. In *Type III* survivorship, the probability of death of the prereproductive creature is very high and survivorship increases with advancing age—frog tadpoles and adults.
15. Density-dependent factors that affect population size—regulatory factors that increase in impact on population size with an increasing proportion of the population affected. These include parasitic infestations, infections and disease, nest parasitism, direct resource competition and aggression, and predation.
16. Density-independent factors—typically abiotic factors, such as severe weather or the limitation of decomposition by permafrost, that results in boom-and-bust population cycles in the tundra.

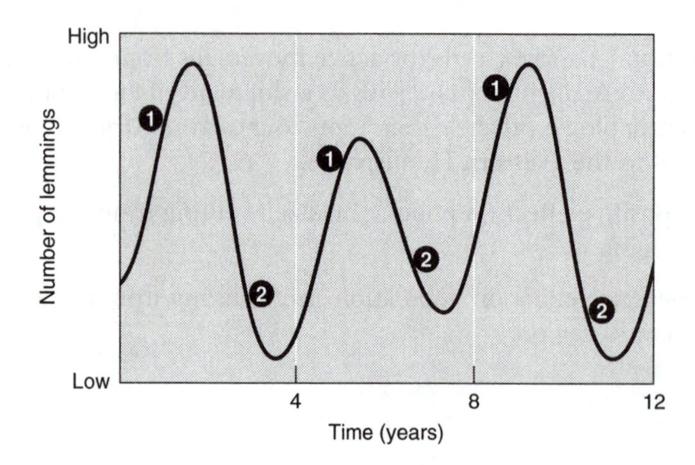

17. Thomas Malthus—The "dismal economist" of the early 19th century who predicted that the consequences of rising population were increasing poverty, starvation, disease, and warfare. Garrett Hardin remarked about malthus, "Anyone who has to be reburied so often cannot be entirely dead."
18. Zero population growth or ZPG—the concept of environmental sustainability is inexorably linked to control of population growth, more specifically, when birthrate minus death rate equals zero—$r = b - d = 0$.
19. Demographics—a branch of sociology dealing with the statistics of population characteristics, such as location, ethnicity, religious preference, educational level, income, sex, and others.
20. Highly developed countries—those countries with high GPI PPPs (GNPs, gross national production), a high amount of industrialization and related service economies, and a great amount of infrastructure supporting the economy, education, defense, and heath care of citizens. Moderately developed and less developed countries are classified accordingly.
21. Stages of demographic transition—as less developed countries become more developed, the following stages are recognized: 1) preindustrial—birth and death rates are high, and there is moderate population growth; 2) transitional stage—as infrastructure develops, birth rate increases and death rate declines; 3) industrial—a decline in birth rate occurs as industrialization progresses; and 4) postindustrial—birth and death rates are low.
22. Age structure diagram—illustrates the number and proportion of people at various ranges of age.
23. Population growth momentum—a positive population growth momentum exists when populations with large numbers of prereproductive individuals become reproductive, the numbers of parents will be larger than those in each succeeding generation.
24. Immigration Reform and Control Act—established priorities for immigrants with essential occupations, those with family members living in the U.S., and refugees from persecution in certain foreign countries.

SEEING THE FOREST

1. Compare the HIV/AIDS infection rates in Nigeria, the Philippines, Canada, Argentina, France, and Australia. Go to the World Factbook: **www.odci.gov/cia/publications/factbook/**

Nation/Continent	**HIV/AIDS infection rate in %**
Nigeria/Africa	
Philippines/Southeast Asia	
Canada	
Argentina/South America	
France/Europe	
Australia	

Questions to Ponder

1. Refer to text Fig. 8–12 and calculate the Mexican crude population growth rate in the years 1975 and 2000.

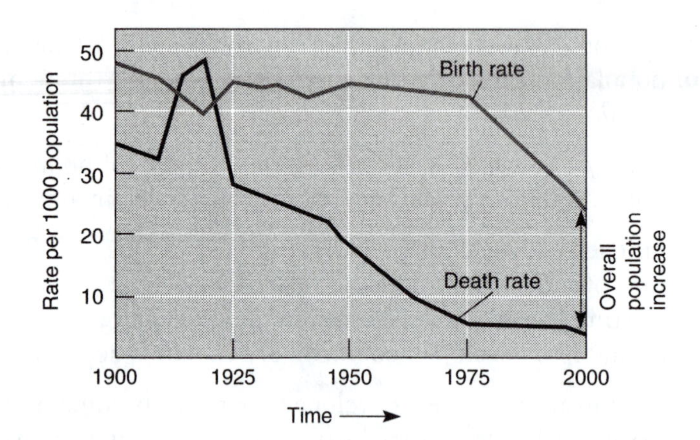

2. Given the following data for the U.S.A.: current birth rate = 0.015, current death rate = 0.009, current population = 287,000,000, current immigration = 800,000, current emigration = 150,000. Calculate the true annual population growth as a decimal.

3. Go to **www.odci.gov/cia/publications/factbook/** and find a country in western Europe that actually has a negative crude population growth rate. What is that negative rate?

4. Do some research on dinosaurs. List three *r* and *K* adaptations that they generally exhibited. See the *Encyclopedia of Dinosaurs*, contribution by Peter Dodson of the U. of Pennsylvania School of Veterinary Medicine: **darwin.apnet.com/dinosaur/dodson.htm** and ZoomDinosaurs.com at **www.allaboutwhales.com/subjects/dinosaurs/**

Adaptation of a specific dinosaur	*r* or *K*?
1.	
2.	
3.	
4.	
5.	
6.	

5. List 4 causes of boom-and-bust population cycles.

   1.

   2.

   3.

   4.

6. Compare the birth rates, death rates, GPI PPPs (GNP), and population growth rates of developed countries, moderately developed countries, and less developed countries by filling in the table below. Compare Canada, Chad, and Costa Rica using **www.odci.gov/cia/publications/factbook/**

	Birth rate	Death rate	GNP per capita in U.S. dollars	$R = b - d$ in % per year
Canada				
Chad				
Costa Rica				

7. Using the figure below, explain why population growth is low in Germany, but high in Nigeria.

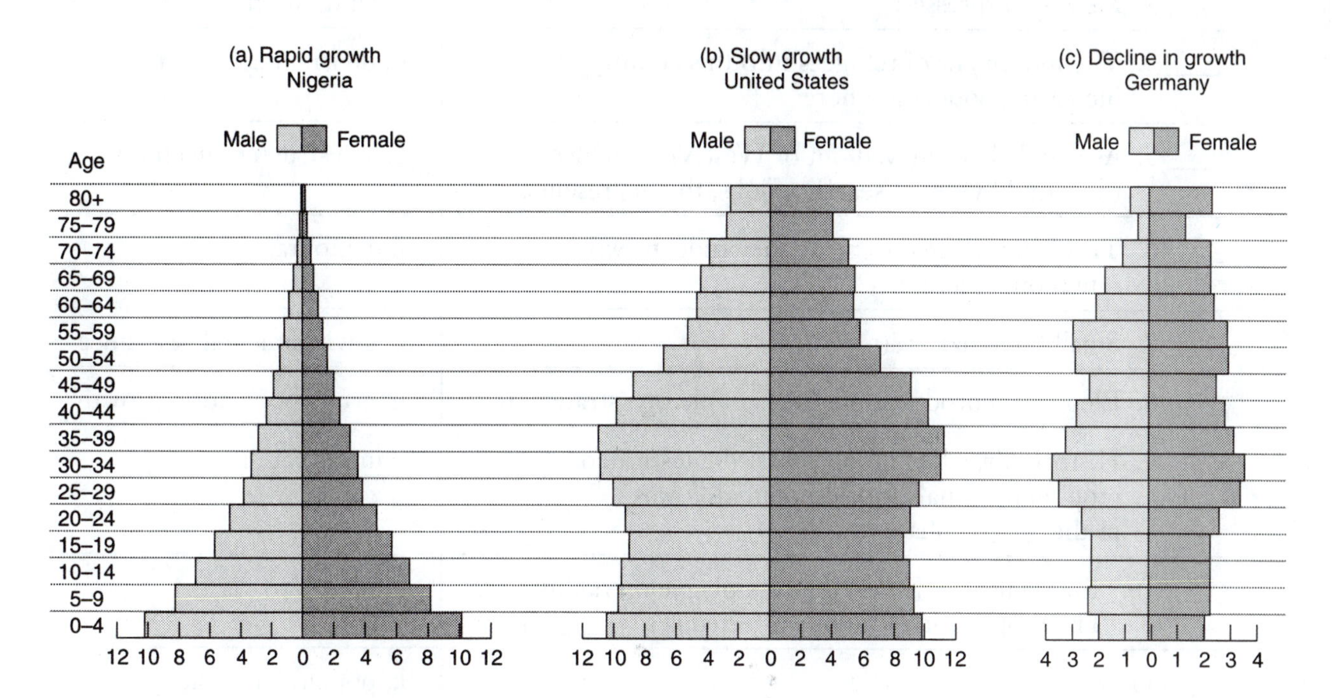

POP QUIZ

Matching

Match the following terms and definitions

____ 1. AIDS may prevent the growth of population in certain African countries. This is an example of:	a. population growth momentum
____ 2. A biologist studies the effects of bird territory size on litter size. This is an activity of the discipline of:	b. keeping population near carrying capacity
____ 3. $R = b - d$ expresses:	c. emigration
____ 4. The movement of refugees into this country, increasing population here:	d. density dependent
____ 5. As people have moved out of West Virginia over the past 40 years in search of jobs, this represents:	e. age structure diagram
____ 6. This factor is used to make the crude growth rate more accurate:	f. demography
____ 7. Small body size is a(n):	g. environmental resistance
____ 8. Elephant reproduction is an example of a strategy to:	h. crude population growth
____ 9. Historically, when humans first domesticated crops and animals, world population was in the ______ phase.	i. immigration
____ 10. Tularemia is a disease that is a plague on rabbits. The proportion of rabbits infected is a ______ factor.	j. *r* strategy
____ 11. $r = b - d = 0$	k. population ecology
____ 12. Weather, an abiotic factor that affects population size is: ______.	l. net dispersal
____ 13. When the U.S. Census Bureau determines the number of households with an income of $30,000 or less, it is doing the activity of this discipline:	m. density independent
____ 14. If each generation of parents becomes larger, this principle is illustrated:	n. lag
____ 15. This device indicates the numbers of prereproductive, reproductive, and postreproductive individuals:	o. zero population growth

Multiple Choice

1. Wildflowers from Europe, Asia, and the Middle East spring up in the grassy fields between runways at airports. This is an example of:
   a. passive emigration
   b. immigration
   c. emigration
   d. passive dispersal
   e. a, b, and c

2. Seeds from an alpine tundra habitat are carried by wind to another mountain where they increase the plant's population. This is an example of:
   a. active migration
   b. immigration
   c. emigration
   d. active dispersal
   e. b and d

3. During an especially cold winter, snowy owls from the tundra migrate to Michigan, decreasing the population in the tundra. This is an example of:
   a. passive migration
   b. immigration
   c. emigration
   d. passive dispersal
   e. b and c

4. Which environmental resistance factor will contribute to the inability of eastern U.S. vesper sparrows to recover from habitat loss?
   a. nest parasitism
   b. competition from chipping sparrows
   c. West Nile virus causing the death of sparrows
   d. crows raiding nests and killing hatchlings
   e. all of the above

5. Refer to the graph below of a population of fruit flies growing in a closed environment with limited resources. In which segment of the graph does exponential growth occur?
   a. A
   b. B
   c. C
   d. D
   e. A and B

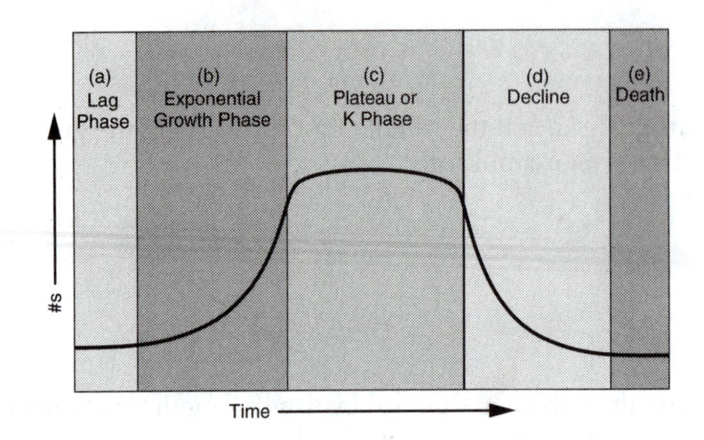

6. Refer to the graph above. In which segment of the graph does the accumulation of toxic wastes and loss of biotic and abiotic resources have great impact?
   a.
   b.
   c.
   d.
   e. b and c

7. Refer to the graph above. In which segment of the graph does the population reach its environmental carrying capacity
   a. A
   b. B
   c. C
   d. D
   e. E

8. Inspect the graph below. In which stage was world human population growth in 2000 AD/CE?
   a. lag
   b. K/plateau
   c. exponential growth
   d. decline
   e. all of the above.

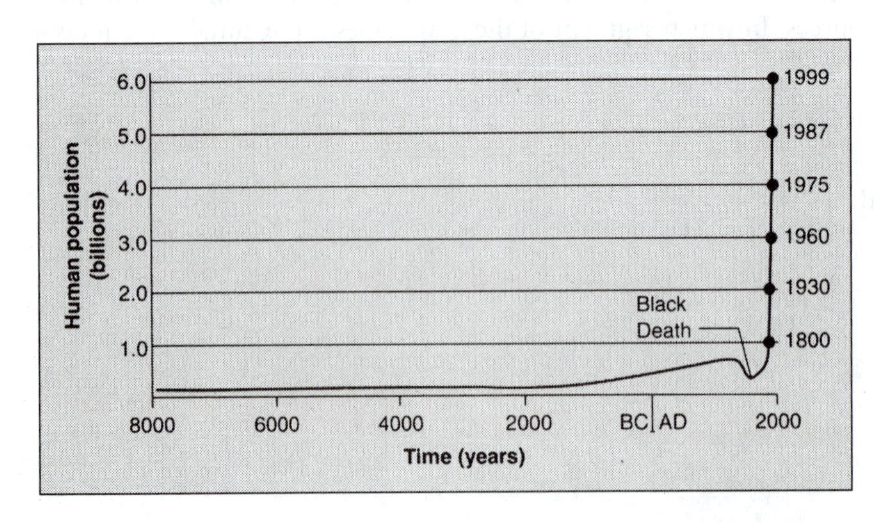

9. Inspect the graph at the bottom of page 124. In which stage was world human population growth in 1000 AD/CE?
   a. lag
   b. K/plateau
   c. exponential growth
   d. decline
   e. all of the above

10. Inspect the graph at the bottom of page 124. In which stage was population growth of developed countries in 2000 AD/CE?
   a. lag
   b. K/plateau
   c. exponential growth
   d. decline
   e. all of the above

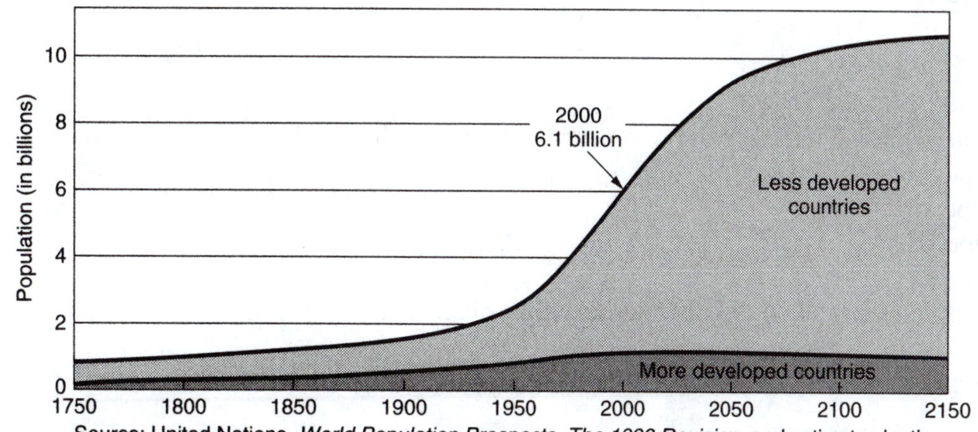

Source: United Nations, *World Population Prospects, The 1998 Revision*; and estimates by the Population Reference Bureau.

11. In 1890, there were approximately 1.55 billion people on Earth; in 1930, approximately 3.1 billion; in 1960, approximately 3.05 billion; in 2000, 6.1 billion. In this time period, what has happened to the amount of time between population doublings?
   a. Population growth rate has not changed.
   b. Zero population growth policies have worked.
   c. The number of years between doublings has decreased.
   d. exponential growth
   e. c and d

12. Sea urchins produce a cloud of eggs that quickly produce larvae that become part of the zooplankton that filter-feeding and other animals consume. This is an example of which type of survivorship?
   a. Type I
   b. Type II
   c. Type III
   d. Type IV
   e. Trick question alert! All of these are applicable.

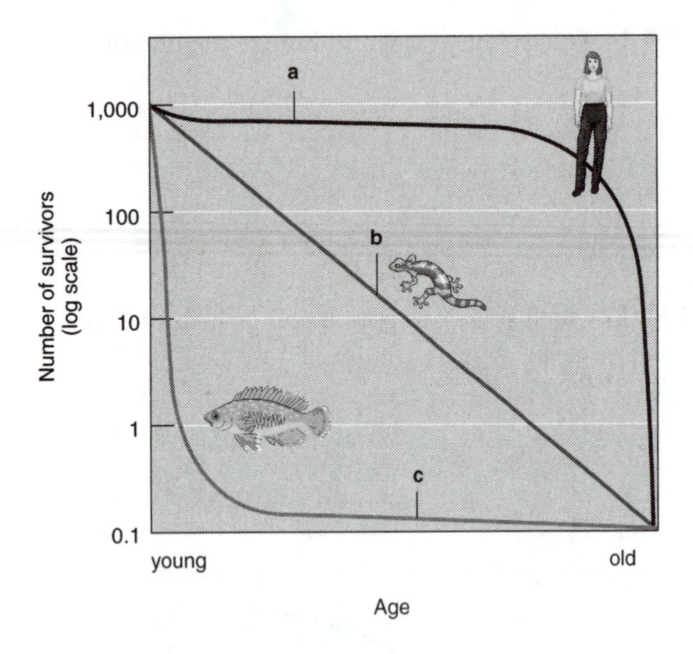

13. Songbird survival typically fits which type of survivorship?
    a. Type I
    b. Type II
    c. Type III
    d. Type IV
    e. Trick question alert! All of these are applicable.

14. In the graph above, which curve represents Type I survivorship?
    a. A
    b. B
    c. C

15. In the year 2000, what percentage of the world's population lived in undeveloped countries?
    a. 40%
    b. 50%
    c. 60%
    d. 70%
    e. 80%

16. Which of the following would represent an *r* strategy of the female human *Ascaris* roundworm?
    a. parental care of young
    b. large body size
    c. short life span
    d. 10,000 eggs made per day
    e. a high and stable population level

17. In the northern coniferous forest, removal of wolves from ecosystems have first caused an explosion of the population of elk, followed by a collapse. Which statement is correct concerning this phenomenon?
    a. environmental resistance to elk population growth decreased.
    b. The elk exceeded the carrying capacity of the environment.
    c. The elk are *K* strategists.
    d. The elk ate most of the plants they depended on and starved.
    e. All are correct.

18. The area with the largest prereproductive population is:
    a. Africa
    b. Asia
    c. Latin America
    d. North America
    e. Europe

19. The country with the largest proportion of immigrants is:
    a. United States
    b. Canada
    c. Germany
    d. Australia
    e. Japan

20. Industralized populations are characterized by:
    a. moderate population growth, but birth rates and death rates are high
    b. increased birth rate
    c. an increase in birth rate occurs as industrialization progresses
    d. low birth and death rates
    e. increased birth rate and increased death rate

CYBER SURFIN'

1. To see most current annual birth and death rates for any country in the world, go to the U.S. Central Intelligence Agency's World Factbook: **www.odci.gov/cia/publications/factbook/**
2. Go to **www.popexpo.net/eMain.html.** How many people were on Earth when you were born? When your mother was born? When your mother's grandmother was born? When your mother's great grandmother was born?
3. How do bacteria quickly produce infections that can kill? See **ampere.scale.uiuc.edu/pb102/02/part1.html.**
4. Do the population exercises found at **www.nap.edu/readingroom/books/evolution98/evol6-h.html.**

'LIKE' BOOKING IT

1. Brown LR, Gardner G, and Halwell B. ***Beyond Malthus: Nineteen Dimensions of the Population Challenge.*** New York: W. W. Norton, 1999.
2. Ehrlich PR. ***The Population Bomb.*** Cutchogue, NY: Buccaneer Books, 1997.
3. Brown LR. ***Who Will Feed China?: Wake-Up Call for a Small Planet (The Worldwatch Environmental Alert).*** New York: W. W. Norton, 1995.
4. Livi-Bacci M. ***A Concise History of World Population: An Introduction to Population Processes,*** 3rd ed. Oxford, U.K.: Blackwell Publishers, 2001.

ANSWERS

Seeing the Forest

Nation/Continent	HIV/AIDS infection rate in %
Nigeria/Africa	5.06, 1999 est.
Singapore/Southeast Asia	4.04
Canada	0.3
Argentina/South America	0.69
France/Europe	0.44
Australia	0.15

Questions to Ponder (Many of these answers are examples of many possible responses, please don't memorize them.)

1. Refer to text Fig. 8-12 and calculate the Mexican population growth in the years 1975 and 2000.

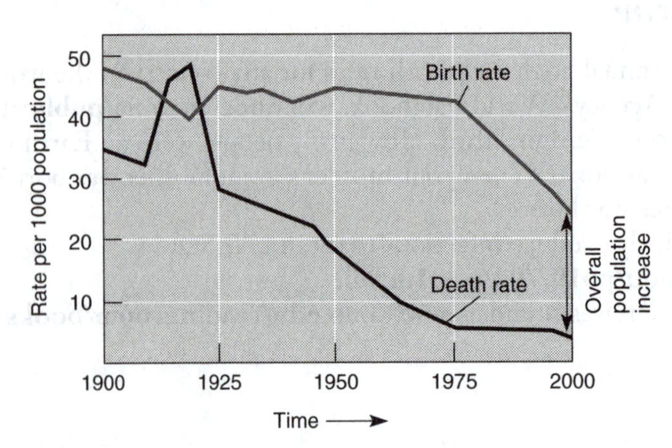

2. Given the following data for the U.S.A.: 1998 birth rate = 0.015, current death date = 0.009, current population = 287,000,000, current immigration = 800,000, current emigration = 150,000.

Calculate the true annual population growth as a decimal. Answer: 0.015 – 0.009 + (800,000 – 150,000/287,000,000) = 0.006 + 0.0023 = 0.0083 or 0.83%.

3. Go to **www.odci.gov/cia/publications/factbook/** and find a country in western Europe that is actually has a negative crude population growth rate. What is that negative rate? Answer: Austria with a birth rate of 9.58/1,000 and a death rate of 9.73 (2002), for a negative 0.15/1000 = 0.00015 or 0.015%.

4. Do some research on dinosaurs. List six r and K adaptations that they generally exhibited. See the *Encyclopedia of Dinosaurs*, contribution by Peter Dodson of the University of Pennsylvania School of Veterinary Medicine: **darwin.apnet.com/dinosaur/dodson.htm,** the BBC's "Walking with Dinosaurs," **www.bbc.co.uk/dinosaurs/dinosaur_world/index.shtml,** and ZoomDinosaurs. com at **www.allaboutwhales.com/subjects/dinosaurs/**

Adaptation of a specific dinosaur	r or K?
1. Large body size of sauropods	K
2. Low survivorship of young sauropods?	r
3. Large number of eggs laid by maiasaurs (22)?	r
4. Parental care of maiasaur offspring	K
5. Older age at first *Diplodocus* reproduction event	K
6. Long life span of sauropods	K

5. Compare the birth rates, death rates, GPI PPPs (GNPs), and population growth rates of developed countries, moderately developed countries, and less developed countries by filling in the table below. Compare Canada, Chad, and Costa Rica using **www.odci.gov/cia/publications/factbook/**

	Birth rate	**Death rate**	**GNP per capita in U.S. dollars**	**$R = b - d$ in % per year**
Canada	11.09	7.54	27,700	0.355
Chad	47.74	15.06	1,030	3.27
Costa Rica	19.83	4.31	8,500	1.5

Pop Quiz

Matching

1. g
2. k
3. h
4. i
5. c
6. l

7. j
8. b
9. n
10. d
11. o
12. m
13. f
14. a
15. e

Multiple Choice

1. d
2. b
3. c
4. e
5. b
6. e
7. b
8. e
9. a
10. b
11. a
12. b
13. e
14. a
15. a
16. d
17. b
18. e
19. e
20. c

9

Facing the Problems of Overpopulation

LEARNING OBJECTIVES

After you have studied this chapter you should be able to:

1. Relate human population size to each of the following: hunger, natural resources and the environment, and economics.
2. Distinguish between people overpopulation and consumption overpopulation.
3. Describe the concept of ecological footprints.
4. Outline some of the complexities associated with the concept of sustainable consumption.
5. Define urbanization and describe trends in the distribution of people in rural and urban areas.
6. Explain how cities are analyzed in ecological terms.
7. Describe some of the problems associated with the rapid growth rates of large urban areas.
8. Relate total fertility rates to each of the following: cultural values, social and economic status of women, and the availability of family planning services.
9. Compare the ways the governments of China, India, Mexico, and Nigeria have tried to slow human population growth.
10. Outline some of the steps that governments and individuals can take to achieve global population stability.

SEEING THE FOREST

HIV/AIDS is the number-one cause of death in sub-Saharan Africa. More than 70% of the world's HIV-positive people live there. The U.S. Agency for International Development (USAID) estimates that 20–35% of children under 15 have lost at least one parent. By the year 2010, there will be more than 33 million children orphaned. Life expectancies for the entire region will fall to nearly 30 years of age. Specific predictions are as follows: Botswana (29), Namibia (33), Swaziland (30), and Zimbabwe (33). In Botswana, Namibia, Swaziland, and Zimbabwe approximately 25% of people aged 15–49 are infected. See the World Factbook:
www.odci.gov/cia/publications/factbook/geos/wa.html

1. Make informed predictions that this epidemic will have an effect on the environment, economy, and society of sub-Saharan Africa.

   1. ______________________________
   2. ______________________________
   3. ______________________________
   4. ______________________________
   5. ______________________________

2. Identify three contrasting views of the problem of population growth and hunger from Chapter 9.

   1. ______________________________
   2. ______________________________
   3. ______________________________

SEEING THE TREES

VOCAB

We do not inherit the earth from our ancestors, we borrow it from our children.
Native American Proverb

1. Consumption—the social and economic act of using and depleting renewable and nonrenewable resources. Social consumption is a behavior that depletes natural resources. Economic consumption largely depletes the natural resources of countries where the raw materials of production originate.

2. Overpopulation—the concept that overpopulation and its accompanying resource demand does damage to the environment. There are two types: 1) *people overpopulation* as is the case in developing/underdeveloped countries, and 2) *consumption overpopulation* as is the case in developed nations in which the consumption of natural resources per country or per individual greatly exceeds that of the average individual in the world.
3. Total material requirement—the total weight of natural resources, not including air or water, that are required to support each person. In 1994, the US TMR was approximately 84 tons per individual; for Portugal, it was 15 tons (source: European Environmental Agency). Natural resources included fossil fuels, metals, minerals, soil erosion, and biomass (food and wood).
4. Ecological footprint—the average amount of all natural resources needed to maintain an individual or a specific technology—a heat pump or a natural gas furnace.
5. Sustainable consumption—consumption that minimizes the use of nonrenewable resources and ensures the availability of renewable resources for future generations.
6. Urbanization—the relative increase of urban populations while rural populations decrease—the process of moving from an agrarian to an industrialized economy and society.
7. Urban ecology—the discipline that studies the trends in population, social structure, environment, and technology of cities.
8. Urban heat island—the effect of heat-producing or radiating technologies or structures making cities generally warmer than surrounding rural areas.
9. Compact development—development that attempts to minimize the ecological footprints of people, their housing, and workplaces.
10. Total fertility rate—the average number of children born per woman in a given population.
11. Rule of 70—70 divided by the population growth rate in % equals the approximate time to double a population. To illustrate, the U.S. has about a 0.6% population growth rate, so the doubling time (in years) of the population size is 70/0.6 = 117 years.

Questions to Ponder

1. What has the government of Egypt done in an attempt to control its population growth?

Governmental method or policy to control population growth
1.
2.

What have other governments done? List six methods or policies and the country that uses each.

Government method or policy to control population growth	Country
1.	
2.	
3.	
4.	

5.	
6.	

2. What does the acronym POET stand for in urban ecology? Give a specific example of each as it affects the environment.

Term: Description of Term	Example
P	
O	
E	
T	

3. According to the Heat Island Research group, the daily maximum, minimum, and mean temperatures of the months of July and August, 1948–1988 in Phoenix, AZ, during which the city tripled in population and area, increased at a rate of 0.13°F/year. The mean increase of the minimum temperature was a higher 0.18°F/year. Give four specific (heat-generating) technologies that may have accounted for this increase.

Heat-producing or radiating technology
1.
2.
3.
4.

4. Offer four major reasons why fertility rate is high in developing/underdeveloped countries.

Factors that tend to increase fertility rate
1.
2.
3.
4.

5. The 1994 Global Summit on Population and Development focused on which three areas for action?

Areas for action

1.

2.

3.

6. Compare the fertility, infant mortality, and HIV/AIDS infection rates of Sweden, Thailand, and Somalia using **www.odci.gov/cia/publications/factbook/**

Country	Fertility rate	Infant mortality rate/1000	HIV/AIDS infection rate
Sweden			
Thailand			
Somalia			

7. a. What are the high and low estimates for China's total fertility rate in the year 2020?

   b. At the higher figure, how much more will the population grow in the number of individuals by the year 2050? See the graph below.

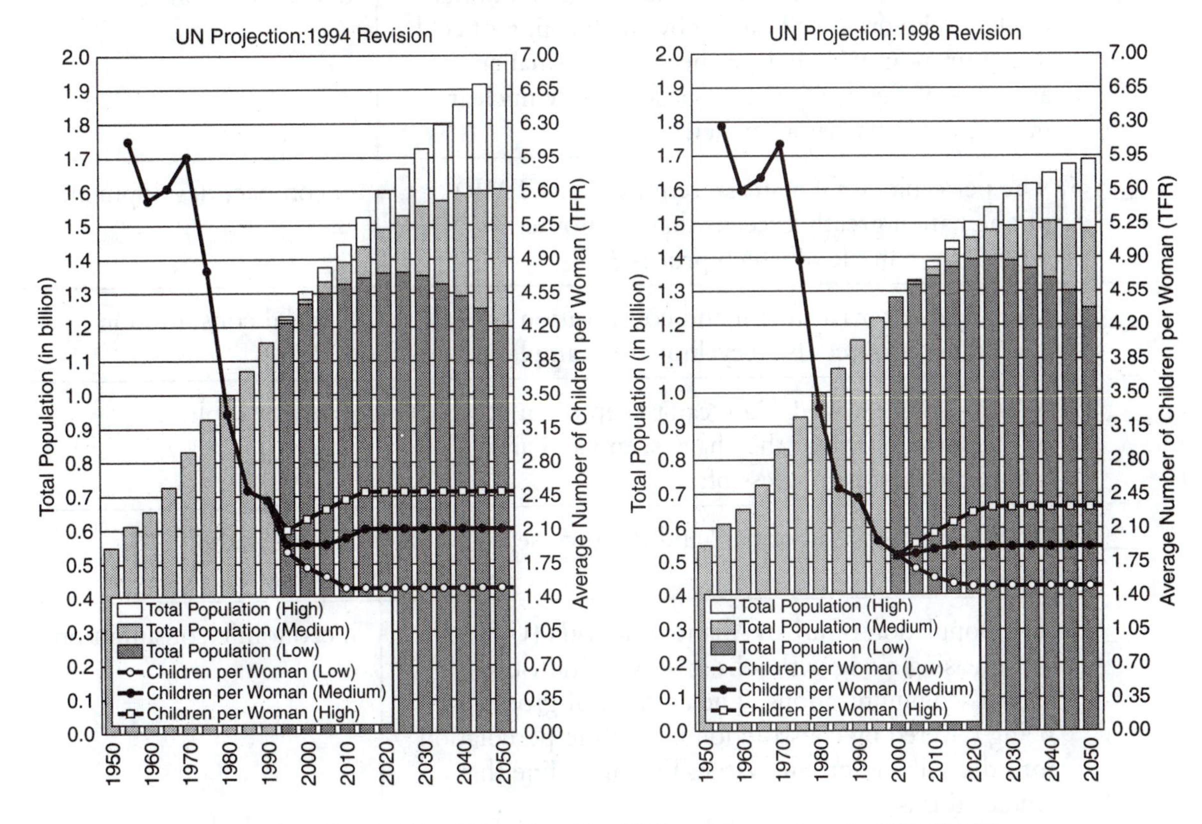

Helig, G.K. (1999): ChinaFood: Can China Feed Itself? Laxenburg, Austria (CD-ROM Publication, International Institute for Applied Systems Analysis), ISBN: 3-7045-0134-4

8. Using the rule of 70, how many years will be required to double Mexico's current population?

POP QUIZ

Matching

Match the following terms and definitions

____1. If more employees get the larger SUVs and demand that parking spaces be made larger, thus reducing available parking space, this is an act of:	a. ecological footprint
____2. Coal mines in western and central Pennsylvania were largely depleted of hard coal or anthracite from 1850–1950 to fuel the development of the U.S. steel industry. This is an example of:	b. pronatalists
____3. The number of tons of natural resources, excluding air and water, needed to support an individual annually is:	c. U.N. Global Summit Conference
____4. The standard U.S. water heater uses large amounts of electricity produced mostly by the burning of coal. The European in-line water heater uses smaller amounts of electricity mostly produced by nuclear means, the latter has a smaller:	d. urbanization
____5. The per capita total material requirement (TMR) of one nation greatly exceeds that of another. This is an example of which process?	e. compact development
____6. Municipal water treatment and conservation of aluminum resources by recycling are examples of:	f. social consumption
____7. A 1920s song extolled, "You can't keep them down on the farm, now that they have seen gay Paris." This expresses the process of:	g. urban ecology
____8. "Building up, instead of out" and the reverse of urban sprawl is:	h. heat island
____9. The population of urban Atlanta, Georgia (U.S.) is increasing along with jobs in service industries, such as communications. The amount of ground being covered by construction is limiting percolation into the soil and groundwater. The discipline that studies this is:	i. economic consumption
____10. In the cities of the desert southwest of the U.S., the difference in day and night temperatures is decreasing. This is likely a ______ effect.	j. consumption overpopulation

____11. In Namibia in 2002, the number of children born per woman was 4.77. This is the ______ rate.	k. sustainable consumption
____12. The issue of infant and mother mortality in childbirth was an issue of:	l. low social status of women
____13. Those who favor population growth for economic reasons:	m. total fertility
____14. An important social reason for the large number of children per family in developing countries is:	n. TMR

Multiple Choice

1. This factor prevents population control in Africa and other developing countries?
   a. religious restrictions against practicing birth control
   b. low status of women
   c. cultural traditions of having large families
   d. having children to help with farming
   e. all of the above

2. Demographers from the Population Institute consider which factor to be of most important concern for sustaining the quality of life of the world's individuals?
   a. the birth rate in developed countries
   b. the high death rate in undeveloped countries
   c. the high death rate in developing countries
   d. the high rate of population growth in developing countries
   e. warfare

3. Of the world's six billion people, approximately which percentage are malnourished or under nourished?
   a. 1%
   b. 1.33%
   c. 13%
   d. 23%
   e. 33%

4. The reason why some reasonably assert that industrialization is a solution to world poverty assume that industrialization results in:
   a. a better standard of health care leading to lower death rates
   b. lower birth rates, then control of population growth
   c. higher birth rates from improved health care
   d. better education and job training
   e. all of the above

5. The most important contribution of the U.S. in the problems resulting from overpopulation of the world is which factor?
   a. immigration leading to internal population growth
   b. pollution from heavily industrialized areas

c. failure to export biological means of population control
d. overconsumption of natural resources
e. all of the above

6. The developed countries represent what percentage of world population, and generate how much pollution and waste, respectively?
   a. 50%, over 50%
   b. 20%, 75%
   c. 80%, 20%
   d. 60%, 40%
   e. 10%, 90%

7. Developing/underdeveloped countries have this concern when dealing with increasing their standard of living:
   a. a high debt to developed countries for goods
   b. many people are poorly educated so that technical workers have to be imported
   c. poorly developed countries have great environmental damage due to resource extraction by developed countries
   d. population stabilization first, then economic development
   e. all of the above

8. What is the ratio of the ecological footprint of each U.S. citizen compared to a citizen of India?
   a. 1:1
   b. 5:1
   c. 10:1
   d. 1:4
   e. 1:1.5

9. The young age structure of cities as compared to rural areas is due to which factor?
   a. a higher birth rate in the city
   b. influx of younger people seeking jobs
   c. mechanization of farming
   d. improved crop yields
   e. all except a

10. Which statement below describes the "heat island" effect?
    a. In cities, the low daily temperature is directly proportional to growth.
    b. In cities, the high daily temperature is directly proportional to growth.
    c. Black-topped roofs and roads release heat into the air above cities.
    d. The high and low temperatures of Albuquerque, NM, are higher than the surrounding desert.
    e. All of the above

11. Cities such as Portland, OR, have growth without the urban sprawl and have limited environmental loss. They accomplished this by:
    a. compact development.
    b. utilizing the "brown fields" of vacant lots or dilapidated buildings.
    c. decreasing parking areas.

d. business encouraging the use of public transportation.
e. all of the above

12. Which city is not accurately matched with its major environmental problem?
a. Mexico City: air pollution
b. Curitiba, Brazil: traffic
c. Bombay, India: population of homeless
d. Los Angeles, CA: air pollution
e. New York City: water pollution

13. Of the factors listed below, which best explains the high total fertility rates in developing/underdeveloped countries?
a. Infant and child death rates are high.
b. Birth rate is high, and death rate is low.
c. Total death rate is high.
d. Immigration is higher than birthrate.
e. Trick question alert! All of these are applicable.

14. In China, an outcome of governmental efforts to control population will be:
a. zero population growth by the year 2010
b. men will outnumber women in 2050
c. women will outnumber men in 2050
d. the total fertility rate will be less than 1 in 2010
e. increased abortions of male babies currently and in the near future

15. Which is most effective means of birth and disease control?
a. condoms
b. surgical sterilization
c. abstinence
d. oral contraceptives
e. IUDs

16. Which country fits the following description: It has a large urban population with 33% of its population under the age of 15?
a. Mexico
b. India
c. China
d. North America
e. Europe

17. Refer to the graph below. Which of the ethnic groups listed represents the largest minority population in the United States?
a. Asians
b. African Americans
c. Native Americans
d. Latin Americans
e. Eastern Europeans

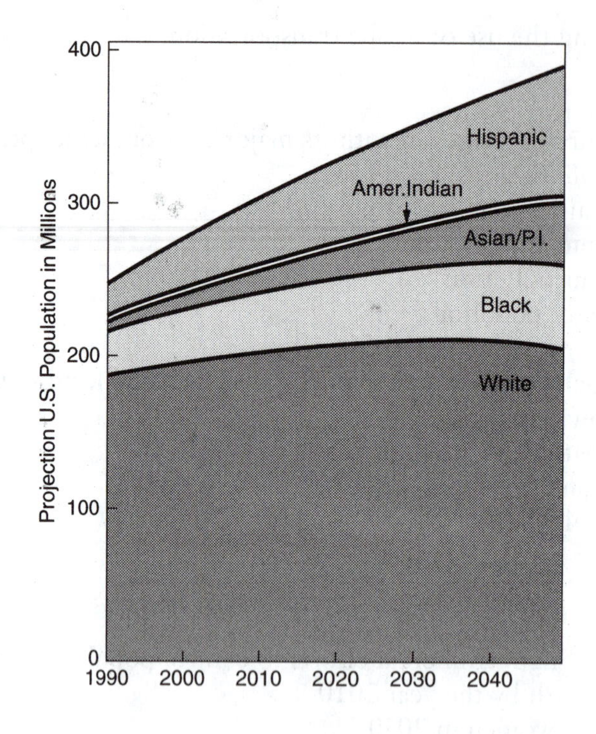

18. The greatest projected increases will occur in which population?
    a. Asians
    b. African Americans
    c. Native Americans
    d. Latin Americans
    e. Eastern Europeans

19. Which of the following was not a specific measure adopted by the U.N. 1994 Global Summit on Population and Development?
    a. reproductive health and disease prevention
    b. improving education for women
    c. family planning
    d. prevention of female infanticide
    e. treaty obligations for adherence

20. Which country is matched correctly with its effort to control population?
    a. Japan: oral contraception with birth control pills
    b. Saudi Arabia: oral contraception
    c. China: surgical sterilization after second pregnancy
    d. Germany: additional pregnancies are discouraged by cash payments
    e. Russia: widespread availability of oral contraceptives

21. Which city is the largest in the Western Hemisphere?
    a. Sao Paulo, Brazil
    b. New York, U.S.
    c. Buenos Aires, Argentina
    d. Mexico City
    e. Los Angeles, CA, U.S.

22. Refer to the graph below, at what time was China's population in exponential growth?
    a. 300 AD/CE
    b. 750 AD/CE
    c. 1500 AD/CE
    d. 1950 AD/CE
    e. The population has never been in exponential growth.

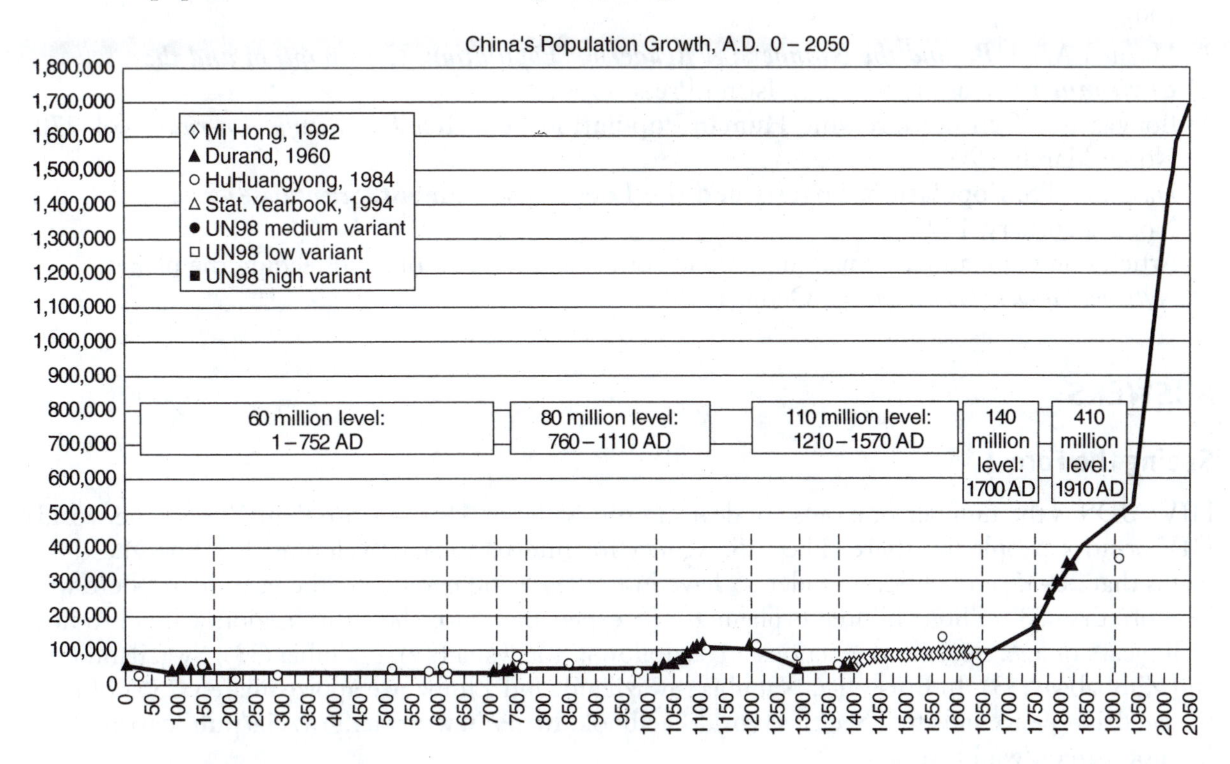

Helig, G.K. (1999): ChinaFood: Can China Feed Itself? Laxenburg, Austria (CD-ROM Publication, International Institute for Applied Systems Analysis), ISBN: 3-7045-0134-4

CYBER SURFIN'

1. To see most current annual total fertility, birth, and death rates for any country in the world, go to the U.S. Central Intelligence Agency factbook: **www.odci.gov/cia/publications/factbook/**
2. For a good general discussion of population growth, go to John W. Kimball's biology pages at **users.erols.com/jkimball.ma.ultranet/BiologyPages/P/Populations.html**
3. For an objective look at world population trends, see *World Population. Prospects: Analyzing the 1996 UN Population Projections.* Gerhard K. Heilig, International Institute for Applied Systems Analysis (IIASA), Schlossplatz 1, A-2361 Laxenburg, Austria, 1996. See **www.iiasa.ac.at/Admin/PUB/0docs/WP-96-146.html.971006/**
4. From the Union of Concerned Scientists, "Frequently asked questions about population growth," go to **www.ucsusa.org/global_environment/connections/page.cfm?pageID=548**
5. See also Chapters 1 and 8 resources.

'LIKE' BOOKING IT

1. Newman LF. ***Hunger in History: Food Shortage, Poverty, and Deprivation.*** New York: Blackwell Publishers, 1990.
2. Ehrlich PR, and Ehrlich A. ***Betrayal of Science and Reason.*** Covelo, CA: Shearwater Books, 1998.
3. Mazur LA, ed. ***Beyond the Numbers: A Reader on Population, Consumption and the Environment.*** Washington, DC: Island Press, 1994.
4. Bongaarts J. **Can the Growing Human Population Feed Itself?** *Scientific American*, Vol. 270, No. 4: March, 1994.
5. Dasgupta PS. **Population, Poverty and the Local Environment.** *Scientific American*, Vol. 272, No. 2, February, 1995.
6. Articles on population, human migration, women and population, and feeding the planet. *National Geographic Magazine*, October, 1998.

ANSWERS

Seeing the Forest

HIV/AIDS is the number-one cause of death in sub-Saharan Africa. More than 70% of the world's HIV-positive people live there. The U.S. Agency for International Development (USAID) estimates that 20–35% of children under 15 have lost at least one parent. By the year 2010, there will be more than 33 million children orphaned. Life expectancies for the entire region will fall to nearly 30 years of age. Specific predictions are as follows: Botswana (29), Namibia (33), Swaziland (30), and Zimbabwe (33). In Botswana, Namibia, Swaziland, and Zimbabwe approximately 25% of people aged 15–49 are infected. See the World Factbook: **http://www.odci.gov/cia/publications/factbook/geos/wa.html**

1. Make informed predictions that this epidemic, in combination with an increase in population, will have an effect on the environment, economy, and society of sub-Saharan Africa.

    1. Deforestation for firewood/fuel and building materials
    2. Loss of species diversity and collapse of populations, such as monkeys and apes
    3. Increasing numbers of homeless children without parents
    4. Increasing numbers of children forced into armies
    5. Increasing poverty

2. Identify three contrasting views of the problem of population growth and hunger from Chapter 9.

    1. Population growth/high fertility rates cause food shortages
    2. Promoting economic development will increase agricultural productivity and educational level so that fertility rate will fall as the nation transitions from developing/underdeveloped to developed/industrialized.
    3. The inequitable distribution of food and resources causes world hunger.

Question to Ponder (Many of these answers are examples of many possible responses, please don't memorize them.)

1. What has the government of Egypt done in an attempt to control its population growth?

Governmental method or policy to control population growth
1. The government sanctioned radio and TV soap operas promoting birth control
2. The government encouraged religious participation and family planning clinics.

What have other governments done? List six methods or policies and the country that uses each.

Governmental method or policy to control population growth	Country
1. Coercion of women to limit children to one with job sanctions and forced sterilization	China
2. Education and publicity campaigns	India
3. Economic/industrial development	Mexico
4. Abortions	Bangladesh

2. What does the acronym POET stand for in urban ecology? Give a specific example of each as it affects the environment.

	Term: Description of term	Example
P	Population: This is the number of people, the factors that change that number and the demographic characteristcs such as ethnicity and sex.	Increase in population increases consumption of natural resources. Look at Mexico City's consumption of fossil fuels and the air pollution resulting.
O	Organization: How is the population governed?	Do local laws or national laws regulate the the environment in a city? Is there any disagreement between the governmental philosophy of local, state and national authorities regarding the protection of the environment? Is your local government pro-development and anti-protection and conservation?
E	Environment: In which biome is the city located? s there a river, lake or ocean nearby?	Does a city such as Phoenix, AZ, withdraw more water from an aquifer than is being recharged?
T	Technology: What technologies have altered the urban environment?	Los Angeles' growth was made possible by the construction of an aqueduct from the Colorado River. Chicago, IL, is investigating reflective building surfaces and greenery covered surfaces (instead of tar black top roofs) as a way of reducing the heat buildup of the city in the summer

3. According to the Heat Island Research group, the daily maximum, minimum, and mean temperatures of the months of July and August, 1948–1988 in Phoenix, AZ, during which the city tripled in population and area, increased at a rate of 0.13°F/year. The mean increase of the minimum temperature was a higher 0.18°F/year. Give four specific (heat-generating) technologies that may have accounted for this increase.

Heat-producing or radiating technology

1. Air conditioning and heat pump compressors.
2. Automobiles, trucks, etc. transportation and construction vehicles.
3. Black tar shingles and coatings for roofing materials and roads (asphalt)
4. Home appliances powered directly or indirectly by the combustion of fossil fuels.

4. Offer four major reasons why fertility rate is high in developing/under-developed countries.

Factors that tend to increase fertility rate

1. Low socio-economic status of women.
2. Lack of education.
3. Availability of birth control technologies or a prohibitive cost of those technologies due to the poor economic conditions there.
4. Ethnic or religious taboo.

5. The 1994 Global Summit on Population and Development focused on which three areas for action?

Areas for action

1. Reproductive rights
2. Empowerment of women
3. Reproductive health

6. Compare the fertility, infant mortality, and HIV/AIDS infection rates of Sweden, Thailand and Somalia using www.odci.gov/cia/publications/factbook/

Country	Fertility rate	Infant mortality rate/1000	HIV/AIDS infection rate
Sweden	1.54	3.44	0.08%
Thailand	1.86	29.5	2.15%
Somalia	7.05	122.15	3%?

7. a. What are the high and low estimates for China's total fertility rate in the year 2020? *High: 2.3 TFR, low 1.6.*

   b. At the higher figure, how much more will the population grow in the number of individuals by the year 2050? See the graph below. *1.4 billion.*

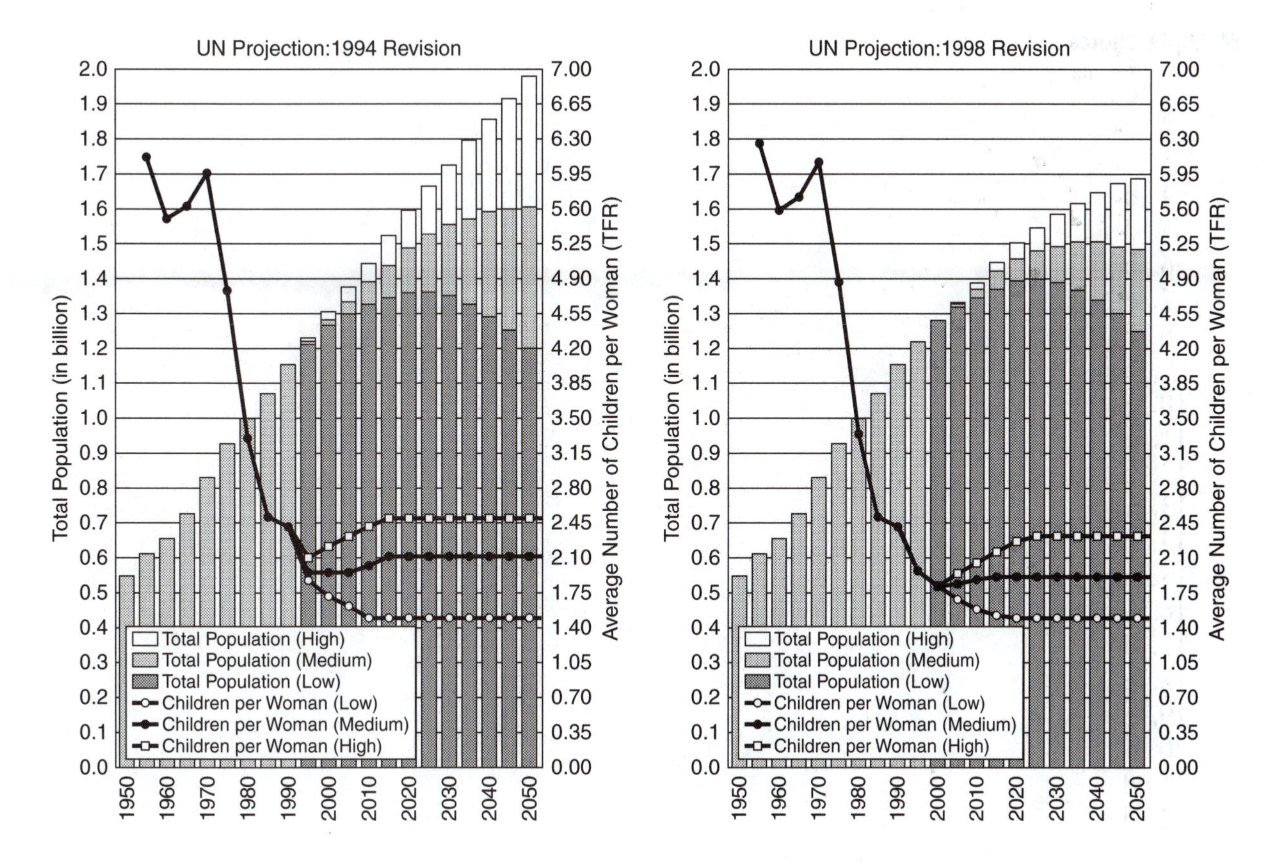

8. Using the rule of 70, how many years will be required to double Mexico's current population
   70/1.47 = 47.6 years.

Pop Quiz

Matching

1. f
2. i
3. n
4. a
5. j
6. k
7. d
8. e
9. g
10. h
11. m
12. c
13. b
14. l

Multiple Choice

1. e
2. d
3. c
4. b
5. d
6. b
7. e
8. c
9. e
10. e
11. e
12. b
13. a
14. b
15. c
16. a
17. d
18. e
19. c
20. a
21. d
22. d

10

Fossil Fuels

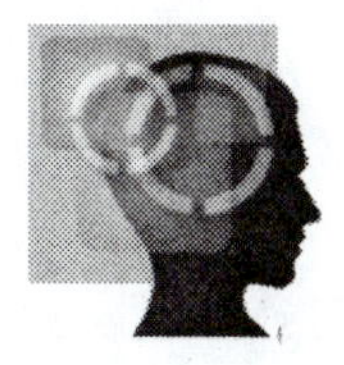

LEARNING OBJECTIVES

After you have studied this chapter you should be able to:

1. Briefly explain U.S. dependence on foreign oil.
2. Compare per capita energy consumption in highly developed and developing/underdeveloped countries.
3. Describe the processes that formed coal, oil, and natural gas.
4. Discuss the advantages and disadvantages (including environmental problems) of using coal.
5. Discuss the advantages and disadvantages (including environmental problems) of using oil and natural gas.
6. Summarize the continuing controversy surrounding the Arctic National Wildlife Refuge.
7. Distinguish among the five types of synfuels (tar sands, oil, shales, gas hydrates, liquid coal, and coal gas) and briefly consider the environmental implications of using synfuels.
8. Relate three reasons that the U.S. needs a comprehensive national energy strategy. Briefly describe the national energy policy of the Bush administration.

SEEING THE FOREST

Since the year 2000, the George W. Bush administration of the U.S. government has made decisions in regards to energy and the environment, such as these listed:

a. failed to ratify the Kyoto Global Warming Treaty limiting the production of carbon dioxide from combustion
b. proposed building 1,300 new coal-burning and nuclear power plants
c. opened up more public lands to logging, mining, and drilling for gas and oil

d. removed restrictions on road building in national forests to allow for increased recreation and harvest of natural resources
e. rejected higher fuel efficiency standards for automobiles. Some say the Bush administration has mostly ignored energy conservation and efficiency technologies
f. changed water pollution regulation to allow for surface-mined overburden (rock and soil) to be pushed into streams at the foot of the mountains
g. proposed oil drilling in the Arctic National Wildlife Refuge

Compare that with the administration's declared National Energy Policy to:

a. modernize conservation.
b. accelerate the protection and improvement of the environment. See Chapter 10.
c. increase energy supplies.

Using information from the text book, fill-in the following observations:

1. How have energy dependence and consumption changed in the U.S. after the OPEC oil embargo of 1973?

	1973	**2002**
Percentage of oil imported		

2. Describe five manners of consumption that have increased per capita energy used per person in the U.S. about 0.8% a year since 1982. Refer to U.S. Per Capita Use of Energy graph.

Consumption characteristic
1.
2.
3.
4.
5.

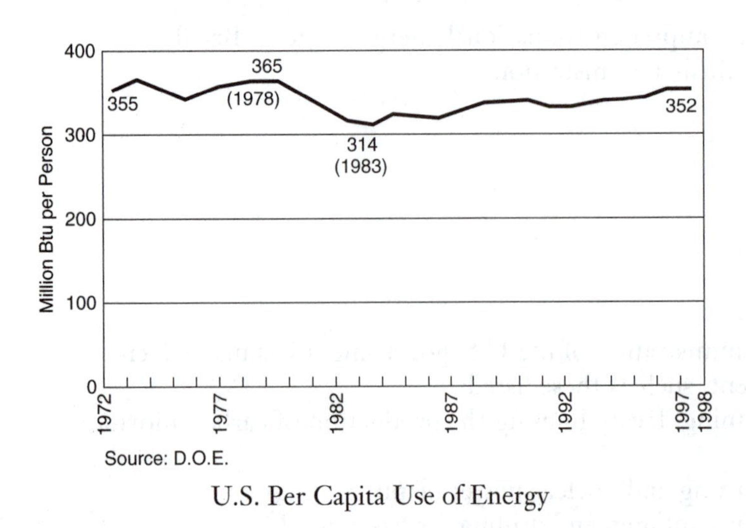

U.S. Per Capita Use of Energy

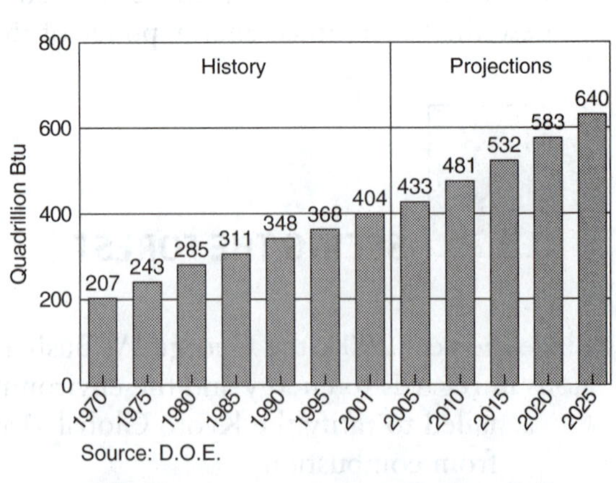

World Enegy Consumption 1970–2050

3. In six years, 1995–2001, world energy consumption per capita increased by ______ %. Refer to the World Enegy Consumption 1970–2050 graph.
4. What measures were taken in the early 1970s and again in the early 1980s to reduce energy consumption?

Energy conservation measures
1.
2.
3.
4.

SEEING THE TREES

VOCAB

We abuse land because we regard it as a commodity belonging to us. When we see land as a community to which we belong, we may begin to use it with love and respect.

Aldo Leopold

1. Fossil fuels—carbon-based fuels (coal, natural gas, and oil) that were created by the decay of buried organisms in anaerobic conditions and high pressure from rock layers above. In a contemporary time frame, they are nonrenewable resources.
2. Surface mining—surface or strip mining is conducted by dragline shovels, blasting, and bulldozers that remove soil and rock covering and harvest seams of coal that encircle mountains or lie beneath the soil of flatter lands.
3. Overburden—the layers of soil and rock that cover valuable deposits of minerals. Minerals such as sulfur, selenium, copper, and iron are leached out by rainwater from overburden deposited on the surface and flow into sediment ponds or natural streams of water. Sulfuric acid is formed by the action of chemosynthetic bacteria to eventually from sulfuric acid that contaminates streams. Rainfall washes soil into streams, causing an increase in turbidity (siltation) that restricts photosynthesis by algae and plants, and nest building by fishes.
4. Subsurface mining—subsurface or deep mining creates shafts and tunnels to expose and remove coal underground. Sulfuric acid commonly drains into streams from mines where water can percolate through sulfur-bearing rocks.
5. Acid deposition—when bituminous coal is burned sulfur, nitrogen and carbon oxides are produced. When these are further oxidized and combined with water in clouds, acids form and are deposited in precipitation on soils and plants, draining streams and lakes. See next page and the chemistry appendix.

Acid rain gas	Reaction to make acid in air	Principal sources
CO_2	$CO_2 + H_2O \rightarrow H_2CO_3$ (carbonic acid)	Combustion of fossil fuels and wood. C_xH_x (Hydrocarbons) + $O_2 \rightarrow CO_2 + H_2O$
SO_2	$SO_2 + H_2O + O_2 \rightarrow H_2SO_4$ (sulfuric acid)	Coal-burning power plants, autos (sulfur and its acids erode exhaust systems over time and decrease the effectiveness of catalytic converters by coating the catalyst) $S + O_{2-} \rightarrow SO_2$.
N_2O, NO, and NO_2	N_2O, NO, and $NO_2 + H_2O + O_2 \rightarrow HNO_3$ (nitric acid)	High-compression diesel and gasoline engines, and coal-burning power generators. $N_2 + O_2 \rightarrow$ N_2O, NO, and NO_{2-}

6. Smoke stack scrubber technology—chemicals, such as calcium carbonate (dry scrubber) or water (wet scrubber), are sprayed into the exhaust gas to remove acids. In addition electrostatic precipitators remove silicate fly ash.
7. Clean Air Act Amendments of 1990—a U.S. law that required major cuts of sulfur dioxide emissions by the year 2000. The requirements were largely revoked by the Bush administration.
8. Clean coal technology—fluidized bed combustion takes place at a lower combustion temperature, reducing nitrogen oxides. Sulfur is removed using the calcium carbonate of limestone. Coal may be transformed into gas, liquid fuels, or oil. Commercials encouraging the use of more coal from Appalachia are appearing, but the clean coal technology being offered is neither economical nor available at this writing in 2002. Historically, coal gasification or liqufication created tar and liquid by-products that were highly carcinogenic and toxic. Some older coal gasification locations are now Superfund sites because of the carcinogenic byproducts deposited there.
9. Petroleum or oil—a mixture of lighter and heavier hydrocarbons from petroleum gas to oils—to gasoline to tars to waxes. Oil is used to make petrochemicals, such as fibers and plastics.
10. Natural gas—a mixture of small hydrocarbon molecules, such as methane, ethane, propane and butane, that is piped into homes and industries for heating and cooking.
11. Liquid petroleum gas—a mixture of small hydrocarbon molecules, such as methane, ethane, propane (mostly), and butane, that can be pressurized into a liquid form and stored in tanks, supplying fuel for heating rural homes.
12. Cogeneration—natural gas combustion can be used to generate steam for heating as well as electricity. This is a good approach for large buildings.
13. Structural traps—geological pockets between porous rock and nonporous rock that can trap gas or oil.
14. Oil Pollution Act of 1990—a congressional law that requires oil and shipping companies to be liable for the environmental damages caused by oil spills. It also requires oil tankers entering U.S. waters after 2015 to have double hulls.
15. Arctic National Wildlife Refuge—the refuge consists of 19 million acres of tundra grassland that has been called "America's Serengeti." Located in northeast Alaska between the Brooks Mountain Range and the Beaufort Sea of the Arctic Ocean, this ecosystem supports thousands of animals, including reindeer, musk oxen, arctic fox, snowy owls, wolves, and polar bears. Unfortunately, there are oil deposits under it.

16. Synfuel—gaseous or liquid fuels that are created from the hydrogenation of coal. This technology was developed and used extensively by Germany in the Second World War.
17. Tar sands—geologic deposits of sand that are saturated with oil. It must be heated with steam underground to fluidize the oil to harvest it.
18. Oil shales—mudstone rock that is saturated with oil. Oil shales must be mined, crushed, and heated to recover the oil. At the present time, this is uneconomical.
19. Gas hydrates—geologic deposits of methane trapped in ice in porous rock.
20. Energy subsidy—a subsidy is a payment, loan, or tax incentive given by governments to industries to encourage the use or consumption of a given service or product—airlines are subsidized through payments for airmail. Fossil fuels and nuclear power industries are subsidized by approximately 32 billion tax dollars a year.
21. Broad form mining deed—the rights to recover buried minerals were sold for pennies an acre over 100 years ago throughout most of the coal-bearing regions of Appalachia. Current landowners have no recourse legally to protect their land from surface mining operations.
22. The Surface Mining Control and Reclamation Act of 1977—required surface mined land to be "reclaimed" by backfill and planting. Nevertheless, recovery is slow, and water pollution abounds.

Questions to Ponder

1. Referring to Table 10-1 and the chapter, compare types of coal and contrast the ways oil, natural gas, and coal are formed by filling in the table below.

Nonrenewable fossil fuel	Source of decaying materials	Conditions of transformation	Generally, where is it found?	Purity of carbon and contaminants
Oil		Anaerobic, high pressure, high temperature		Somewhat impure, high sulfur content, burns with some smoke
Natural gas			Above oil deposits in oceans and land	
Peat	Moss bogs	Somewhat anaerobic but near the surface	U.K.	Very impure, burns with a lot of smoke. Low in sulfur, high in nitrogen
Lignite				Lower sulfur content, carbon content is low (15%), burns better than peat
Bituminous (soft coal)				

Subbituminous				High in nitrogen but low in ______. Carbon content is ______
Anthracite (hard coal)				

Table 10.1 **A Comparison of Different Kinds of Coal**

Type of Coal	*Color*	*Water Content (%)*	*Relative Sulfur Content*	*Carbon Content (%)*	*Average Heat Value (BTU/pound)*	*2000 Cost at Mine for 2,000 lbs of Coal*
Lignite	Dark brown	45	Medium	30	6,000	$11.41
Subbituminous coal	Dull black	20–30	Low	40	9,000	$7.12
Bituminous coal	Black	5–15	High	50–70	13,000	$24.15
Anthracite	Black	4	Low	90	14,000	$40.90

2. Name two types of coal mining, the environmental or health effects associated with them, and the percentages of coal mined by each.

	Method	Environmental or health effect	Percentage coal mined
1.	Subsurface (deep) mining	a.	______
		b.	______
		c.	______
2.	Surface (strip) mining	a.	______
		b.	______
		c.	______

3. Many believe that the residents of the coal mining regions of Appalachia and their local and state governments unintentionally subsidize the mining of coal. Speculate on three ways.

Unintentional local and state subsidy
1.
2.
3.

4. Using Raven-Berg 4e Table 10-A, extrapolate the percent increase in approximate annual gasoline consumption for the U.S. and per person from the year 2000 to 2010 when the population reaches approximately 290 million and gasoline is consumed at 10 million barrels/day.

5. a. Research and explain how cars traveling at 75 mph get up to 25% less gas mileage than when traveling at 55 mph. Figure about 25% less gas mileage at 75 mph.

b. Find your car's optimum EPA gas mileage and compute a 400-mile trip at 75 mph and again at 55 mph. Go to **www.fueleconomy.gov/**

6. According to projections made by the U.S. Department of Energy, how much more (in %) coal, natural gas, and oil energy will the world consume in 2020 as compared to 2002? Refer to the graph below.

	Percentage increase in 2020
Coal	
Natural gas	
Oil	

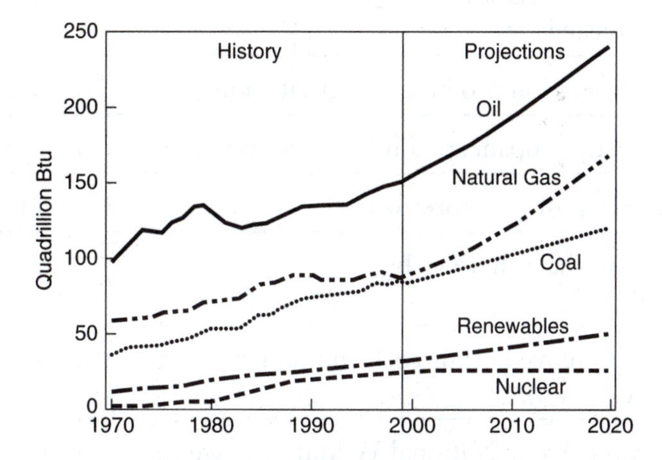

7. Using the graph above, estimate the increase in carbon dioxide emissions from 2000 to 2020.

8. Which acid rain and global warming gas does the U.S. government *not* see as a problem in their *Clear Skies* program of economic incentives to control air pollution? See **www.epa.gov.**

POP QUIZ

Matching

Match the following terms and definitions

____1. A geologic pool of carbon in a gaseous, liquid, or solid state that at one time was in the bodies of organisms:	a. double-hulled tankers
____2. A type of mining that causes water pollution through siltation and acid drainage is:	b. liquid petroleum gas
____3. The most pure form of geologic carbon fuel is:	c. heated

____4. This form of coal burns with a lot of heat but produces large amounts of sulfur and nitrogen oxides:	d. acid rain
____5. This material is formed by the action of chemosynthetic bacteria in the wet overburden of surface mines and in deep mines:	e. anthracite
____6. Smoke stack scrubber technology uses this to remove sulfur dioxide:	f. synfuel
____7. When sulfur, nitrogen, and carbon oxide gases are combined with water, further oxidized in the air and rainfall occurs. This is called:	g. tundra
____8. Clean coal technology removes ______ from coal before it is combusted:	h. sulfur
____9. Natural gas cogeneration produces electricity and ______.	i. strip mining
____10. This fuel is mostly propane, and it is stored in tanks:	j. bituminous
____11. Gas hydrates are geologic stores of:	k. sulfuric acid
____12. Tar sands and oil shales have to be ______ to release the oil.	l. fossil fuel
____13. Where oil forms above porous rock and under nonporous rock is a(n):	m. steam for heating
____14. The biome of the Arctic National Wildlife Refuge is:	n. structural trap
____15. The Oil Pollution Act of 1990 requires:	o. methane
____16. Gasoline or naptha made from coal would be ______:	p. calcium carbonate

Multiple Choice

1. The long lines at gasoline stations in 1973–1974 were due to what event?
   a. the advent of large vans and SUVs
   b. the 1973 war between Egypt and Israel
   c. a large explosion at an oil refinery
   d. an embargo by the Persian Gulf and other oil producers
   e. a mandated switch from leaded to unleaded gasoline

2. In 1979, President Jimmy Carter said that dealing with the ______ crisis was the moral equivalent of war.
   a. war between the Arab states and Israel
   b. energy
   c. Three Mile Island reactor meltdown
   d. the Iranian hostage taking
   e. striking miners

3. Most of the electrical power used in the U.S. is generated by what process?
   a. hydroelectric
   b. burning of coal
   c. burning of natural gas
   d. nuclear fission
   e. solar

4. It could be said that the power generation process that produces the most air pollution is ______?
   a. hydroelectric
   b. hydrogen fuel cells
   c. combustion of natural gas
   d. nuclear fission
   e. burning of coal

5. A form of power generation that is currently underutilized is which of the following?
   a. hydroelectric dams
   b. solar
   c. wind
   d. geothermal
   e. all of the above

6. This fossil fuel is found in geologic ices and structural traps:
   a. methane
   b. oil
   c. LPG
   d. tar sand
   e. synfuel naptha

7. King Edward I of England forbade, under the penalty of death, the burning of a tar-filled "sea coal" in London in the year 1306. Most likely the coal being burned was:
   a. anthracite
   b. bituminous
   c. subbituminous
   d. peat
   e. all of the above

8. In 1974, the national interstate speed limit was reduced to 55 miles per hour in the U.S. At the former speed limit of 70 mph, approximately how much more gas was burned at the higher speed before the law was enacted?
   a. 10%
   b. 25%
   c. 35%
   d. 45%
   e. There was no change in fuel consumption after 1974.

9. Hybrid fuel engines in automobiles have this positive effect:
   a. eliminating gasoline use
   b. making the car's power totally electric

c. combining solar power with electric
d. conserving gasoline
e. all except a

10. This coal is mined in the western U.S., and its use represents the chief efforts made by power generating companies to reduce sulfur dioxide emissions in the 1980's and 1990's:
    a. lignite or brown coal
    b. peat
    c. bituminous or soft coal
    d. subbituminous
    e. anthracite or hard coal

11. Strip or surface mining causes the following environmental problem:
    a. mine acid drainage
    b. increased water turbidity
    c. acids and minerals in the water damage metal plumbing
    d. destruction of mountains
    e. all of the above

12. Which means of power generation is mismatched with its chief pollutant released?
    a. coal burning: sulfur dioxide
    b. natural gas combustion: carbon dioxide
    c. nuclear: radioactive gas emissions
    d. nuclear: solid radioactive waste
    e. geothermal: hot steam

13. Which pollutant is voluminous and common to the use of synfuels, natural gas, LPG, gasoline, and coal?
    a. carbon monoxide
    b. carbon dioxide
    c. sulfur dioxide
    d. nitrogen oxides
    e. Trick question alert! All of these are equally voluminous.

14. Which nation has the largest supply of natural gas?
    a. Iran
    b. Saudi Arabia
    c. Russia
    d. United States
    e. Quatar

15. Which area has the greatest oil reserves?
    a. Persian Gulf states
    b. North Africa
    c. South America
    d. North Sea
    e. Russia

16. Which area has the greatest coal reserves?
    a. Mexico
    b. UK, France, and Germany
    c. United States and Canada
    d. Venezuela, Colombia, and Brazil
    e. North Africa

17. Which material occurs naturally in some soil and is mined and sprayed in smoke stack scrubbers to neutralize sulfuric acid?
    a. humus
    b. limestone/calcium carbonate
    c. water and lye
    d. sulfur
    e. silicate fly ash

18. The principal problem with the U.S. Surface Mining Control Act is ______?
    a. mine acid drainage control
    b. siltation control
    c. breaking of sediment ponds and resulting flooding
    d. reclamation of areas mined previous to 1977
    e. dumping toxic wastes into mines

19. Which is a pollutant from coal-burning power plants that results in acid rain?
    a. sulfur dioxide
    b. nitrogen dioxide
    c. carbon dioxide
    d. nitric oxide (NO), and dinitrogen oxide
    e. all of the above

20. Which natural resource has been used by the U.S. and Germany to produce large amounts of plastics, methane, gasoline, and oil?
    a. oil
    b. coal
    c. wood
    d. natural gas
    e. a and b only

21. In North and Central America, which country has the highest per capita consumption of energy?
    a. Canada
    b. United States
    c. Mexico
    d. Bermuda

22. Excepting eastern Europe and the former Soviet Union, which statement is an accurate prediction of the energy use and uncontrolled pollution generated in industrialized and developing countries in 2020 as compared to 2002?
    a. More energy will be used in industrialized countries, producing an equal amount of pollution.
    b. Developing countries will use more energy and produce more pollution.

c. The energy consumption and pollution produced by developing and industrialized countries will be about the same.
d. The energy consumption of industrialized and developing countries will nearly the same, but the pollution generated by developing countries will be greater.

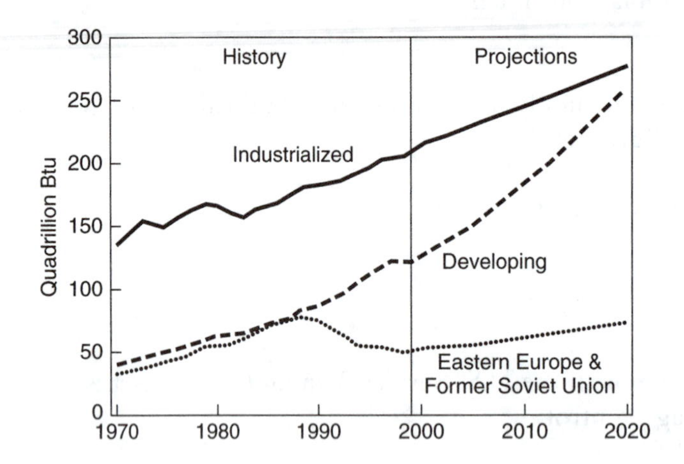

CYBER SURFIN'

1. For a good general discussion of the politics of energy generation and pollution, see the U.S. National Public Radio site: **www.pbs.org/newshour/bb/environment/energy/#**
2. To see the administration's proposals on air pollution, go to **www.epa.gov/**
3. An interesting series of article on natural resource consumption and pollution can be seen at the British Broadcasting Corporation (BBC) site: **news.bbc.co.uk/hi/english/static/in_depth/world/2002/disposable_planet/**
4. To see the author's aerial photos of strip mining, go to **www.gpc.edu/~jaliff/slidesho.htm**
5. For the environmentalist perspectives of energy consumption, pollution, and politics go to the Sierra Club site: **www.sierraclub.org/energy/** At the same site, learn more about mountaintop removal in Appalachia: **www.northstar.sierraclub.org/corps_mountain_tops.htm.** Then take a virtual tour of the Arctic National Wildlife Refuge: **www.sierraclub.org/wildlands/arctic/**
6. The University of Connecticut has a special report on ANWR, see **arcticcircle.uconn.edu/ANWR/**
7. To see a current map of acid rain measurements in the U.S. (pH below 7 = acid, with each number less equal to 10× increase), go to the U.S. Geological Survey, **http://nadp.sws.uiuc.edu/isopleths**

'LIKE' BOOKING IT

1. Wilson EO. ***The Future of Life.*** New York: Knopf Publishing Group, 2002.
2. McNeil JR, McNeil JR, and Kennedy P. ***Something New Under the Sun: An Environmental History of the Twentieth-Century World (Global Century Series).*** New York: W.W. Norton, 2001.
3. Montrie C. ***To Save the Land & People: A History of Opposition to Surface Coal Mining in Appalachia.*** Chapel Hill, NC: University of North Carolina Press, 2003.
4. Worldwatch Institute, ed. ***State of the World 2002.*** New York: W.W. Norton, 2002.

5. Worldwatch Institute, Renner M, and Sampat P. ed. ***Vital Signs 2002: The Trends That Are Shaping Our Future (Vital Signs, 2002).*** New York: W.W. Norton, 2002.

ANSWERS

Seeing the Forest

1. How have energy dependence and consumption changed in the United States after the OPEC oil embargo of 1973?

	1973	2002
Percentage of oil imported	35%	50%

2. Describe five manners of consumption that have increased per capita energy used per person in the U.S. about 0.8% a year since 1982. Refer to the graph on page 148 from the U.S. Department of Energy.

Consumption Characteristic
1. An increased number of vehicles per capita
2. An increased number of miles driven to commute to work
3. Increased number of home appliances, especially larger water heaters and hot tubs
4. Increased lighting and heating in homes
5. Energy used for plastics and aluminum production

3. In six years, 1995–2001, world energy consumption per capita increased by 9.8%. Refer to the graph on page 148.
4. What measures were taken in the early 1970s and again in the early 1980s to reduce energy consumption?

Energy conservation measure
1. Reduction of U.S. national speed limit to 55 mph
2. Improved fuel efficiency of automobile engines
3. Improved fuel efficiency of automobile tires (radials) and oils
4. Recycling aluminum cans and plastic

Questions to Ponder (Many of these answers are examples of many possible responses, please don't memorize them.)

1. Referring to Table 10-1 and the chapter, compare and contrast the ways oil, natural gas, and coal are formed by filling in the table below.

Nonrenewable fossil fuel	Source of decaying materials	Conditions of transformation	Generally, where is it found?	Purity of carbon and contaminants
Oil	Aquatic organisms	Anaerobic, high pressure, high temperature between rock layers	Under oceans and the land, in the Persian Gulf states, the North Sea, Texas, etc.	Somewhat impure, high sulfur content, burns with some smoke
Natural gas	Aquatic organisms	Anaerobic, high pressure, high temperature between rock layers	Above oil deposits in oceans and land in Russia, Iran, etc.	Burns cleanly but produces carbon dioxide
Peat	Moss bogs	Somewhat anaerobic but near the surface	Scotland, U.K.	Very impure, burns with a lot of smoke
Lignite	Buried swamps	Anaerobic, high pressure, high temperature between rock layers, but not as much as blacker, harder forms	Colorado, Montana, and Wyoming	Moderate sulfur content, carbon content is 30%, burns better than peat
Bituminous (soft coal)	Buried swamps	Anaerobic, high pressure, high temperature between rock layers	Appalachia, Europe, China	May have high sulfur, nitrogen, and silica levels. Carbon content is 50–70%
Subbituminous	Buried swamps	Anaerobic, high pressure, high temperature between rock layers	Colorado, Montana, and Wyoming; British Columbia	High in nitrogen but low in sulfur. Carbon content 35–45%.
Anthracite (hard coal)	Buried swamps	Anaerobic, high pressure, high temperature between rock layers, but more so as compared to bituminous	Eastern Pennsylvania	Greatest purity and least in contaminants. Carbon content 90%

Table 10.1 A Comparison of Different Kinds of Coal

Type of Coal	*Color*	*Water Content (%)*	*Relative Sulfur Content*	*Carbon Content (%)*	*Average Heat Value (BTU/pound)*	*2000 Cost at Mine for 2,000 lbs of Coal*
Lignite	Dark brown	45	Medium	30	6,000	$11.41
Subbituminous coal	Dull black	20–30	Low	40	9,000	$7.12
Bituminous coal	Black	5–15	High	50–70	13,000	$24.15
Anthracite	Black	4	Low	90	14,000	$40.90

2. Name two types of coal mining, the environmental or health effects associated with them, and the percentages of coal mined by each.

	Method	Environmental or health effect	Percentage coal mined
1.	Subsurface (deep) mining	a. mine acid drainage	40%
		b. mineral and coal dust caused lung diseases, respectively, silicosis and black lung or anthracosis	
		c. deep mining is one of the most dangerous occupations	
2.	Surface (strip) mining	a. mine acid damage	60%
		b. destruction and displacement of wildlife and landscape	
		c. siltation and mineralization of streams	

3. Many believe that the residents of the coal mining regions of Appalachia and their local and state governments unintentionally subsidize the mining of coal. Speculate on three ways.

Unintentional local and state subsidy
1. Repair of smaller local roads used to haul coal
2. Destruction of farms and natural areas
3. Increased failure of plumbing. Increased expenses in water treatment

4. Using Raven-Berg 4e Table 10-A, extrapolate the percent increase in approximate annual gasoline consumption for the U.S. and per person from the year 2000 to 2010 when the population reaches an approximately 290 million and gasoline is consumed at 10 million barrels/day.
The consumption for the U.S. should be about 10 million/barrels/day × 42 gallons of gasoline/barrel = 420 million gallons/day × 365 days/year = 153,300 million gallons/year. The year 2000 figure was 128,160 million gallons/year (8.36 million barrels/day × 42 gallons/barrel × 365 days/year). Therefore, the projected increase for the U.S. in 2010 will be 25,140 million gallons/year: 153,300-128,200 million gallons/year is an increase of 19.6% (25,100/128,200 × 100 = %). In 2010 each person will consume about 529 gallons of gasoline/year (420 million gallons/day divided by 290 million people × 365 days/year) as compared to the 2000 figure of 494 gallons per year (351.1 million gallons/day divided by 287.4 million people × 365 days/year), an increase of 7% consumption per person, 35/494 = 0.07 or 7%.

5. a. Research and explain how cars traveling at 75 mph get up to 50% less gas mileage than when traveling at 55 mph. *As you go faster, air piles up in front of the car, providing resistance to its movement.*

   b. Find your car's optimum EPA gas mileage and compute a 400-mile trip at 75 mph and again at 55 mph. Figure that the 400-mile trip at 75 mph will take 25% more gas/mile. Go to www.fueleconomy.gov. *My Ford gets 26.5 mpg at 55 mph and it would get 26.5 × .25 at 75 mph, or 20 mpg. I would use 15.1 gl at 55 mph and 20 gl at 75 mph. At $1.50 per gallon, I would save almost $6.50.*

6. According to projections made by the U.S. Department of Energy, how much more (in %) coal, natural gas, and oil energy (BTU) will the world consume in 2020 as compared to 2002? Refer to the graph below.

	Percentage increase in 2020
Coal	Coal will increase from 80 quadrillion BTU in 2002 to 110 q. BTU in 2020, an increase of 30/80 or 37.5%.
Natural gas	Natural gas will increase from 80 q BTU to 165 q. BTU, an increase of 85/80 or 106.3%.
Oil	Oil will increase from 150 q. BTU to 245 q BTU, an increase of 95/150 or 63.3%.

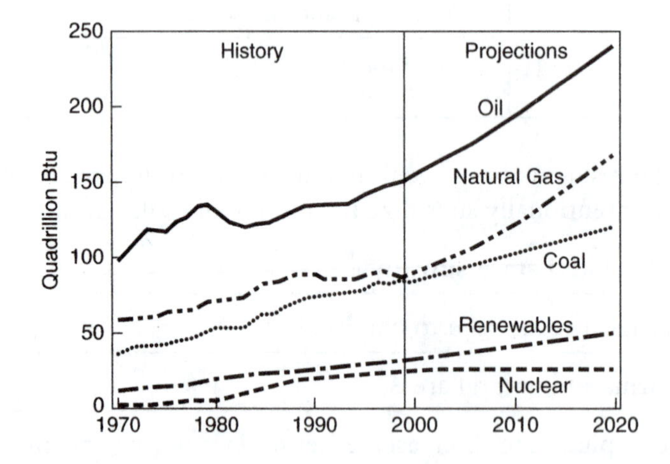

7. Using the graph above, estimate the increase in carbon dioxide emissions from 2000 to 2020. *The total BTUs produced by the consumption of fossil fuels will increase from 310q BTU (80 q from coal, 80 q from natural gas, and 150 q from oil) to a grand total of 520 q BTU (110 q from coal, 165 q from natural gas and 245 q from oil). 210 increase/310 in 2000 × 100 = 68% increase.*
8. To see the administration's proposals on air pollution, go to www.epa.gov/.

Which acid rain and global warming gas does the U.S. government *not* see as a problem in their Clear Skies program of economic incentives to control air pollution? *Carbon dioxide.*

Pop Quiz

Matching

1. l
2. i
3. e
4. j
5. k
6. p

7. d
8. h
9. m
10. d
11. e
12. c
13. n
14. g
15. a
16. f

Multiple Choice

1. d
2. b
3. b
4. e
5. e
6. a
7. b
8. b
9. d
10. d
11. a
12. c
13. b
14. c
15. a
16. c
17. b
18. d
19. e
20. a
21. a
22. d

11

Nuclear Energy

LEARNING OBJECTIVES

After you have studied this chapter you should be able to:

1. Distinguish between nuclear energy and chemical energy and define radioactive elements and radioactive decay.
2. Summarize how mined uranium becomes fuel for nuclear power plants.
3. Diagram a pressurized water reactor and show its three water circuits.
4. Contrast conventional nuclear fission, breeder nuclear fission, and fusion.
5. Describe the two major nuclear power plant accidents discussed in the text—Three Mile Island and Chernobyl.
6. Distinguish between low-level and high-level radioactive waste.
7. Relate the pros and cons of permanent storage of high-level radioactive waste at Yucca Mountain.
8. Explain what happens to nuclear power plants after they are closed.

SEEING THE FOREST

The phenomenon of radiation emitted by certain materials was first discovered approximately 100 years ago. Interestingly, many of the nuclear physicists of the 1920s and 1930s were German or eastern European Jews or Italians who fled fascist persecution.—Lucky for us!

For the scientists listed below, describe their contributions to the discovery of nuclear power.

Scientist	Contribution or discovery
Wilhelm Röntgen	
Henri Becqueril	
Marie Curie	
Albert Einstein	
Ernest Rutherford	
Otto Hahn and Fritz Strassmann	
Enrico Fermi	
Edward Teller	

SEEING THE TREES

VOCAB

There is no evil in the atom—only in mens' souls.

Adlai Stevenson, U.S. senator, presidential candidate and U.N. ambassador.

1. Atomic energy—Energy is the ability to do work in physics terms. It has several forms. Atomic energy holds the nucleus of an atom together. Atoms may be split or fused to release that energy. The sun converts atomic energy to radiant energy that includes light. Green plants have chlorophyll and accessory pigments that trap the light and convert it into energy stored in the chemical bonds of glucose and other chemical compounds. Therefore, photosynthesis converts electromagnetic light energy into chemical energy.
2. Nuclear fission—atoms of heavy elements (uranium-233, uranium-235, and plutonium-239) are split into smaller atoms (cobalt-60, strontium-90, cesium-137, and iodine-131). The first atomic bombs were fission bombs.

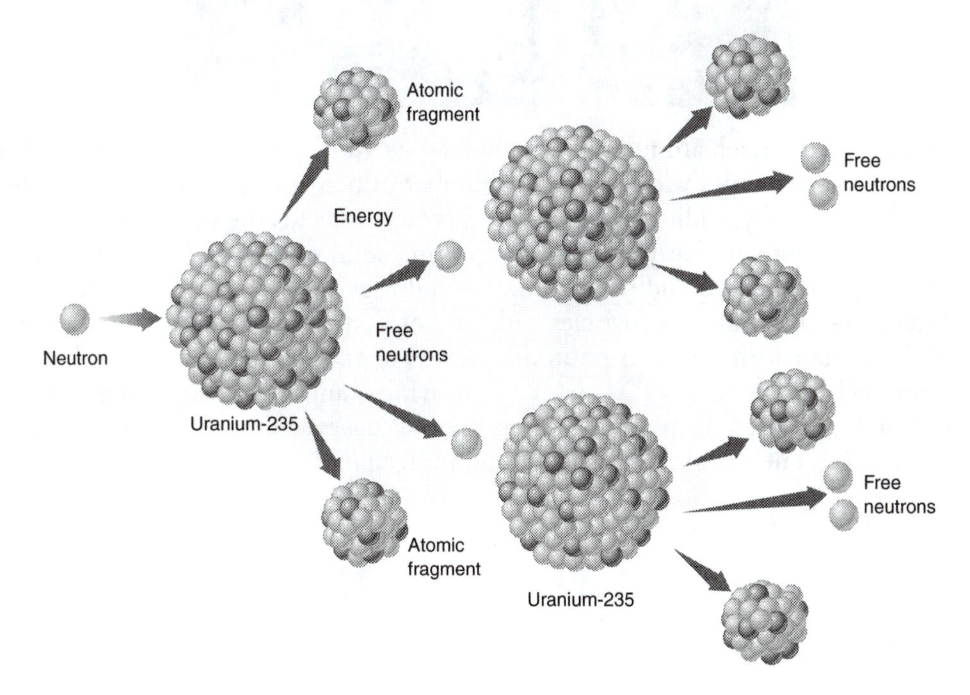

3. Nuclear fusion—simple hydrogen atoms are fused with neutrons to make heavier isotopes of hydrogen that, under very high temperatures and pressures, are fused into a heavier element, mostly helium. There is an accompanying loss of measurable mass that is converted into a large amount of energy (mass defect principle). The fusion process was first observed in the sun and later applied to bomb production. The Hungarian expatriate Edward Teller argued unsuccessfully for the first atomic bombs to be made this way.
4. $E = mc^2$—Einstein's famous equation that hinted at the vast amount of energy produced in a nuclear reaction. Because the speed of light is a large number—300,000,000 meters/sec—this squared is even larger, multiplied by the mass of the material used. E in joules = mass in kilograms × 300,000,000 m/sec squared.
5. Atomic mass—the number of protons plus the number of neutrons. Simple hydrogen does not have neutrons, but heavier isotopes of hydrogen and other heavier atoms have them—aluminum has an atomic weight of 27—it has 13 protons and 14 neutrons.
6. Atomic number—the atomic number is the same as the number of positively charged protons, which, in turn, is equal to the number of negatively charged electrons in an atom. Therefore, all solitary atoms have an electrical charge of zero—aluminum is atomic number 13: it has 13 protons and 13 electrons. See **www.webelements.com**

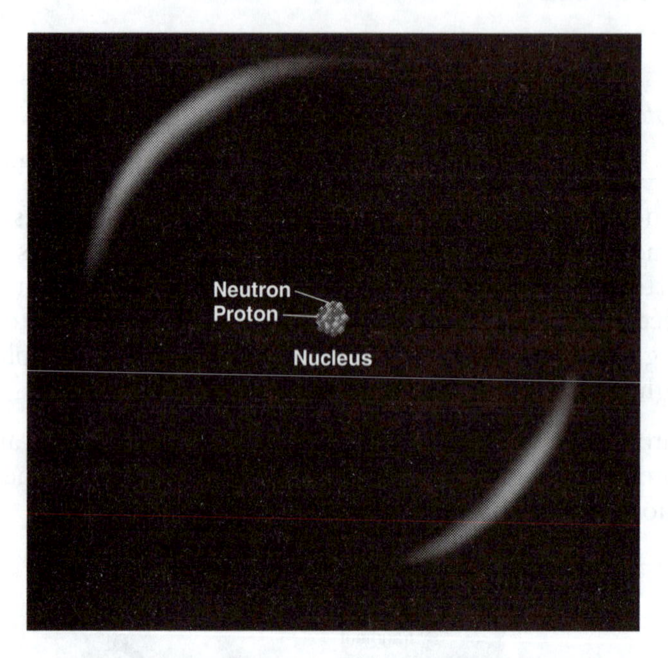

7. Radioactive isotopes—Each atom, as determined by its atomic number or number of protons, has several forms or isotopes with a different number of neutrons and, therefore, different atomic weight. Generally, adding neutrons to an atomic nucleus (neutron capture) makes it unstable. Then it begins to decay its mass by emitting solid particles with measurable mass—alpha particles (two neutrons and two protons bound together), beta radiation (electrons, protons and neutrons) and waves of particles with masses that are not measurable (electromagnetic radiation). The latter form includes photons (packets of energy) of shorter wavelength and more energy such as gamma rays, X-rays, and ultraviolet; and longer wavelengths of less energetic visible light, infrared (heat), microwaves, and radio waves. In the spectrum from gamma rays to radio waves, energy content decreases and wavelength increases.

Table 11.1 Some Common Radioactive Isotopes Associated with the Fission of Uranium

Radioisotope	*Half-Life (years)*
Iodine-131	0.02 (8.1 days)
Xenon-133	0.04 (15.3 days)
Cerium-144	0.80
Ruthenium-106	1.00
Krypton-85	10.40
Tritium	12.30
Strontium-90	28.00
Cesium-137	30.00
Radium-226	1,600.00
Plutonium-240	6,600.00
Plutonium-239	24,400.00
Neptunium-237	2,130,000.00

8. Radioactive half-life—the amount of time required for the decay of one-half of the mass of a given radioactive isotope and have one-half of its mass remaining. The half-life of tritium (the isotope of hydrogen with a mass of 3) is 12.3 years. That means fusion bomb warheads in storage lose half their fusion mass in 12.3 years. Government "nuclear reservations" at the Savannah River, SC, and Hanford, WA, sites have made tritium to refurbish warheads.

9. Nuclear reactor—a device that starts and controls nuclear fission reactions. Reactors are composed of fissionable material in the form of nuclear fuel pellets that are placed in rods. The nuclear fuel rods are inserted so that they obtain a critical mass—a sufficient mass of fissionable material must accumulate in one location for there to be enough neutrons to be generated to cause fission or splitting of the heavy uranium atoms.

10. Moderators and control rods—deuterium oxide (heavy water), carbon (graphite), and boron can slow or absorb neutrons and control the criticality of the mass so that a bomb-like fission does not occur. The control rods are placed so that they regulate the critical mass of the reactor by making it smaller or larger in concentration. Nuclear reactor cores have water or liquid sodium to transfer heat so that steam can be generated to drive electricity-generating turbines. Sodium core reactors are very dangerous. If pipes carrying liquid sodium leak and come into contact with secondary circuit water, a steam explosion occurs. This happened in Russia.

11. Water reactor circuits—The *primary water circuit* contains water in a closed pipe at high pressure that is heated by the nuclear reactor. The pressurization prevents the superheated liquid water (293°C, water boils at 100°C) from turning into steam. The water in the primary circuit heats water in the *secondary water circuit,* producing steam that drives electricity-generating turbines. This water is cooled in a condenser by cooling water provided by the *tertiary water circuit.* When that water is heated by the condenser, it is put into a cooling pond or tower beside the generating plant to return heat to the atmosphere. Coal and nuclear plants may return water to large lakes or rivers, causing thermal pollution in aquatic ecosystems. See the figure on top of next page.

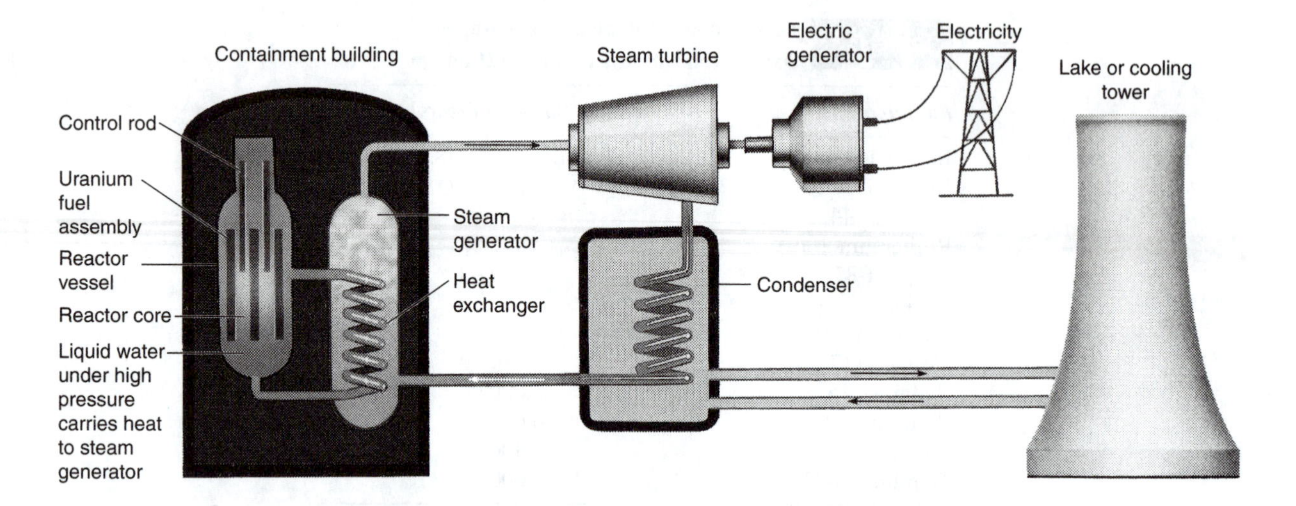

12. Enriched uranium—"natural" uranium is uranium-238 that decays very slowly to lead. U-235 is fissionable, it generates neutrons that collide with U-238 to eventually change the material into plutonium-239, which is also fissionable. The enriching on natural U-238, with U-235 and Pu-239, to about 3% increases the heat generated in nuclear reactors. Bomb-grade uranium is 20% U-235.

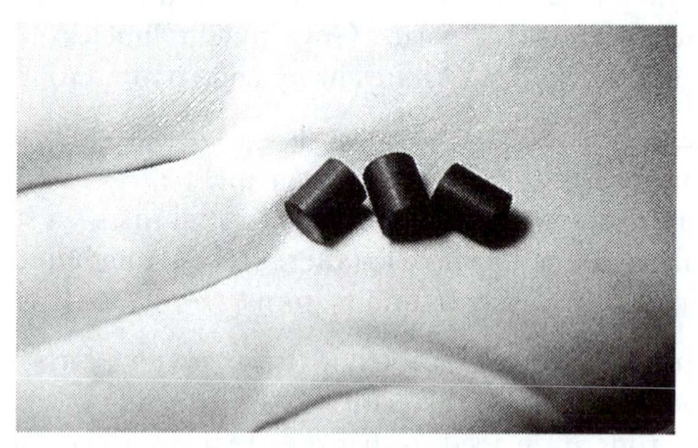

Nuclear fuel pellets

13. Breeder reactors—these convert U-238 to Pu-239 or thorium-232. In this process, they actually generate more fissionable material than they use. The U-238 from nuclear power plant waste could be reprocessed to use in breeder reactors, thus eliminating the need to mine and process more uranium for quite some time, maybe 100 years. Unfortunately, it is still cheaper to mine the uranium, and many of the plants have been closed due to safety concerns. North Korea has a breeder reactor.
14. Radioactive fallout—if fissionable materials explode, the radioactive isotope atoms produced by fission are carried by wind to other locations where the contaminants can enter food chains or cause direct damage to humans and other organisms. Gamma rays and X-rays can mutate DNA—producing cancer—or cause the direct death of cells—causing radiation sickness or death—depending on the amount of exposure. The explosion of the Chernobyl, Ukraine, nuclear power plant spread radioactive fallout first to northwest Europe (Belarus, Russia, and Scandinavia), but eventually to France, Greece, and Siberia.
15. Meltdown—if the reactor rods become too hot due to a loss of coolant or inadequate moderation, they may melt. In melting, they may penetrate the reactor lining and sink into the ground.

A 50% meltdown occurred at Three Mile Island, PA, in 1979. Also, the superheated metal may come into contact with coolant water, producing an explosion. This is what happened at Chernobyl, Ukraine. See the simulation in **Cyber Surfin'** below.

16. Radioactive waste—fissionable materials produce radioactive waste that can leak into the soil and groundwater to other locations. The radioisotopes can enter food chains or cause direct damage to humans and other organisms. There has been some groundwater contamination at the Savannah River, SC, and Hanford, WA, bomb materials production sites. Radioactive wastes are classified as *low level* if they give off small amounts of gamma and X-radiation (ionizing, because their photons can knock electrons from atoms), or high level. Generally, low-level wastes are contaminated clothing and equipment, and high-level wastes are spent reactor rods, cooling fluids, or radioactive gases, such as tritium. Liquid waste may be solidified into *glass logs* for underground storage.
17. 1982 U.S. Nuclear Waste Policy Act—required that the federal government identify and provide permanent storage of nuclear waste underground. Salt beds near Carlsbad Caverns, NM, and Yucca Mountain, NV, are active storage sites, the latter activated over the protests of residents and elected officials of Las Vegas and the state of Nevada—illustrating the NIMBY (not in my backyard) and NIMTOO (not in my term of office) reactions, respectively.
18. Decommissioning nuclear power plants and bomb-producing facilities—unsafe or uneconomical nuclear power plants and bomb-producing facilities have to be "stored" or protected while the radioactive materials decay. Also, nuclear reactors are "entombed" in concrete (as was Chernobyl's exploded reactor). Some liquid waste materials, now solidified, and their above-ground storage containers, have to be maintained and protected for hundreds of years while their radioisotopes decay sufficiently.

Questions to Ponder

1. Identify, define, and describe the three types of energy involved in the chain of energy from the Sun to plants to animals.

Energy form	Definition and description

2. Consult a periodic table of the elements or go to www.webelements.com. Consider the following questions:
   a. Why is uranium atomic number 92?
   b. How many electrons do hydrogen and its isotopes have?
   c. The isotopes of hydrogen are protium-1, deuterium-2, and tritium-3. What do the numbers represent?
   d. Common and dangerous radioisotopes that result from atomic fission are cesium-137, strontium-90, and iodine-131. In the periodic table, elements in a vertical column share chemical characteristics. Also, certain radioactive isotopes gain entrance into plants and on into food chains. Therefore, propose a reason for the dangers of cesium-137 and strontium-90.

Radioisotope	Reason why it is particularly dangerous in a food chain

e. Explain how iodine-131 has a higher atomic mass number as compared to iodine-127.
f. Explain how iodine-131 is dangerous.

3. Compare and contrast nuclear fission and nuclear fusion reactions as to the types and supplies of fuel available, and nuclear wastes produced. See the links in Cyber Surfin', below.

Nuclear reaction	Type and supply of fuel	Waste products
Fission		
Fusion		

4. Compare how critical mass is obtained in a reactor and in a fission bomb.
5. Using Table 11-2, compare the types, relative amounts, and environmental or health effects of pollution from coal-burning power plants and nuclear power generators by filling in the table below.

Table 11.2 **Comparison of Environmental Impacts of 1,000-MWe Coal and Nuclear Power Plants***

Impact	*Coal*	*Nuclear (Conventional Fission)*
Land use	17,000 acres	1,900 acres
Daily fuel requirement	9,000 tons/day	3 kg/day
Availability of fuel, based on present economics	A few hundred years	100 years, maybe longer (much longer with breeder fission)
Air pollution	Moderate to severe, depending on pollution controls	Low
Climate change risk (carbon dioxide emissions)	Severe	No risk
Radioactive emissions, routine	1 curie	28,000 curies
Water pollution	Often severe at mines	Potentially severe at nuclear waste disposal sites
Risk from catastrophic accidents	Short-term local risk	Long-term risk over large areas
Link to nuclear weapons	No	Yes
Annual occupational deaths	0.5 to 5	0.1 to 1

* Impacts include extraction, processing, transportation, and conversion. Assumes coal is strip-mined. (A 1,000-MWe utility, at a 60% load factor, produces enough electricity for a city of 1 million people.)

	Pollutant or land use effect	Rate the amount as very large, large, or small	Health or environmental consequence
Coal burning	Land use for mining		
	Sulfur dioxide, nitrogen oxides emissions		
	Mine acid and siltation		

	Risk from catastrophic accident		
Nuclear	Land use for mining		
	Nuclear waste generation		
	Local radiation, water and air pollution		
	Risk of catastrophic accident		

6. Compare and contrast the accidents at Chernobyl and Three Mile Island *See the text* at the World Nuclear Association: **www.world-nuclear.org/info/chernobyl/inf07.htm** and **www.world-nuclear.org/info/inf36.htm.**

7. Describe two problems associated with the gamma-ray sterilization of food. (Food is exposed to gamma radiation, and the bacteria and fungi on it are killed.)

	Problem associated with gamma ray sterilization of food
1.	
2.	

8. The energy requirement for each U.S. citizen is approximately 5×10^{11} joules. Using $E = mc^2$, compute the number of joules produced by 0.0056 grams of fissionable atoms.

POP QUIZ

Matching

Match the following terms and definitions

____ 1. The energy stored in the food molecule glucose is ______ energy.	a. isotopes
____2. The energy that holds protons and neutrons together in the nucleus of an atom is ______ energy.	b. Nuclear Waste Policy Act
____ 3. Hydrogen fusion occurs to make ______ (type) energy that plants use in photosynthesis.	c. breeder
____ 4. The process of splitting U-235 atoms is called:	d. half-life
____ 5. Atoms that have the same atomic number, but different atomic weights are called:	e. nuclear fuel pellets

____ 6. Adding extra neutrons to the stable nucleus of an atom makes it ______.	f. atomic
____ 7. Radiation can consist of solid particles that include(s):	g. high-level radioactive waste
____ 8. The most dangerous form of radiation is the electromagnetic wave called:	h. chemical
____ 9. The length of time required to decay three grams of carbon-14, with 1.5 grams remaining is called ______.	i. critical mass
____ 10. The amount of fissionable material that has to be concentrated in one spot to start a nuclear fission chain reaction is called:	j. light/radiant
____ 11. The stream that drives turbines that generate electricity is produced from water in the ______.	k. nuclear fission
____ 12. Water that allows for the steam to be condensed to liquid water comes from the:	l. gamma rays
____ 13. Many in the world fear that this type of reactor can make more fissionable material that is used for making bombs. This is a ______ reactor.	m. emit radiation
____ 14. The presence of strontium-90, iodine-131, and cesium-137 in the atmospheric dust is:	n. electrons, neutrons, helium nuclei
____ 15. Spent reactor rods are:	o. tertiary circuit
____ 16. "Out of sight, out of mind" is the theme of the:	p. secondary circuit
____ 17. Enriched uranium that is placed into fuel rods:	q. radioactive fallout

Multiple Choice

1. Nuclear fusion reactions on the Sun convert atomic energy to:
   a. light.
   b. gamma rays.
   c. X-rays.
   d. heat.
   e. all of the above.

2. The first atomic bombs and reactors used this process:
   a. fusion of hydrogen
   b. fission of uranium
   c. fusion of uranium
   d. fission of tritium
   e. all of the above

3. Which of the following technologies generates the least electrical power in the U.S.?
   a. hydroelectric
   b. burning of coal
   c. burning of natural gas
   d. nuclear fission
   e. solar

4. The power generation process that produces the least air pollution is ______.
   a. hydrogen fuel cells assisted by carbon fuel combustion
   b. combustion of natural gas
   c. nuclear fission
   d. clean coal technology
   e. combustion of oil

5. This material is used to make the most powerful atomic bombs:
   a. tritium-3
   b. uranium-238
   c. uranium-235
   d. hydrogen/protium-1
   e. all of the above

6. Enriched uranium has more ______ and less ______.
   a. U-238, U-235
   b. U-235, U-238
   c. iodine-131, strontium-90
   d. boron, tritium
   e. metal, radiation

7. This material can absorb or slow down neutrons:
   a. graphite
   b. boron
   c. deuterium oxide
   d. heavy water
   e. all of the above

8. Which country has attained a good safety record of nuclear power generation and relatively low cost by standardizing their facilities?
   a. U.S.
   b. Russia
   c. U.K.
   d. France
   e. Germany

9. The water that is sent to cooling towers in a nuclear power plants comes from here.
   a. primary water circuit
   b. water heated by the reactor
   c. water that was converted to steam
   d. water that was used to condense the secondary circuit water
   e. all except a

10. Which of the following radioisotopes in nuclear fallout has the longest half-life?
    a. tritium-3
    b. strontium-90
    c. cesium-137
    d. krypton-85
    e. radium-226

11. Because of its similarity to calcium, this radioisotope accumulates in bones:
    a. cesium-137
    b. strontium-90
    c. tritium-3
    d. radon-226
    e. radium-226

12. Atomic fusion of deuterium and tritium, makes which product?
    a. helium
    b. uranium
    c. gamma rays only
    d. carbon-14
    e. steam

13. In World War II, the Nazis built a facility in Norway for extracting this material from seawater, perhaps intending to use it in a nuclear reactor or bomb.
    a. hydrogen
    b. tritium
    c. deuterium
    d. uranium
    e. neptunium

14. The Chernobyl accident occurred here:
    a. Ukraine
    b. Poland
    c. Russia
    d. Lithuania
    e. Siberia

15. Which area received the most fallout from the Chernobyl explosion?
    a. Poland
    b. United Kingdom
    c. Lithuania
    d. Belarus
    e. Germany

16. Which form of radiation from nuclear reactions has the greatest energy and the shortest wavelength?
    a. alpha
    b. beta
    c. gamma
    d. X
    e. UV-C

17. Which region has the most nuclear reactors with safety problems?
    a. France
    b. former Soviet Union and eastern European bloc
    c. United States
    d. Middle East
    e. U.K.
18. Of those listed below, which is the greatest source of ionizing radiation exposure to the general public?
    a. X-rays
    b. radon-226 leakage into basements from certain geologic formations
    c. sunlight
    d. gamma radiation from radioisotopes in the Earth's core
    e. nuclear waste in groundwater
19. Which method can generate the high temperatures needed for nuclear fusion reactors?
    a. explosion of TNT
    b. solar radiation focused by mirrors
    c. combustion of natural gas
    d. lasers
    e. all of the above
20. Many cancers in Japanese exposed to the radiation produced by the atomic bombs in World War II occurred years after. Which of the statements below is an accurate explanation of this fact?
    a. Severely damaged cells could not divide and multiply.
    b. Genes were damaged by ionizing radiation.
    c. Once tissues are exposed to gamma rays, they emit gamma rays for years.
    d. Normal chemicals in their bodies were transformed into radioisotopes from the gamma rays they were initially exposed to.
    e. All of the above are accurate.
21. Which country has many instances of radioisotopes leaking into groundwater and streams?
    a. Canada
    b. U.S.
    c. Russia
    d. France
    e. Trick question alert! All of theses countries have major problems.
22. At Chernobyl, trees were cut down and buried for this reason:
    a. There was radioactive dust fused into the bark by the explosion.
    b. Tree roots had soaked up radioactive isotopes in the soil.
    c. Gamma rays had mutated the cells into bizarre shapes.
    d. Thirty-foot tall trees suddenly grew into 60-foot trees.
    e. All of the above.
23. The principal concern about storing nuclear waste at the Yucca Mountain site is:
    a. leakage into ground water.
    b. radiation levels at the surface on the mountain.
    c. transportation to the Nevada site.
    d. formation of radioisotopes in the rock layers above the hot waste.
    e. all of the above.

CYBER SURFIN'

1. From the American Institute of Physics and the J. Arthur Rank Organization, go to **www.aip.org/history/einstein/voice1.htm** to hear Einstein explain his equation.
2. For a general background information of radioactive isotopes, chemistry, and radiation, bring up **www.gpc.edu/~jaliff/envschem.htm**
3. To see a simulation of nuclear fission from Michigan State University's Lectures Online, go to **lectureonline.cl.msu.edu/~mmp/applist/chain/chain.htm**
4. To see a simulation of nuclear fusion from *How Stuff Works, How Nuclear Bombs Work*, by Craig Feuudenrich, see **science.howstuffworks.com/nuclear-bomb1htm**
5. From Linköping University, Sweden by Henrik Eriksson, see how good you are at preventing a Chernobyl nuclear plant meltdown and explosion at **www.ida.liu.se/~her/npp/demo.html.**
6. Learn how to operate your own Tokamak nuclear fusion reactor from the Princeton University Internet Plasma Physics Education eXperience **w3.pppl.gov/~dstotler/SSFD/**
7. To see satellite images of the Chernobyl nuclear plant disaster, scope out the U.S. Geological Survey's Earthshots website: **edcwww.cr.usgs.gov/earthshots/slow/Chernobyl/Chernobyl**
8. The U.S. Public Broadcasting Corporation has produced an informative presentation on the pros and cons of nuclear power, *Nuclear Reactions: Why do Americans Fear Nuclear Power*, go to **www.pbs.org/wgbh/pages/frontline/shows/reaction/**
9. From the World Nuclear Association, an excellent single site for information on nuclear power generation and safety: **www.world-nuclear.org**

'LIKE' BOOKING IT

1. Rhoades R. ***The Making of the Atomic Bomb.*** Carmichael, CA: Touchstone Books, 1995.
2. Fusco P and Caris M. ***Chernobyl Legacy.*** Millbrook, NY: de.MO, 2001.
3. Morris RC. ***The Environmental Case for Nuclear Power: Economic, Medical, and Political Considerations.*** St. Paul, MN: Paragon, 2000.
4. Collman JP. ***Naturally Dangerous: Surprising Facts About Food, Health, and the Environment.*** Herndon, VA: University Science Books, 2001.

ANSWERS

Seeing the Forest

For the scientists listed below, describe their contributions to the discovery of nuclear power.

Scientist	Contribution or Discovery
Wilhelm Röntgen	Discovered X-rays by observing the radiation emitted when a high voltage and current of electrons strikes a metal plate in a vacuum tube. X-rays were first called cathode rays.
Henri Becquerel	Observed radiation emitted by crystals of uranium using photographic (silver) film.

Marie Curie	The first woman in 650 years to hold a professorship at the famous Sorbonne University in Paris. Won the Nobel Prize for her discoveries of radium, polonium, and her study of the radioactivity produced by radium.
Albert Einstein	$E = mc^2$ was the foundation for understanding the atomic power observed by nuclear physicists when they split atoms.
Ernest Rutherford	The first scientist to demonstrate radioactive decay and calculate half-life.
Otto Hahn and Fritz Strassmann	Conducted the first nuclear fission of uranium producing barium isotopes that were about 1/2 the size of the uranium atoms.
Enrico Fermi	Constructed the first nuclear reactor in a squash court under the west stands of the football stadium at the University of Chicago.
Edward Teller	The principal designer of the hydrogen bomb. He proposed using nuclear fusion instead of fission for the first atomic bombs.

Questions to Ponder (Many of these answers are examples of many possible responses, please don't memorize them.)

1. Identify, define, and describe the three types of energy involved in the chain of energy from the Sun to plants to animals.

Energy form	Definition and description
Atomic	The energy that holds the particles of the atomic nucleus together.
Electromagnetic waves	Chlorophyll and accessory pigments are called antenna molecules because they can catch the energy of visible light photons, VIBGY-OR (the rainbow, violet-indigo-blue-green-yellow-orange-red).
Chemical	Electromagnetic energy energizes electrons that make chemical bonds in food molecules.

2. Consult a periodic table of the elements or go to **www.webelements.com.** Consider the following questions:

   a. Why is uranium atomic number 92? *It has 92 positively charged protons.*
   b. How many electrons do hydrogen and its isotopes have? *One*
   c. The isotopes of hydrogen are protium-1, deuterium-2, and tritium-3. What do the numbers represent? *The atomic mass (protons + neutrons).*
   d. Common and dangerous radioisotopes that result from atomic fission are cesium-137, strontium-90, and iodine-131. In the periodic table, elements in a vertical column share chemical characteristics. Also certain radioactive isotopes gain entrance into plants and into food chains. Therefore, propose a reason for the dangers of cesium-137 and strontium-90.

Radioisotope	Reason why it is particularly dangerous in a food chain
Cesium (Caesium)	It is the same family as sodium and potassium, which are the most important electrolytes in our bodies.
Strontium	It is a relative of calcium and taken into bones. Cows eating grass with strontium-90, give milk with elevated levels of radioactivity.

e. Explain how iodine-131 has a higher atomic mass number as compared to iodine-127. *I-131 has 4 more neutrons as compared to I-127.*

f. Explain why iodine-131 is dangerous. *It is a radioactive isotope that is incorporated into the thyroid gland where it can cause cancer. Taking potassium iodide (KI) tablets can limit the amount of I-131 taken into the thyroid.*

3. Compare and contrast nuclear fission and nuclear fusion reactions as to the types and supplies of fuel available, and nuclear wastes produced.

Nuclear reaction	Type and supply of fuel	Waste products
Fission	uranium-238 enriched with U-235	cerium-144, cesium-137, iodine-131, plutonium-239, radium-106, ruthenium-106, strontium-90, tritium-3 (see Table 11-1).
Fusion	deuterium-2 and tritium-3	helium that is an inert gas and gamma ray radiation control.

4. Compare how critical mass is obtained in a reactor and in a fission bomb. *In a reactor, fuel rods that contain enriched uranium pellets are brought in close enough proximity to generate enough neutrons to split other atoms of uranium. In a fission bomb, a conventional explosive charge drives a smaller mass of uranium or plutonium into a larger mass of the same, instantaneously creating a critical mass to produce a nuclear explosion.*

5. Using Table 11-2, compare the types, relative amounts, and environmental or health effects of pollution from coal-burning power plants and nuclear power generators by filling in the table below.

	Pollutant or land use effect	Rate the amount as Very Large, Large or Small	Health or environmental consequence
Coal burning	Land use for mining	Nearly 10 times larger than nuclear at 17.000 acres	Displacement or elimination of birds, trees, etc.
	Sulfur dioxide, nitrogen oxides emissions	Large, but has decreased in recent years due to burning coal with low sulfur and removing sulfur from gasoline	Acid rain, global warming gas, and lung disease

	Mine acid and siltation	Large	Poisoning aquatic ecosystems and inhibiting photosynthesis of producers in food chains, respectively
	Risk from catastrophic accident	Small	Little effect
Nuclear	Land use for mining	Small relative to the energy produced	Poisoning or inhibiting photosynthesis in aquatic ecosystems
	Nuclear waste generation	Large	Waste escape into groundwater, nuclear waste used for bomb making, radiation level increases, increasing cancer risk
	Local radiation, water and air pollution	Small	Some increased exposure to radiation, generally no air pollution, thermal pollution of aquatic ecosystems
	Risk of catastrophic accident	Long-term risk is significant	Radioisotope escape into groundwater and atmospheric fallout, radiation level increases, increasing cancer risk

6. Compare and contrast the accidents at Chernobyl and Three Mile Island. See the text at the World Nuclear Association: **www.world-nuclear.org/info/chernobyl/inf07.htm** and **www.world-nuclear.org/info/inf36.htm.** *Chernobyl: an experiment was being conducted, and poorly trained technicians and poor procedures decreased the flow of reactor coolant water. A peculiarity of the reactor design caused a surge of power, and the reactor overheated, causing water in the primary circuit and around the reactor to explode into steam. The reactor blew up through the roof and ejected parts of the reactor and radioisotopes into the air—particularly cesium-137 and iodine-131. The fallout spread first to northwest Europe, later to southern Europe, and northern Asia. At Three Mile Island, due to faulty valve open/closed and pressure indications, technicians accidentally overfilled the primary water circuit, then decreased the water in the primary circuit, which caused the reactor to overheat and a meltdown. The nuclear core did not escape its containment. The reactor room was filled with hydrogen that could have exploded but did not.*

7. Describe two problems associated with the gamma-ray sterilization of food. (Food is exposed to gamma radiation and the bacteria and fungi on it are killed.)

	Problem associated with gamma ray sterilization of food
1.	Transport of radioactive materials and waste, accidents resulting from the transport
2.	Radiation exposure to those working in sterilizing facilities. Food sterilized by gamma radiation does not become radioactive or retain radioactivity.

8. The energy requirement for each U.S. citizen is approximately 5×10^{11} joules. Using $E = mc^2$, compute the number of joules produced by 0.0056 grams of fissionable atoms. *(0.0056 × 10^{-3}* [to convert to kg]) × (9×10^{16}) = 5.04×10^{11} joules.

Pop Quiz

Matching

1. h
2. f
3. j
4. k
5. a
6. m
7. n
8. l
9. d
10. i
11. p
12. o
13. c
14. q
15. g
16. b
17. e

Multiple Choice

1. e
2. b
3. e
4. c
5. a
6. b
7. e
8. d
9. d
10. c
11. b
12. a
13. c
14. a
15. d
16. c
17. b
18. a
19. d
20. b
21. c
22. b
23. c

12

Renewable Energy and Conservation

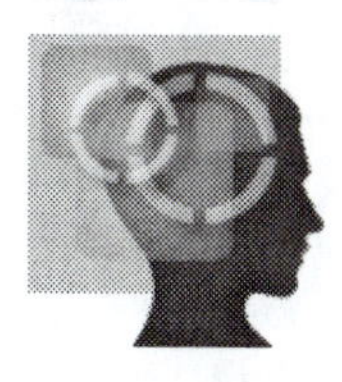

LEARNING OBJECTIVES

After you have studied this chapter you should be able to:

1. Distinguish between active and passive solar energy and describe how each is used.
2. Contrast the advantages and disadvantages of solar thermal electric generation and photovoltaic cells in converting solar energy into electricity.
3. Define biomass, explain why it is an example of indirect solar energy, and outline its disadvantages as a source of energy.
4. Describe the locations that can make optimum use of wind energy and hydropower. Compare the potential of wind energy and hydropower.
5. Describe two renewable energy sources that are not direct or indirect results of solar energy.
6. Distinguish between energy conservation and energy efficiency and give examples of each.
7. Define cogeneration and give an example of a large-scale cogeneration system.

SEEING THE FOREST

There are alternatives to using nonrenewable fossil fuels and uranium to generate electrical power. List 10 alternative power resources that are renewable.

	Renewable power resource
1.	
2.	
3.	
4.	
5.	
6.	
7.	
8.	
9.	
10.	

SEEING THE TREES

VOCAB

It is really quite amazing by what margins competent but conservative scientists and engineers can miss the mark, when they start with the preconceived idea that what they are investigating is impossible. When this happens, the most well-informed men become blinded by their prejudices and are unable to see what lies directly ahead of them.

Arthur C. Clarke, 1963

1. Renewable energy resources—sustainable energy resources that can be replenished by natural processes.
2. Direct solar energy—sunlight can be used in solar photovoltaic panels to generate electricity, it can heat buildings with southern facing windows (see Chapter 1), and it can heat water for use for washing and heating.
3. Indirect solar energy—the energy that is accumulated in the bodies of plants, and hydropower created by the action of the sun causing evaporation of water, which then runs downhill and releases its potential energy.
4. Nonsolar renewable energy resources—geothermal power uses the heat of the Earth's molten core in geologically active areas. Lunar hydroelectric uses the gravitational pull of the moon.
5. Infrared radiation—photons of electromagnetic radiation that are longer in wavelength and of lower energy content compared to visible red photons.
6. Passive solar heating—in the U.S. and southern Canada, southern exposure glass windows or "sky light" panels heat interior spaces using the greenhouse effect. Solar photons pass through

the glass and heat interior surfaces, creating infrared radiation. Infrared photons are trapped inside the glass, being unable to move through it.

7. Active solar heating—solar panels or panels of black metal trap solar radiation and heat a liquid in a primary heating circuit. That liquid flows to a heat exchanger with a secondary water circuit. The heated water then flows into a tank, where it is available for hot water use.

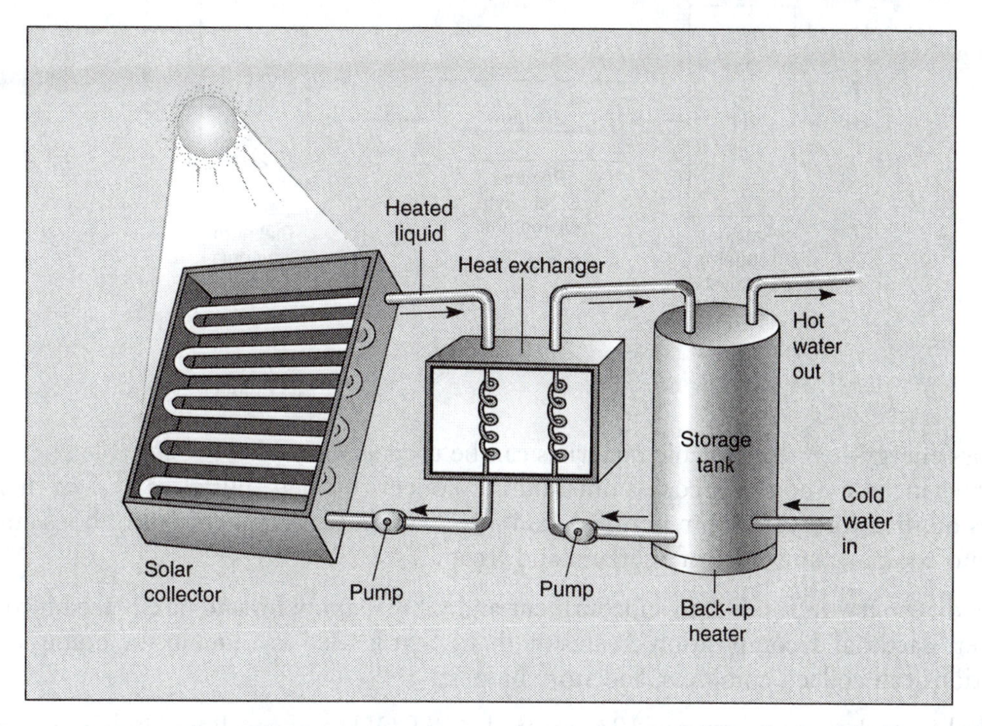

8. Solar thermal electric generation—mirrors, aimed at the Sun, trap and focus solar radiation on oil-filled pipes in a primary heating circuit. The oil then heats water in a secondary circuit and creates steam that drives electricity-generating turbines.

9. Photovoltaic (PV) solar cells—when light strikes certain chemicals, such as silicon and gallium arsenide crystals, electrons are moved away from the crystals, creating a direct current (DC). The cost per watt of electricity generated is high considering the high costs of equipment and the low efficiency of the PV cells, and the electricity must be used as it is produced. This process is similar to that in plants where chlorophyll traps solar radiation and energizes electrons.

10. Solar-generated hydrogen fuel cells—electricity from PV cells can break down water into hydrogen and oxygen (electrolysis). At the anode (negative pole), the hydrogen molecules are split into hydrogen ions or protons (H^+) and electrons (e^-). The flow of electrons or electricity drives an electric motor, heat pump compressor, lights, etc. At the cathode (positive pole), the "spent" electrons are recombined with oxygen atoms and the protons/hydrogen ions from the polymer electrolyte membrane to create water.

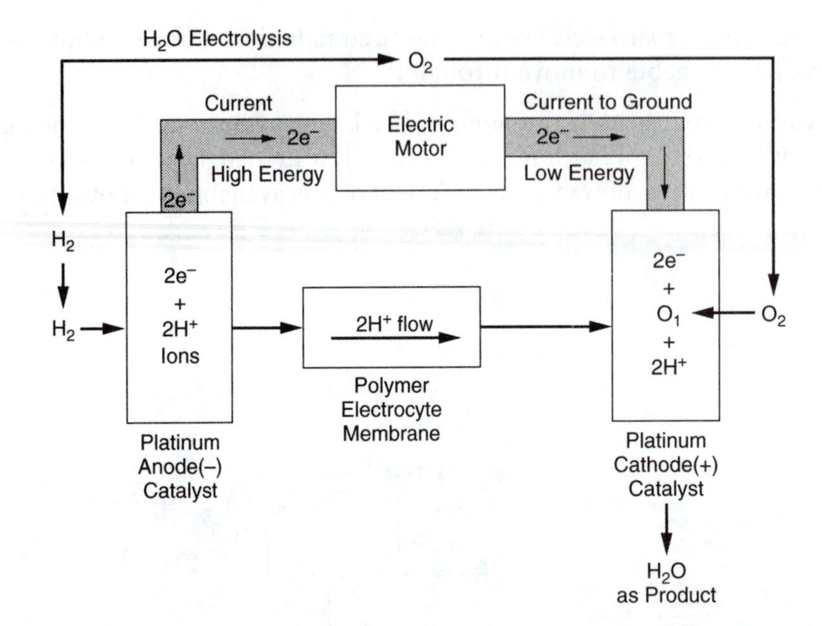

11. Biomass energy—certain organic materials can be used as fuel. These renewable energy resources include wood, charcoal, animal dung (pioneers burned "buffalo chips" in their sod houses on the North American prairies), and peat. The latter is still cut out of bogs and used to heat and cook in some areas of Scotland and Ireland.

12. Biogas digesters—in the Indian subcontinent and China, animal wastes are placed into a tank or digester. Bacterial decomposition creates methane that is used as a fuel in the home. Large operations can collect, compress, and store biogas.

13. Gasohol—gasohol is a mixture of 10% ethanol (C_2H_5OH) and gasoline. Higher concentrations won't work because they vaporize in the gas lines under the high heat around engines, creating vapor lock that stops the flow of fuel. Because alcohols have oxygen in their formula, they are cleaner burning than standard gasoline. Certain *crop residues* have been used to make ethanol, but farmers value residues for adding humus and nutrients back to soils.

14. Wind energy—many countries in the world are developing "wind farms," for harnessing the power of currents of moving air. Certain geographic areas are ideal. Seacoasts have on-shore wind in the morning and off-shore winds at night. The slopes of the Sierra, Rocky, and Appalachian mountains have a fairly consistent wind flow from the west.

15. Hydropower—the sun heats surface water on oceans and other bodies of liquid water, vaporizing it and causing it to rise in the atmosphere. Then, water has potential energy because it has risen against the force of gravity. As the water falls from the clouds, collects into streams, and flows toward the sea, it can be channeled through dams that contain water-driven turbines. The gravitational energy has been transformed into electrical energy. The glittering, gaudy lights of Las Vegas, NV, are powered by energy from the Colorado River falling through Hoover Dam. Lunar hydropower harnesses the pull of the moon on ocean water creating tides that drive turbines.

16. Ocean thermal energy conversion (OTEC)—except under ice, there is a temperature gradient from the warmer surface water to the colder, deeper water of oceans. The warm water is pumped into a primary circuit to boil ammonia in a secondary circuit. The liquid ammonia,

having a low boiling point (–33°C) does not need much heat to make it boil. The gaseous ammonia then drives an electricity-generating turbine. The liquid ammonia is recondensed by the colder water that is deeper (4°C).

17. Hydrothermal reservoir—hydrothermal reservoirs are underground chambers of water heated by the proximity of molten rock of the Earth's core. Yellowstone National Park is one area with hydrothermal reservoirs that form steam geysers and springs. Geologically active areas can produce steam to generate electricity.
18. Geothermal heat pump—heat pumps exchange heat from the warmer soil and rock below the surface with cool water brought from the surface. The deeper the heat exchanger is put into the ground, the more energy is available for heating buildings.
19. Energy conservation—Reducing the use of various forms of energy or eliminating the wasteful use of energy—driving slower or turning the winter thermostat down—conserves energy.
20. Energy efficiency—the mathematical comparison between the amount of energy available and the amount that can actually be used for work. Improved synthetic motor oils reduce engine friction, improving engine efficiency so that the energy stored in gasoline can be more efficiently applied to moving the car.
21. Energy intensity—a comparison of economic output and energy use—total energy consumption divided by gross national product.
22. National Appliance Energy Conservation Act—This U.S. law requires improved energy efficiency of household appliances. Refrigerators manufactured today use 75% less energy than those in the 1970s.
23. Cogeneration of power—the production of two forms of useful energy at one time. When water is heated to produce steam to drive electricity-generating turbines, the hot condensed water can also be used to heat buildings.
24. Demand-side management—to delay costs in building new power-generating facilities, commercial power companies have encouraged conservation by subsidizing improved insulation and more efficient home appliances, air conditioners, and heat pumps.

Questions to Ponder

1. Identify, define, and describe the three types of energy involved in the chain of energy from the Sun to plants to animals.

Energy form	Definition and description
1.	
2.	
3.	

2. Compare and contrast the advantages and disadvantages of passive and active solar heating.

3. Imagine you are the architect designing a "green" energy-conserving, high-efficiency classroom building with technologies available today.

   a. What features would you design? What technologies would you use?
   b. If you had to pick between photovoltaic and heat-collecting panels, which would you use? Please explain your decision.

4. One reason why large centralized power generation stations are inefficient is the long distances traversed by transmission lines and the power losses that occur in overcoming the electrical resistance of the long lines. You have literally heard that lost power when your AM radio in your car is near high-tension power lines. How does cogeneration deal with that problem?

5. Offer four disadvantages of solar power.

	Disadvantage of solar power
1.	
2.	
3.	
4.	

6. For the following energy resources offer two advantages and disadvantages of each.

	Advantage	**Disadvantage**
Wind		
Biomass/wood		
Biomass/animal waste		
Biomass/plant residues		
Freshwater hydropower		
Geothermal		

7. Debate this proposition pro and con: "Geothermal power is not reusable."

POP QUIZ

Matching

Match the following terms and definitions

____1. Designing southern exposure windows in Canadian buildings is an example of:	a. energy efficiency
____2. Solar radiation transferred to a liquid in a primary heating circuit is an example of:	b. ocean thermal energy conversion
____3. Sunlight displaces and moves electrons from silicon/phosphorus chips in a ______ .	c. biogas digester
____4. Solar electrolysis of ______ makes hydrogen for fuel cells.	d. hydrothermal reservoir
____5. Hog farms in the southeastern U.S., infamous for their water pollution and odors, could recycle animal waste using a(n):	e. active solar heating
____6. Corn can be fermented to produce this oxygenating gasoline additive:	f. water
____7. The process of photosynthesis makes carbohydrates in wood. This is ______ energy.	g. energy conservation
____8. This is a preferred area for "wind farms":	h. Home Energy Generation Act
____9. This process extracts heat from seawater:	i. ammonia
____10. The periodic eruptions of "Old Faithful" geyser are due to a(n):	j. passive solar heating
____11. Making sure that your office or room light is turned off when you are not there is an example of:	k. photovoltaic solar cell
____12. In the 1970s, steel-belted radial ply tires replaced the bias ply tires that created more side-wise movement as they contacted the pavement. This is an example of improved:	l. Aswan
____13. *Energy Star*^{R} ratings resulted from this measure:	m. mountain passes
____14. Electricity can be generated at home and the excess sold to local power lines:	n. demand-side management

____15. This hydroelectric dam spread the disease of schistosomiasis in irrigation canals:	o. ethanol
____16. It takes little energy to boil this liquid to create a gas that will drive a turbine:	p. National Appliance Energy Conservation Act
____17. The Southern Company gave a rebate to put additional insulation in my house. Their attempt to reduce their costs by reducing consumption is called ______:	q. biomass

Multiple Choice

1. Which of the following is a disadvantage of solar energy technology?
   a. Large areas of solar panels would cover landscapes
   b. Current technology is not efficient in converting sunlight to electricity.
   c. day and night length
   d. cost of equipment
   e. all of the above

2. Offshore areas in oceans are ideal for which type of power generation?
   a. lunar/tidal hydropower
   b. wind farms
   c. temperature gradient energy exchanges
   d. surface water heating to boil liquid ammonia to drive turbines
   e. all of the above

3. Which of the following appliances has an efficiency that has not been improved appreciably by new technology since 1970?
   a. refrigerators
   b. air conditioners
   c. convection ovens
   d. heat pump
   e. televisions

4. Which technology of power generation produces the least air pollution?
   a. hydrogen fuel cells
   b. combustion of methane
   c. biogas combustion
   d. natural gas cogeneration
   e. electric/gasoline hybrid engines

5. The use of biomass energy has resulted in severe desertification and deforestation in which area?
   a. Arizona
   b. Sudan
   c. Egypt
   d. Kenya
   e. British Columbia

6. Which location would be best for locating solar energy collectors in the U.S. and Canada?
   a. Death Valley, CA
   b. East Texas
   c. Appalachia
   d. Edmonton, Alberta
   e. Seattle, WA

7. Hydrogen fuel cells and the process of photosynthesis have which process in common?
   a. Water is split into hydrogen and oxygen gases.
   b. Hydrogen ions and electrons are combined with oxygen to make water.
   c. They both make electricity that can be used in homes.
   d. They produce chemical energy.
   e. a and b

8. Which material cannot be used as a raw material to make methane or biogas?
   a. feces
   b. plant residues
   c. wood
   d. water
   e. Trick question alert! All these can be used to make biogas.

9. In rural areas of the Indian subcontinent and China, energy for heating and cooking is commonly supplied by:
   a. solar-generated electricity.
   b. geothermal power from the Deccan shield.
   c. solar electrolysis of water.
   d. fermentation and digestion of organic refuse, feces, and urine.
   e. all except a

10. Which of the following energy resources requires the greatest allocation of land surface area?
    a. geothermal
    b. active solar
    c. wind
    d. cogeneration
    e. passive solar

11. Which location would be best for a wind farm in terms of disruption to the environment and consistency of wind flow?
    a. mountain pass
    b. plains of Oklahoma
    c. offshore ocean
    d. deserts
    e. tropical rain forest

12. In California, geothermal power generators use this energy source:
    a. heat generated by friction of rocks at the San Andreas fault
    b. heat generated by volcanic activity

c. gravitational pull on surface plates of the Earth's crust
d. heat from the molten core of the Earth

13. Which process does not produce carbon dioxide as a product of combustion?
a. hydrogen fuel cell
b. solar-generated hydrogen
c. natural gas cogeneration
d. biogas
e. biomass

14. Biogas is mostly:
a. oxygen
b. hydrogen
c. carbon dioxide
d. methane
e. nitrogen

15. If you want to do wintertime passive solar heating in New Zealand, your windows should face which way?
a. north
b. south
c. east
d. west
e. up!

16. Which material forms the catalyst in an electrolytic hydrogen fuel cell?
a. silicon
b. gallium arsenide
c. platinum
d. copper
e. electrolytic membrane

17. Photovoltaic cells use this material in which electrons can the dislodged and moved to conducting wires:
a. silicon
b. gallium arsenide
c. platinum
d. copper
e. electrolytic membrane

18. The sun evaporates water from an ocean, and the water condenses in clouds. The condensed water then has this type of energy.
a. hydroelectric
b. potential
c. kinetic
d. atomic
e. hydrogen bonding

19. Which problem is associated with the building of hydroelectric dams?
    a. flooding of valuable farm lands
    b. elimination of flood plain soil enrichment upstream and downstream
    c. siltation of the upstream portion of the dam
    d. blocking the paths of migratory fishes
    e. all of the above

20. Which of the following biomass materials would produce the least air pollution when combusted?
    a. biogas
    b. hardwood
    c. peat
    d. animal feces
    e. sugar cane residue

21. Which energy technology has the least efficiency?
    a. hydroelectric
    b. passive solar heating
    c. active solar heating
    d. photovoltaic electrical
    e. wind

22. Which property of ammonia makes it an ideal gas to extract heat from water?
    a. As a gas, it can drive a turbine better than steam.
    b. When the liquid is turned into a gas, the gas can be used to heat homes.
    c. As a liquid, it has a low boiling point.
    d. The liquid, when compressed, cools the water returning to the ocean.
    e. All of the above are correct.

23. In developing countries, which is the most available and used renewable energy resource?
    a. biogas
    b. biomass/wood
    c. hydroelectric
    d. solar
    e. biomass/peat

24. When conserving energy in the home, which single measure saves the most energy?
    a. using fluorescent bulbs
    b. turning the thermostat to 70 in the winter and 78 in the summer
    c. thick insulation
    d. turning the water temperature down in the water heater
    e. taking shorter showers

CYBER SURFIN'

1. The U.S. Department of Energy has an animation on "Turning Sunlight into Electricity." See **www.eere.energy.gov/pv/pvmenu.cgi?site=pv&idx=0&body=video.html**
2. Check out at the U.S.D.O.E. pages on Zero Energy Buildings and Solar Buildings Technologies, go to **www.eere.energy.gov/solarbuildings/**
3. To see animation of a wind power generator and information on the subject, surf to the U.S.D.O.E. site, ƐNERGY.Gov, **www.nrel.gov/clean_energy/wind.html**
4. The UnoCal site has igeneral information about geothermal power, go to **www.unocal.com/geopower/power.htm**
5. From Geoexchange: Geothermal Heat Pump Consortium, see a movie describing Geothermal heat pumps: **www.ghpc.org/about/movie.htm**
6. Go to How Stuff Works and explore hydrogen fuel cells: **www.howstuffworks.com/fuel-cell.htm** and how hybrid cars work: **www.howstuffworks.com/hybrid-car.htm**
7. The Alternative Energy Institute has an informative site with additional links listed and evaluated, see **www.altenergy.org/Non-Renewable/non-renewable.html**

'LIKE' BOOKING IT

1. Gipe P. ***Wind Power for Home & Business: Renewable Energy for the 1990s and Beyond (Real Goods Independent Living Book).*** White River Junction, VT: Chelsea Green Pub. Co., 1993.
2. Strong SJ, and Scheller WG. ***The Solar Electric House: Energy for the Environmentally-Responsive, Energy-Independent Home.*** White River Junction, VT: Chelsea Green Pub. Co., 1994.
3. Hoffman P, and Harkin T (foreword). ***Tomorrow's Energy: Hydrogen, Fuel Cells, and the Prospects for a Cleaner Planet.*** Cambridge, MA: MIT Press, 2001.
4. Lerner S. ***Eco-pioneers: Practical Visionaries Solving Today's Environmental Problems.*** Cambridge, MA: MIT Press, 1997.

ANSWERS

Seeing the Forest

1. There are alternatives to using nonrenewable fossil fuels and uranium to generate electrical power. List 10 specific alternative power resources that are renewable.

	Renewable power resource
1.	active solar heating
2.	passive solar heating
3.	photovoltaic electric
4.	freshwater hydroelectric
5.	lunar/tidal hydroelectric

6.	biomass digester/biogas
7.	solar hydrogen generation
8.	biomass- wood
9.	wind
10.	ocean thermal energy conversion

Questions to Ponder (Many of these answers are examples of many possible responses, please don't memorize them.)

1. Identify, define, and describe the three types of energy involved in the chain of energy from the Sun to plants to animals.

Energy form	Definition and description
1. Atomic	Energy that holds atomic nuclei together that is liberated when atomic fusion converts mass to energy
2. Solar radiation	Electromagnetic: energy in the form of Ultraviolet-A and visible light photons traveling in waves
3. Chemical	The energy stored in the chemical bonds of a food molecule

2. Compare and contrast the advantages and disadvantages of passive and active solar heating. *In concert with improved insulation of roofs, walls, and windows, passive solar heating offers great savings in energy costs and nonrenewable resources. Oil, gas, and coal (used to generate the electricity) can be conserved. The disadvantage is the variability of the strength of sunlight from a sunny day to a cloudy day, and the total amount of cloud cover per year. Active solar heating is a sound idea theoretically, but vast improvements in efficiency need to be made, particularly in heat-conducting liquid technology. It has an inherent advantage over passive solar in that water, which heats and cools slowly, can store some heat. It shares the variability of sunlight strength with passive solar heating.*

3. Imagine you are the architect designing a "green" energy-conserving, high-efficiency classroom building with technologies available today.

   a. What features would you design? What technologies would you use?
   I would use solar electrolysis of water, producing hydrogen. That technology would require photovoltaic cells on the roof. The hydrogen would be compressed and stored. Hydrogen fuel cells would power electrical generation for the building. I would also have a wind generator on the roof that would supply electricity for the electrolysis of water. I would have a biogas digester to process human and plant waste products from the building. The combustion of biogas will heat steam to drive turbines, producing electricity. We will sell our excess power to the local power company as allowed by U.S. law. A geothermal heat pump will provide heating and cooling year round.

   The building will have southern exposure triple-paned window glass with translucent reflective covers applied in the summer. The building will have no black tar roof materials. Except for the solar panels, all surfaces will be reflective. The building will have the best insulation that can be obtained. This building will cost a lot, but figure the monetary savings and keeping a mountaintop in Kentucky from being lopped off! My vision is that large buildings and houses, to some extent, would be energy sufficient.

b. If you had to pick between photovoltaic and heat collecting panels, which would you use? Please explain your decision. *See my answer above.*

4. One reason why large centralized power generation stations are inefficient is the long distance traversed by transmission lines and the power losses that occur in overcoming the electrical resistance of the long lines. You have literally heard that lost power when your AM radio in your car is near high-tension power lines. How does cogeneration deal with that problem? *Cogeneration of power and steam for heating centralizes electrical power generation and greatly reduces the power lost in transmission over large power grids by having a short distance between the generator and end user.*

5. Offer four disadvantages of solar power.

	Disadvantage of solar power
1.	Variability of sunlight according to season/day length and daily weather patterns
2.	Inefficiency of power generation technology
3.	Large areas of surface ecosystems are covered, killing or displacing plants and animals
4.	The sun goes down at night, therefore, water electrolysis would be a better approach than generating electricity only

6. For the following energy resources offer two advantages and disadvantages of each.

	Advantage	Disadvantage
Wind	Wind is free!	Bird kills.
	Wind always blows, particularly at higher altitudes	Only certain locations have constant wind.
Biomass/wood	Renewable and as long as the ecosystem is maintained	Produces significant amounts of sulfur and nitrogen oxides and greenhouse gases.
	Low technological requirements, usually readily available	Deforestation and desertification occur in overpopulated areas.
Biomass/animal waste	Readily available, particularly in agricultural areas	Peee-eew! The odor problem can be overcome.
	Easy to harvest. Biogas can be compressed and stored.	Makes greenhouse gases when combusted.
Biomass/plant residues	Readily available, particularly in agricultural areas	Reduction of humus for soil.
	Easy to harvest. Biogas can be compressed and stored.	Makes greenhouse gases when combusted.
Freshwater hydropower	Available in streams from creeks to rivers.	Disrupts migration of fishes to spawning areas.

	Renewed by the water cycle. No greenhouse gases emitted.	Dams cause the best farming areas to be permanently covered. Elimination of the natural flood plains that are ideal for agriculture. Also dams silt up.
Geothermal	Renewable as long as the Sun's gravitational, the Earth's rotational, and nuclear fission forces provide heat.	Available only in certain geologically active areas where molten magma intrudes into the Earth's crust.
	Consistent supplies of heat are available.	Power must be distributed in wasteful grids from a centralized location.

7. Debate this proposition pro and con: "Geothermal power is not reusable." *Geothermal power is renewable as long as the Sun's gravitational energy, the Earth's rotational energy, and nuclear fission forces provide heat. In human time frames, it is renewable. A problem exists with the water brought up from hydrothermal reservoirs, because some is lost in the process of electrical generation—the reservoir has to be replenished.* (Why not use geothermal heat transferred to boil ammonia to drive turbines. Yes, large ammonia plants would be somewhat dangerous to operate. Ammonia gas is toxic and explosive. Sixty years ago, ammonia compressors were making ice for refrigeration.)

Pop Quiz

Matching

1. j
2. e
3. k
4. f
5. c
6. o
7. q
8. m
9. b
10. d
11. g
12. a
13. p
14. h
15. l
16. i
17. n

Multiple Choice

1. e
2. e
3. c
4. a
5. b
6. a
7. b
8. d
9. d
10. b
11. c
12. d
13. a
14. d
15. a
16. c
17. a
18. b
19. e
20. a
21. d
22. c
23. b
24. c

13

Water: A Fragile Resource

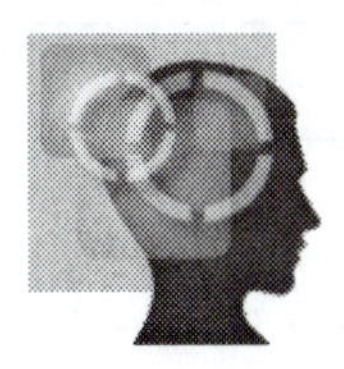

LEARNING OBJECTIVES

After you have studied this chapter you should be able to:

1. Draw a simple diagram of a water molecule, indicating the regions of partial positive and negative charges and how hydrogen bonds form between adjacent water molecules.
2. Describe surface water and groundwater using the following terms in your descriptions: wetland, runoff, drainage basin, unconfined and confined, aquifer, and water table.
3. Explain how humans exacerbate property damage caused by floods, using the upper Mississippi River basin as an example.
4. Relate some of the problems caused by overdrawing surface water, aquifer depletion, and salinization of irrigated soil.
5. Relate the background behind each of the following international water problems: Mono Lake, the Colorado River basin, and the Ogallala Aquifer.
6. Briefly describe each of the following international water problems: drinking water problems, population growth and water problems, the Rhine River basin, the Aral Sea, and potentially volatile international situations over water rights.
7. Contrast the benefits and drawbacks of dams and reservoirs, using the Columbia River to provide specific examples.
8. Give examples of how water can be conserved by agriculture, industry, and individual homes and buildings.

SEEING THE FOREST

Consider the problems that have arisen in regard to water quality and the environment. Fill-in the table below.

	Action	Environmental impact
1.	Diversion of river delta water for municipal use in coastal areas	
2.	Filling-in estuarine marshes	
3.	Irrigation of arid or semiarid land	
4.	Depletion of inland aquifers	
5.	Dam building	
6.	Flood control through stream channelization and the building of levees	
7.	Diversion of river or lake water for inland irrigation (Aral Sea)	
8.	Water pollution from sewage, sewage treatment, and fertilizer runoff	
9.	Industrial toxic water pollution	

SEEING THE TREES

VOCAB

Anything else you're interested in is not going to happen if you can't breathe the air and drink the water. Don't sit this one out. Do something! You are by accident of fate alive at an absolutely critical moment in the history of our planet.

Carl Sagan

1. Reservoir—an artificial impoundment of streams and rivers of water for the use of industries and cities.
2. Hydrogen bonding of water—adjacent liquid water molecules are pulled together by bonds of electrical attraction. These extend from an electronegative oxygen atom of one water molecule to the positive protons in the hydrogen atoms of another molecule. This accounts for the high boiling point of water; its high specific heat (it heats up slowly and cools slowly) and the fact that it is a liquid that covers most of the Earth's crust.

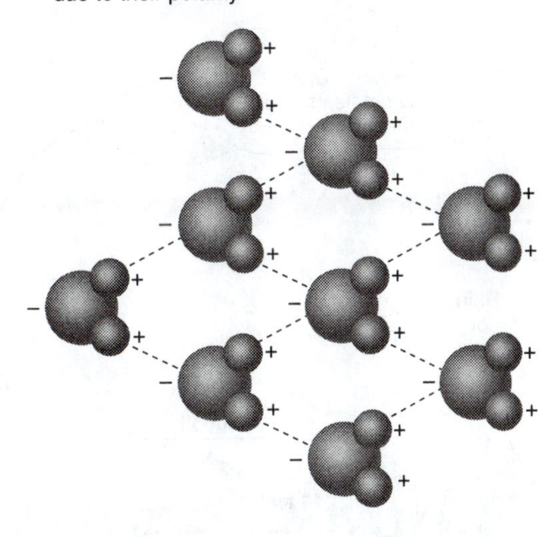

3. Vaporization of water—when a liquid changes into a gas or vaporizes, heat is taken into the liquid to break the bonds between the molecules. The heat comes from the surface from whence the evaporation occurs, and it goes into the water to make it a gas. Therefore, the surface is cooled. If you reverse this process by compressing the gas, heat is produced as the liquid is formed.

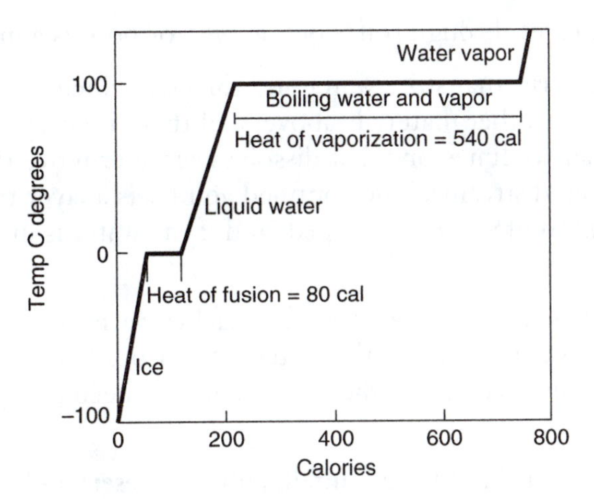

4. Sublimation—a solid such as ice or frozen carbon dioxide can change directly from a solid to a gas. This process is used in the freeze drying of foods.
5. Hydrologic cycle—water cycles from the gaseous state, to liquid and solid, to gaseous again. Most of the biosphere's water is in the liquid state due to hydrogen bonding. Water evaporates or sublimates from ocean and polar ice into the air. There water molecules begin to adhere to each other around a nucleus of dust. This cohesion allows for the water droplets to enlarge to the point that they are carried by gravity to the ground as rain or snow. Rainwater or snowmelt takes two paths back to the ocean: as runoff and by percolation into the soil and into deeper porous rock called aquifers.

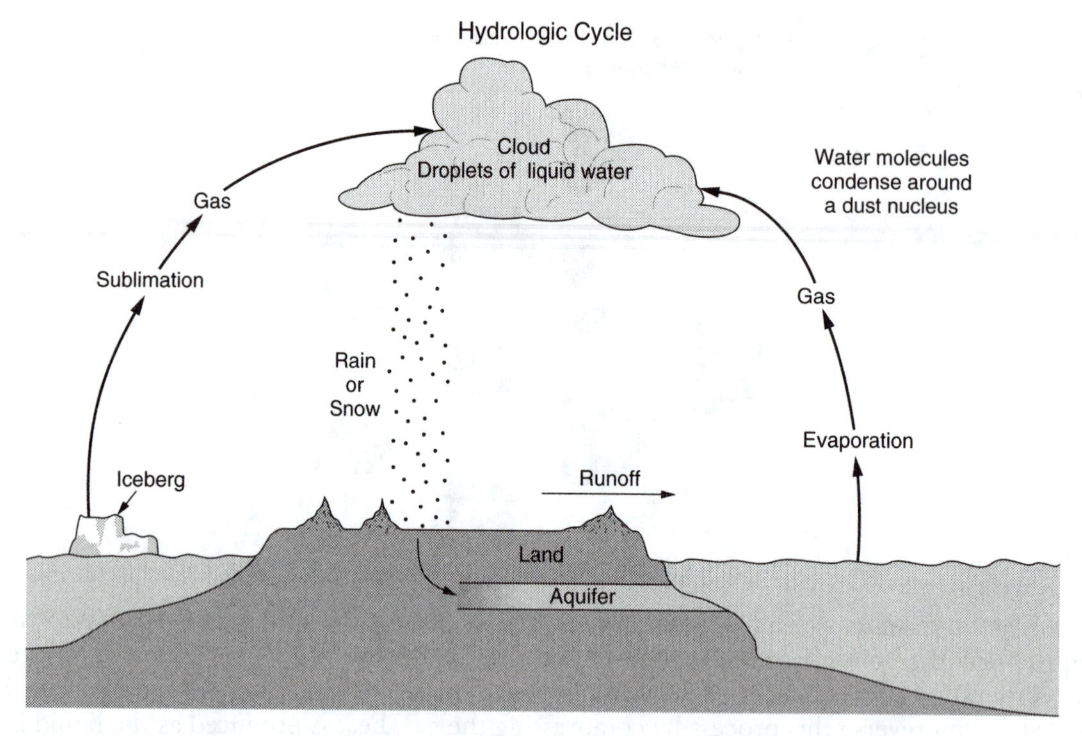

6. Drainage basin—the area of surface runoff of a specific river and its tributaries.
7. Groundwater—water that moves through soil, sand, gravel, or porous aquifers.
8. Aquifers—porous rock or caverns that store water and through which water moves. The *unconfined aquifer* has porous rock or other materials above, and there is always an impenetrable layer below. Most aquifers are made of limestone that dissolves as water move through, producing caves, caverns and underground streams. The confined aquifer is a layer trapped between two impenetrable layers of rock. Aquifers can be tapped, and if the water is under pressure, an artesian well will spring up to the surface.
9. Water table—the depth at which groundwater can be reached varies widely from the desert to the boreal forest. Wells have gone dry when the water table drops below the level of the pump inlet. This can happen in droughts or when the water table has been pumped dry for municipal or agricultural use.
10. Arid and semiarid lands—Deserts have little rainfall, and semidesert and dry grasslands more. See Chapter 7.
11. Flood plains—flood plains are covered by water overflowing from streams. The floods are typically annually in the spring. Major, perhaps catastrophic, floods occur every 100 years (roughly). The ancient Egyptian civilization developed around the Nile River flood plain. Floods carry organic and mineral enriched soil from upstream (the upper Nile originates in tropical areas), depositing the enriched soil on the flood plains. Flood plains are ideal agricultural lands.
12. Aquifer depletion—if the recharging of aquifers does not equal the withdrawal of water, aquifer depletion occurs. Water tables drop, wells have to be drilled deeper, and land may subside as it has in the Imperial Valley of California. In coastal areas, saltwater intrusion occurs as aquifer freshwater pressure drops to a level below seawater pressure. See the figure above.
13. Salinization of soil—irrigation can be done 2 ways: 1) *shallow water irrigation* in arid or semiarid areas leaves dissolved salts behind when the water evaporates. The salt then accumulates to the

preventing plant growth. This happened in ancient Mesopotamia. 2) *Deep irrigation* applies a large amount of water to the soil, which percolates salts away from the root level. Nevertheless, the water that returns to the stream from the runoff or groundwater percolation will make the stream more highly concentrated in dissolved minerals. This is the case of the Colorado River when it reaches Mexico.

14. Xeriscaping—in Phoenix, AZ, one can choose to have a small grass lawn or desert landscape. The xeriscaping uses local cacti (there is a problem of cactus poaching) and yucca, with rock or gravel in between. The folks with grass pay big water bills.
15. Reclaimed water—treated wastewater and water used in the home for washing (gray water) can be used for irrigation as long as it does not have toxic minerals in it.
16. Ogallala Aquifer—the high plains of the Midwest U.S. from Nebraska to North Texas are semiarid areas underlain by the largest aquifer in the world. Water is being withdrawn for irrigation and municipal use at a rate 40 times faster than it is being recharged.
17. Stable runoff—the amount of runoff water that is available throughout the year. In India during the monsoon, most of the water quickly runs off into rivers and on to the Indian Ocean.
18. Sustainable water use—water use rates that can be sustained or replenished by nature or technology, thus, ensuring the availability of clean water for the future.
19. Desalinization of water—coastal cities in desert biomes can remove salt from sea water by evaporation, distillation by boiling, freezing (when water freezes, the salt stays behind), or reverse osmosis. The latter uses a piston pressure to exceed the osmotic pressure of water when it moves toward higher concentration of dissolved chemicals. Using this pressure, the water can be forced through a filter toward an area where the dissolved materials concentrations are lower—the fresh water.
20. Microirrigation—if porous pipes are set at root level, water can be delivered directly there, avoiding the large amounts of evaporation that accompany surface water irrigation.

Questions to Ponder

1. Identify the major problems of water use and quality at the following locations described in the chapter.

Location of body of water or drainage	Problems
1. Puget Sound (see Chapter 2)	
2. Columbia River	
3. San Francisco Bay	
4. Colorado River	
5. Mississippi River	
6. Mono Lake	
7. Ogallala Aquifer	
8. Aral Sea	
9. Rhine River Valley, Germany	

2. Describe the hydrologic cycle.
3. Imagine you are the engineer designing a "green" water system for a coastal city.
   a. What features would you design? What technologies would you use?
   b. If you had to pick between building reservoirs or allowing water to run in streams, which would you use for a water source? Please explain your decision by considering the advantages and disadvantages.
4. Imagine that you are a "green" politician representing farmers in Nebraska using Ogallala Aquifer well water. What would you propose to replenish and sustain this water source, so that their great grandchildren could farm there? Offer 3 measures.

	Measure to sustain or replenish Ogallala water
1.	
2.	
3.	

5. Offer four disadvantages of solar power.

	Disadvantage of solar power
1.	
2.	
3.	
4.	

6. Describe ten ways to conserve water at home.

	Action to conserve water at home
1.	
2.	
3.	
4.	
5.	
6.	
7.	
8.	
9.	
10.	

POP QUIZ

Matching

Match the following terms and definitions

____ 1. This characteristic of water explains its high boiling point and cohesiveness:	a. drainage basin
____2. Most of the dihydrogen oxide (H_2O) that is on the Earth is in this state of matter:	b. Colorado
____3. Coastal cities are in danger of future flooding for this reason:	c. artesian wells
____4. Three western U.S. states and Mexico share this river basin:	d. shallow water desert irrigation
____5. The building of dams on the Columbia River interferes with ______ migration.	e. sublimation
____6. These areas are best left for agriculture instead of real estate development:	f. liquid
____7. One of the first civilizations developed in the ______ River Valley.	g. water table
____8. Salinization of soil is a consequence of:	h. reverse osmosis
____9. This level has been falling in the Imperial Valley, CA:	i. Germany
____10. If confined, groundwater sources are tapped, these result:	j. fish
____11. Water evaporating from ice is termed ______.	k. saltwater intrusion
____12. This results when coastal aquifers are not being recharged at sufficient rates:	l. melting of polar ice
____13. Most of the industrial water pollution in the Rhine River comes from here:	m. microirrigation
____14. This process uses contaminated water pressurized against a filtering membrane for purification:	n. flood plains
____15. Water that is used in the home for washing and recycled as garden water is called:	o. Nile

____16. An area drained by a specific river is called a(n) ______.	p. hydrogen bonding
____17. Placing water at root level is ______.	q. gray water

Multiple Choice

1. The heat of vaporization of water is 540 calories/gram. That energy is needed to do which process?
   a. break bonds between oxygen and hydrogen atoms within a molecule
   b. break bonds between oxygen and hydrogen atoms of adjacent molecules
   c. convert water gas to a liquid
   d. heat the surface water
   e. all of the above

2. When water condenses in a cloud, this happens:
   a. Heat is taken into the liquid water.
   b. The heat of vaporization turns water gas into a liquid.
   c. Heat is released into the cloud.
   d. Hydrogen bonds break.
   e. All of the above are correct.

3. Which part of the world is most interested in desalinization of water?
   a. desert coastal cities
   b. the Arctic
   c. U.S.
   d. tropical regions with high stable runoff
   e. Russia

4. Water can be purified by which process?
   a. Freezing water creates pure ice crystals that can be scraped off and melted.
   b. Boiling water leaves contaminates behind, leaving a pure vapor that can be condensed.
   c. A piston can drive mineral-laden water against a semipermeable membrane, leaving pure water on the other side.
   d. Water can be evaporated by the sun, the vapor condensed, and salt eaten.
   e. All of the above are correct.

5. Water is most dense at which temperature?
   a. +4°C
   b. –4°C
   c. +100°C
   d. –100°C
   e. Water has the same density of 1 gram/cubic centimeter at all temperatures.

6. Which statement describes aquifers accurately?
   a. They include porous limestone.
   b. They contain caverns hollowed out by flowing water.
   c. They carry water into larger bodies of water.

d. They contain water confined by lower impenetrable rock layers.
e. All of the above are correct.

7. When more water is withdrawn from an aquifer than is recharged, this can happen:
   a. saltwater intrusion
   b. wells have to be drilled deeper
   c. subsidence of land
   d. lowering of the water table
   e. all of the above

8. Dam building on the Columbia River has disrupted the migration of which fish?
   a. smelt
   b. sea bass
   c. salmon
   d. flounder
   e. cod

9. The U.S. and Mexico have disputed the use of which river?
   a. Rio Grande
   b. Colorado
   c. Mississippi as it flows into the Gulf of Mexico
   d. a and b
   e. all except a

10. The building of dams has which negative environmental effect?
    a. siltation of the upstream wall of the dam
    b. preventing annual flood plain deposition of enriched soil
    c. interfering with fish migration
    d. scouring of stream beds downstream
    e. all of the above

11. Which area of the biosphere has the most freshwater?
    a. polar ice
    b. tropical rivers
    c. lakes
    d. aquifers
    e. stable surface runoff

12. The Johns Hopkins University School of Public Heath predicted that by 2025 what percentage of world population will be living in areas where there is not enough water for drinking and irrigation?
    a. 80%
    b. 66%
    c. 33%
    d. 20%
    e. 10%

13. This country has high total water runoff, but low stable runoff for use:
    a. Saudi Arabia

b. India
c. Chile
d. Northern Mexico
e. Canada

14. Which practice destroys ideal farm land and costs insurance companies millions of dollars yearly?
 a. dam construction
 b. artesian well drilling
 c. irrigation
 d. building on flood plains
 e. stream channelization

15. Which program of flood control did the U.S. Corps of Engineers conduct in farming areas that increased flooding downstream and lowered water tables around the affected streams?
 a. dam construction
 b. well drilling
 c. irrigation
 d. transportation canal building
 e. stream channelization

16. In New Orleans and similar lowland cities, siltation of the river bottom has forced the building of higher ______ so that in some places the surface of the stream is above the street level.
 a. dams
 b. levees
 c. canals
 d. wetlands
 e. all of the above

17. In 1993, flooding of this drainage caused heavy amounts of pesticides and fertilizer runoff to accumulate in the Gulf of Mexico.
 a. Ohio River
 b. Missouri River
 c. Rio Grande
 d. Mississippi
 e. Colorado

18. In the spring, this river carries snow melt water from the western U.S. to the Gulf of California:
 a. Los Angeles
 b. Merced
 c. Colorado
 d. Columbia
 e. Missouri

19. The pioneers said the north central areas of this river, the subject of much dam building by the U.S. Corps of Engineers, was "too thick to drink, and too thin to plow," by reason of its heavy load of sediment. Identify it.
 a. Mississippi

b. Ohio
c. Colorado
d. Columbia
e. Missouri

20. Which university reduced its water consumption by 50% between 1987 and 1994?
   a. University of Michigan
   b. Harvard
   c. MIT
   d. University of California Santa Barbara
   e. Cal Poly Technical

21. Which geographical area has the worst problem with soil salinization due to shallow irrigation?
   a. desert
   b. semiarid grassland
   c. coniferous forest
   d. temperate grassland
   e. chaparral

22. Which two nations are *mismatched* in their disputes over water resources?
   a. Mexico/U.S.
   b. Turkey/Iraq
   c. Israel/Jordan
   d. Switzerland/Germany
   e. Trick question alert! All these are matched correctly.

23. Which statement accurately describes the trend in U.S. water use from 1980 to 1995?
   a. The population rose 16%, and the water use rose 16%
   b. Water use declined in households but increased in industries.
   c. Water use declined in homes and industries.
   d. Water use declined in industries, but not homes.
   e. Water use remained unchanged.

24. Which measure would provide for sustainable water use?
   a. reduction of surface water evaporation from large shallow reservoirs
   b. legally restricting water extraction from aquifers to that amount recharged by rainfall
   c. cleaning water that has been used
   d. supplying water-conserving appliances to everyone
   e. a, b, and c

CYBER SURFIN'

1. The U.S. Environmental Protection Agency has information on the following subjects:
   a. Basic concepts of watersheds: **www.epa.gov/volunteer/stream/vms21.html**
   b. The Mississippi (including the Missouri River) River watershed: **www.epa.gov/msbasin**
2. To see how Puget Sound Nearshore Project is planning to restore their estuaries and wildlife, go to **www.prism.washington.edu/lc/PSNERP** See also how San Francisco Bay is faring at Save The Bay: **www.savesfbay.org/protect.html**

3. From the University of Illinois, Department of Atmospheric Sciences, an exploration of the hydrologic cycle, see **www2010.atmos.uiuc.edu/(Gh)/guides/mtr/hyd/home.rxml**
4. From the U.S. Geological Survey, see primers on aquifers and groundwater: **sr6capp.er.usgs.gov/aquiferBasics/index.html**, and the chemistry of water **wwwga.usgs.gov/edu/**
5. For an overview of the State of Environment of the Aral Sea Basin Regional report of the Central Asian States' 2000, see **www.grida.no/aral/aralsea/index.htm**
6. To see a critique of the U.S. Corps of Engineers water projects, see the American Rivers.org site at **www.amrivers.org/missouririver/flexibleflow.htm**
7. Check out *H_2Ouse* from the California Urban Water Conservation Council, for do-it-yourself water saving tips: **www.h2ouse.org**
8. Join the U.S.E.P.A. Volunteer Water Monitoring team, go to **http://www.epa.gov/volunteer/stream/stream.pdf**

'LIKE' BOOKING IT

1. Abramowitz JN. ***Imperiled Waters, Improverished Future: The Decline of Freshwater Ecosystems*** (Worldwatch Paper No. 128). Washington, DC: Worldwatch Institute, 1996.
2. de Villiers M. ***Water: The Fate of our Most Precious Resource.*** Boston: Houghton-Mifflin/Mariner Books, 2001.
3. Rothfeder J, and Rothfeder J. ***Every Drop for Sale: Our Desperate Battle Over Water.*** New York: J. P. Tarcher/Penguin Putnam, 2001.
4. Postel S. ***Pillar of Sand: Can the Irrigation Miracle Last?*** New York: W. W. Norton, 1999.
5. Fradkin PL. ***A River No More: The Colorado River and the West.*** Berkeley: University of California Press, 1996.

ANSWERS

Seeing the Forest

	Action	Environmental impact
1.	Diversion of river delta water for municipal use in coastal areas	Decrease of fresh water for estuarine habitats. Salt water intrusion into aquifers.
2.	Filling-in estuarine marshes	Drastic reduction of off-shore fisheries. Reduction of the ecosystem service of sewage reduction and nutrient utilization for plant growth and the decomposition food chains resulting.
3.	Irrigation of arid or semiarid land	Salizination of soils due to the deposition of salts as evaporation occurs.
4.	Depletion of inland aquifers	Reduction of water for farm and municipal use. Land subsidence. Saltwater intrusion in coastal areas.
5.	Dam building	Flooding of prime flood plain farmland and removal from production. Salinization of soils. Movement of lake water into porous rock. Scouring of stream beds downstream of the dam.

6. Flood control through stream channelization and the building of levees	Reduction of groundwater percolation and charging of aquifers. Increased flooding downstream.
7. Diversion of river or lake water for inland irrigation (Aral Sea)	Increased accumulation of salts due to evaporation and inadequate fresh water flow. Accumulation of sediments, pesticide chemical residues, and fertilizer runoff.
8. Water pollution from sewage, sewage treatment, and fertilizer runoff	Increased nutrient loads in water resulting in increased bacterial populations, algal blooms, and odors
9. Industrial toxic water pollution	Toxic water pollution kills the organisms that clean water naturally.

Questions to Ponder

1. Identify the major problems of water use and quality at the following locations described in the chapter.

Location of body of water or drainage	Problems
1. Puget Sound (see Chapter 2)	Industrial and municipal water pollution. 75% of the tidal estuaries have been lost since 1800. Salmon and Orca populations have declined.
2. Columbia River	It is threatened by development, dam construction, navigation, dredging, channelization, and irrigation. Fewer salmon can migrate upstream to spawn and downstream to begin their growth in the ocean.
3. San Francisco Bay	Industrial and municipal water pollution. Saltwater intrusion into groundwater. 94% of the tidal estuaries have been lost since 1850.
4. Colorado River	Dam building has resulted in increased mineralization/salinization of the water as it flows downstream. There are significant losses of water due to evaporation in its drainage, almost entirely desert and semidesert, and seepage into sandstone rock formation particularly along Lake Powell and Lake Mead. The sparse water that reaches Mexico is contaminated with salts and pesticide residues. The cities of Phoenix, AZ, and Los Angeles, CA, divert large amounts of water. In 1988, the city of Scottsdale, AZ, bought land along a tributary, the Bill Williams River in Central northwest Arizona, and then claimed rights to the amount of canal water that the Bill Williams River contributed to the Colorado River. Some engineers even proposed cutting down the pinyon pine-juniper forests so that

	more runoff would appear in Phoenix' reservoirs. (Source, *The Arizona Republic*).
5. Mississippi River	The historic floods of 1927 and 1993 have led many to the conclusion that the levees along the drainage should be torn down, cities moved to higher ground, and the land be allowed to flood naturally. Building in flood plains is a bad idea. Pesticide residues and fertilizer runoff have created a dead zone of the coast of Louisiana in the Gulf of Mexico.
6. Mono Lake, CA	Freshwater that fed Mono Lake was diverted to the city of Los Angeles in 1941. Because evaporation exceeds filling, the lake became saltier and filled with toxic runoff from farms. Wildlife populations declined.
7. Ogallala Aquifer	Withdrawal greatly exceeds the recharging of the aquifer due to irrigation and municipal use. Wells have to be drilled deeper.
8. Aral Sea	Similar to Mono Lake in that it has no outlet, this inland sea in southwestern Asia has begun to fill with salt and toxic residues. The Soviet Union even buried anthrax spores on one island. Desert irrigation has diverted freshwater and added salt and pesticides. The residents of the area are plagued by toxic salt storms that cause or worsen respiratory diseases.
9. Rhine River Valley, Germany	The Rhine River drainage is affected by industrial and municipal pollution from Austria, Lichtenstein, and Switzerland to Germany to France and finally to the Netherlands. The water must be decontaminated for use in the Netherlands.

2. Describe the hydrologic cycle. *Sunlight causes the hydrogen bonds between water molecules to break as the water transforms from a liquid to a gas. When wind blows over ice, water sublimates into the air. The water condenses into droplets that form clouds. Fine ice crystals also form clouds at higher altitudes. Water molecules stick to and cohere around dust particles. When enough water molecules have accumulated, the heavy drops fall as rain or form snow. Rainfall on land masses percolates into the ground, particularly in forested areas and wetlands. The groundwater then flows toward a large body of water where the layer of aquifer intersects with a larger body of fresh or salt water. Surface runoff may be stable when it has nearly constant annual flow or unstable when the flow peaks in a monsoon or dips in a dry season (remember the tropical deciduous forest?).*

3. Imagine you are the engineer designing a "green" water system for a coastal city.

   a. What features would you design? What technologies would you use? *Let's give it a try: Cisterns will collect rainwater for cities and homes. Simple cisterns can be covered barrels with an inlet from your roof gutter. It can be used for "victory gardens" and, when filtered, for other household uses. The small reservoirs will be covered, placed underground, to reduce evaporation. Large wasteful dams and reservoirs will be eliminated, and the land occupied can be turned into prime farmland. Levees will be eliminated, and flooding will replenish farm soils. It will be illegal to build within a 50-year flood plain. It will be illegal to build over wetlands and forests that serve as groundwater charging areas. The water withdrawn from aquifers must be equal to that recharged, but only after the aquifers have been restored. The U.S. Corps of Engineers will only work for the military. Instead the U.S. Department of Agriculture, Natural Resources Conservation Service will take over their civilian responsibilities. Join the Earth Team at* ***www.nrcs.usda.gov/feature/volunteers.*** *In the home, we would establish water conservation labeling and U.S. goals for conservation of appliances. All large industries will be required to recycle and purify their water. All golf courses will use drought-tolerant grasses and be limited in their use of water (particularly in Santa Fe, NM) or eliminated.*

b. If you had to pick between building reservoirs or allowing water to run in streams, which would you use for a water source? Please explain your decision by considering the advantages and disadvantages. *The advantages of the large aboveground reservoirs are in their multiuse for recreation and water for large cities. Large reservoirs have capacity to continue water supply during seasonal or irregular droughts. The disadvantages are large amounts of water lost to evaporation. The lake homes typically are not connected to sewer lines, and therefore the septic tanks leak nutrients into the lake, contributing to high fecal coliform bacteria counts and algal blooms. There is recharging of groundwater by large reservoirs, depending on the geology.*

4. Imagine that you are a "green" politician representing farmers in Nebraska using Ogallala Aquifer well water. What would you propose to replenish and sustain this water source, so that their great grandchildren could farm there? Offer 3 measures.

	Measure to sustain or replenish Ogallala water
1.	Installation of rain-catching cisterns
2.	Limiting withdrawals of water until the aquifer recharge rate exceeds the withdrawal rate in the short term (15 years). Then allowing for withdrawal equal to that recharged.
3.	Eliminating dams and reservoirs. Instead piping a limited amount of water by covered aqueduct to irrigation centers. Then using microirrigation. Both measures will conserve water lost to evaporation

5. Describe ten ways to conserve water at home.

	Action to conserve water at home
1.	Install water-conserving toilets, such as presently used in the U.K.
2.	Install showers and small bath tubs instead of large tubs. Plant flowers in your hot tub.
3.	Install water-conserving, aerating faucets.
4.	Microirrigate vegetable and flower gardens using rain water cisterns.
5.	Irrigate at night and use mulches. Even newspaper or shredded paper can be applied. Use vegetable material destined for the trash as a compost soil builder.
6.	Use filtered rainwater for washing.
7.	Recycle gray water for irrigation. Separating organic waste water (sewage) from gray water used once for washing.
8.	In desert or dry grassland areas, xeriscape, gray water, and drought tolerant plants.

9. Use a water-conserving dishwasher.

10. Use a water-conserving clothes washer.

Pop Quiz

Matching

1. p
2. f
3. l
4. b
5. j
6. n
7. o
8. d
9. g
10. c
11. e
12. k
13. i
14. h
15. q
16. a
17. m

Multiple Choice

1. b
2. c
3. a
4. e
5. a
6. e
7. e
8. c
9. d
10. e
11. a
12. c
13. b
14. d
15. e
16. b
17. d
18. c
19. e
20. d
21. a
22. e
23. c
24. e

14

Soils and Their Preservation

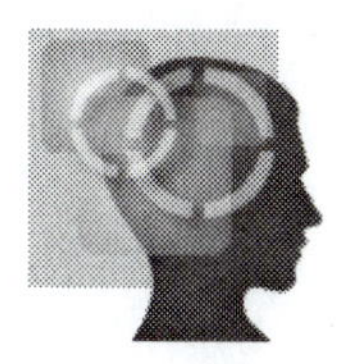

LEARNING OBJECTIVES

After you have studied this chapter you should be able to:

1. Identify the factors involved in soil formation.
2. List the components of soil and give the ecological significance of each.
3. Briefly describe soil texture and soil acidity.
4. Explain the impacts of soil erosion and mineral depletion on plant growth.
5. Describe the American Dust Bowl and explain how a combination of natural and human-induced factors caused this disaster.
6. Define *sustainable soil use* and summarize how conservation tillage, crop rotation, contour plowing, strip cropping, terracing, and shelterbelts help to minimize erosion and mineral depletion of the soil.
7. Discuss the basic process of soil reclamation.
8. Briefly describe the provisions of the Farm Bill regarding the Conservation Reserve Program and the Grasslands Reserve Program.

SEEING THE FOREST

Consider the problems that have arisen in regard to soil quality and the environment. Fill-in the table below.

Area	Soil problems and conservation measures that have been or could be used
1. Green Wall of China	
2. The Everglades of central and south Florida	
3. The high grassland plains of North America	
4. The lower "breadbasket" states along the central Mississippi River and southern Missouri River drainages	
5. The lower Colorado River Basin	
6. Sub-Saharan Africa	

SEEING THE TREES

VOCAB

You will die but the carbon will not; its career does not end with you ... It will return to the soil, and there a plant may take it up again in time, sending it once more on a cycle of plant and animal life.

Jacob Bronowski

1. Desertification—the process of the degradation of arid and semiarid grasslands or chaparral to create desert. Overgrazing, groundwater depletion, and excessive harvesting of wood or brush for fuel can cause desertification.
2. Soil—the thin covering of coarser rock materials that includes minerals derived from the rock and humus that results from the breakdown of plant and animal debris.
3. Soil formation—soil is formed by several processes including sand blowing over and grinding rocks; ice-formation in fine cracks that breaks rocks apart; lichen breaking rock surfaces and adding humus to make soil; heavy rainfall dissolving minerals that accumulate in soil; plant roots breaking up rocks; carbonic acid from rain dissolving rocks and liberating mineral nutrients; bacteria and fungi breaking up plant materials that make humus and releasing minerals into the soil; and earthworms eating soil microorganisms and mixing in organic material as they tunnel through it.
4. Topography—the surface of an area, including mountains, hills, and valleys that vary in altitude and slope.
5. Humus—a dark, amorphous, somewhat gummy substance when highly concentrated that results from the breakdown of plant and animal debris. It increases the water and ionic soil nutrient holding capacity of soils. In cooler soils, it may persist for years. In warmer soils, bacterial action quickly breaks humus down into carbon dioxide and water.

6. Leaching—heavy rains leach or percolate chemical nutrients and other minerals, notably iron and aluminum, down through the soil horizons (illuviation) and into the groundwater aquifer.
7. Soil horizons—soil forms vertical layers from the surface to "parent" rock material underneath as follows:
   a. *O-horizon:* the surface is covered with decomposing plant and animal litter; many arthropods graze on the plant material.
   b. *A-horizon:* a darker layer of soil with humus and minerals leaching out of it quickly (the O-horizon and upper A-horizon form a zone of humification).
   c. *E-horizon:* a highly leached area between the A- and B-horizons; *B-horizon:* the soil color lightens as humus disappears and leached minerals accumulate (the *B-horizon* represents a zone of mineralization).
   d. *C-horizon:* a layer of larger pieces of weathered rock derived from the parent bedrock underneath.
8. Soil microorganisms—soil algae live in the upper soil where light is available; fungi, cyanobacteria, and bacteria decompose the plant and animal litter. Protozoa may be either predatory or photosynthetic.
9. Soil invertebrates—grubs feed on plant roots, leaving feces behind that are decomposed by bacteria and fungi; nematodes (small roundworms) graze on soil algae, bacteria, and fungi (one fungus even preys on nematodes). Earthworms. wood-eating termites, and ants tunnel through the soil distributing organic and inorganic nutrients throughout the *O*- and upper *A*-horizons. Soil porosity is increased. When they are being the proverbial "night crawlers," earthworms leave fecal castings on the surface, bringing nutrients up from lower layers, to be redistributed downward.

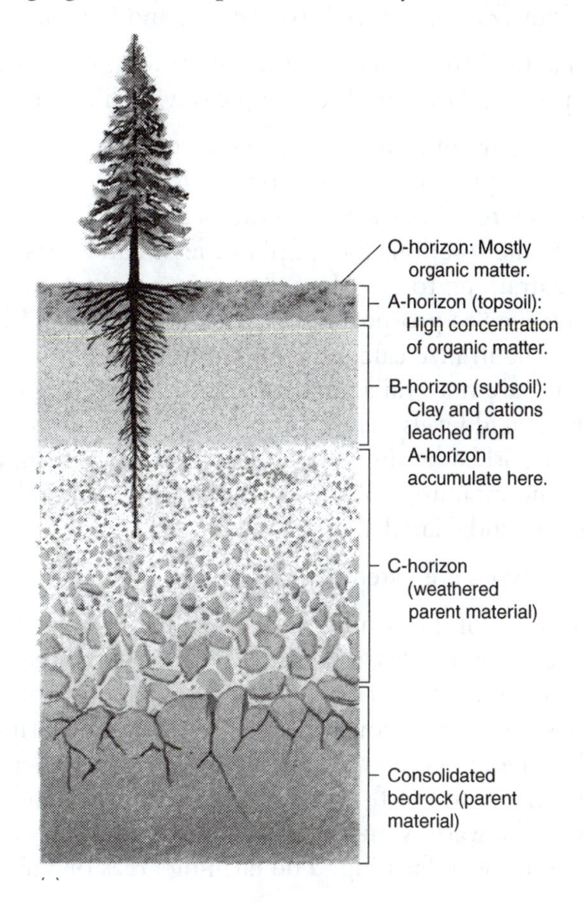

10. Characteristics of a good soil—Also see loam below. Soil good for agriculture consists of the following materials:
    a. 93% minerals (see loam below), 7% organic substances consisting of 85% humus, 10% roots, and 5% organisms.
    b. The organisms in soil consist of:
        1) 40% fungi and algae
        2) 40% bacteria
        3) 12% earthworms
        4) 5% moles and ants
        5) 3% microarthropods
11. Mycorrhizae—literally "fungus roots." Certain fungal mutualists of roots increase the ability of plants to absorb water from the soil. There is a substantial difference in pine tree growth: those with mycorrhizae grow faster and are larger, than those without.
12. The structure or texture of soil—Soil particles are classified according to size of the particle as it affects texture or "feel" as follows: *Sands* are gritty, large particles from 0.05–2 mm in diameter; *Silts* are medium-size particles, 0.002–0.05 mm in diameter; *Clays* are less than 0.002 mm in diameter and have a flat structure—by reason of their small size, smaller objects mathematically have more surface area than larger ones of the same geometric structure, clay particles have more surface area available for holding water and inorganic materials. By comparison, sand particles are like basketballs, silt like tennis balls, and clay like BBs.
13. Soil mineral ions—electrically charged atoms or groups of atoms that are plant nutrients include nitrate (NO_3^-), phosphate (PO_4^{-3}), potassium (K^+), calcium (Ca^{+2}), magnesium (Mg^{+2}), and smaller amounts of zinc(Zn^{+2}), reduced iron (Fe^{+2}), and Copper (Cu^{+1}).
14. Loam—this is the ideal agricultural soil consisting of 40% sand and silt, 20% clay, and 40% in organic materials. It is porous, dark, and breaks up easily, unlike soils with much more clay.
15. Soil acidity—pH is the measure of acidity (H^+ ion concentration in water) with the lower numbers representing more acid (pH = 0 = 10^0 H^+ ions or 1 mole, see the chemistry appendix), and the succeeding number each ten times less than the preceding lower number (pH = 1 = 0.1 mole or 10^{-1} H^+ ions). Acidic solutions have a pH of less than 7, basic or alkaline solutions have a pH of more than 7 (neutral), up to 14. Most plants grow best in soils with a pH of 6–7, however fewer plants have evolved to prosper in more acidic or basic pH levels. Generally, lower pH levels/high acid levels neutralize calcium and magnesium into insoluble forms; acids dissolve minerals so that they are not available for absorption by the roots and are carried away by leaching. Conversely, high acid levels dissolve more toxic aluminum and manganese, so that these toxic minerals are absorbed by the roots. The latter mechanism explains the effects of acid rain on the trees of mountaintops of the northeastern U.S. and lower elevation regions of northeastern New England and Canada.
16. Soil types—the major soil types are listed below:
    a. Spodosols—the typical soil of the coniferous forest that is covered with an acidic layer of decomposing evergreen tree needles, a highly leached ash gray E-horizon and a dark brown illuvial B-horizon. The high levels of precipitation result in leaching, and low temperatures slow needle decomposition. The acidic topsoil is not good for farming.
    b. Alfisol—the topsoil of the temperate deciduous forest. The moderate rainfall washes mineral into the minerals and clay into the B-horizon. When forests are cleared, the plant litter decomposes quickly in the warm summers, and the soil must be chemically or organically fertilized to support long-term farming. The farming areas of Virginia and the Atlantic

Coast were nearly exhausted by cotton farming in the 18th and 19th centuries. The advent of chemical fertilizers allowed for these lands to become productive.

c. Mollisols—the rich soil of temperate grasslands, including "America's breadbasket." They have a dark brown A-horizon rich with humus. Very cold winters freeze the topsoil and slow litter and humus decomposition. Light rainfall tends to wash mineral nutrients to root level where the standard grains or grass roots absorb them.
d. Aridisols—desert soils that lack the layering caused by leaching. Unless they are alkaline, aridisols can be used for a water-intensive agriculture. Many crops can be grown in the cooler temperatures of late fall, winter, and into mid-spring.
e. Oxisols—the soil typical of tropical and semitropical rain forests has a thin organic layer due to warmer temperatures that enhance decomposition of litter. Most of the nutrients of the biome are aboveground in the bodies of the trees.

17. Soil erosion—not only do rainfall and irrigation carry minerals and fertilizers into streams, but soil particles as well. This goes hand in hand with standard European methods of plowing—soil is cut and turned over, that has been used for a thousand years. Because plant roots do not hold the soil year round, fine soil particles wash away toward the oceans. Methods to control soil erosion are defined below.

    a. Conservation tillage—crop residues, such as the stems and roots, are left in the ground to cover and hold the soil over the winter until the next planting. This is also called *no-tillage.*
    b. Crop rotation—alfalfa was discovered and cultivated about 600 BC/BCE by the Persians, who then introduced it to the Greeks as a horse feed. The ancients noted that farm fertility was improved by growing alfalfa. Subsequently, crop rotation applied to western farming when a field was allowed to "lie fallow" for a year, while alfalfa grew and restored nitrogen levels and humus to depleted soils. The alfalfa is a legume plant that is used as an animal feed (although those alfalfa sprouts in salads are yummy). It has mutualistic *rhizobia* bacteria that grow on the roots in nodules. These bacteria can convert atmospheric nitrogen to ammonia (nitrogen fixation), fertilizing the plants and improving their protein content. When the plants die, the nitrogen from the roots is donated to the soil. Soybeans, peanuts, and other beans or peas are important legume food crops.
    c. Contour plowing—using an altimeter, fields can be plowed in curves that follow topographic lines of constant altitude. This reduces the downhill runoff that carries soil into streams.
    d. Strip cropping—a type of contour plowing that increases crop diversity that discourages insect pests and decreases soil erosion by placing a closely sown crop (wheat) under a loosely sown crop (maize/corn).
    e. Terracing—on steeper slopes level areas are placed so as to catch downhill runoff. *Rice paddies* are highly productive ponds that can be made by terracing or located in low-lying wetland areas. The rice paddy grows fish (carp) and rice.
    f. Shelterbelts—to control the blowing dust of the semiarid plains, the U.S. government planted rows of trees around farm fields to control the wind.
    g. Greenbelts—greenbelts of shrubs and trees can be used as boundaries between crops, increasing plant diversity and offering cover for the predators of crop pests. Both increased crop diversity, and greenbelt cover reduces soil erosion and the use of pesticides.

18. Soil Conservation Act of 1935—a U.S. law that organized the Soil Conservation Service, now called the Natural Resources Council. See **Cyber Surfin'.**

19. Food Security Act or Farm Bill of 1985—this U.S. law required farmers to control erosion on land that was subject to large amounts of soil loss. If they failed to do so, they lost governmental price support.

20. Conservation Reserve Program—a voluntary U.S. program that pays farmers not to grow crops on land subject to high rates of erosion.

Questions to Ponder

1. Describe the agricultural, environmental, and sociological conditions that led to the "American Dust Bowl" described by John Steinbeck in *The Grapes of Wrath.* Go to the U.S. National Public Radio website for more information:
www.npr.org/programs/morning/features/patc/grapesofwrath/ and
www.pbs.org/wgbh/amex/dustbowl/

2. Offer three reasons to plant trees and crops together in arid and semiarid areas.

	Agricultural benefit
1.	
2.	
3.	

3. Historically the discovery of alfalfa by the Persians, a "Darwinian" selection of crop seeds, and the invention of the scouring plow revolutionized medieval farming in Europe. Explain the contribution of each.

	Contribution
Alfalfa	
Selection of crop seeds	
Scouring plow	

4. Using common materials around the house or available at little expense at a hardware store, list 5 items you would need to start a compost pile. See the site from Virginia Tech's Department of Environmental Horticulture: **www.ext.vt.edu/pubs/envirohort/426-703/426-703.html.**

	Needed to construct an inexpensive compost pile
1.	
2.	
3.	
4.	
5.	

5. Imagine that you are in charge of one of media magnate Ted Turner's "green farms." What practices would you use to make the large farm ecologically sound and productive? Assume a southern Missouri Valley location and that one part of the farm raises chickens for eggs and dairy cattle.

POP QUIZ

Matching

Match the following terms and definitions

____1. Droughts and overpopulation have led to deforestation and ______ in Sub-Saharan Africa.	a. *A*
____2. When mountains are eroded, this is formed:	b. *B*
____3. A map that indicates slopes, hills, and valleys is called:	c. illuviation
____4. The decomposition of leaf litter forms:	d. contour plowing
____5. Humification occurs in this soil horizon:	e. micorrhizae
____6. In the deciduous forest, soil minerals are leached to this layer and accumulate there:	f. topographic
____7. Below the soil horizons lie:	g. high temperatures and desiccation
____8. Leaching and the accumulation of minerals in lower soil horizons is called:	h. acidic
____9. Soil minerals in this form can be absorbed by plant roots:	i. loam
____10. The soil of "America's breadbasket" is a mixture of smaller and larger particles, and humus called:	j. soil
____11. Mutualistic organism of pine roots:	k. shelterbelts
____12. The topsoils of the northern coniferous forest are ______ because of the slow decay of needles.	1. desertification
____13. Earthworms take nutrients from the *A-horizon* and bring them to the ______ horizon.	m. ionic
____14. Humus quickly disappears from the soils of the Everglades because:	n. humus
____15. In the tropical rainforest, more nutrients are here rather than in the soil:	o. *O*
____16. These were planted to control dust storms in the western plains during the 1930s:	p. parent bedrock
____17. This method of farming follows levels of constant altitude:	q. evergreen broadleaf trees and vines

Multiple Choice

1. Which factor has contributed to the process of desertification in the Sudan and China?
   a. excessive harvesting of wood for fuel
   b. using European-style agriculture that is well-water intensive
   c. excessive grazing of domestic livestock
   d. population growth exceeds environmental carrying capacity
   e. all of the above

2. Which of the following processes does not contribute to the making of soil?
   a. leaching and illuviation
   b. grinding of rocks by wind-blown sand
   c. breakdown of litter at the surface
   d. ice freezing between cracks in rock
   e. erosion of rock

3. Percolation of water through soil horizons leads to which result?
   a. removal of minerals from the *A*- and *E*-horizons
   b. accumulation of minerals in the *B-horizon*
   c. illuviation of minerals to a level below that of grass roots
   d. leaching of water and minerals into groundwater
   e. all of the above

4. An ideal soil includes which percentages of materials?
   a. 100% humus
   b. 40% sand, 40% silt, and 20% clay
   c. 60% clay, 30% silt, and 10% sand
   d. 60% humus and 40% silt
   e. 60% sand, 20% clay, and 20% humus

5. Which factor explains the fertility of "America's breadbasket?"
   a. Sparse rainfall washes calcium and magnesium to root level.
   b. Topsoil freezes for most of the winter.
   c. Loam soils are present.
   d. Corn, rye, and wheat are grasses that tolerate low rainfall.
   e. All of the above are correct.

6. Which soil material is mismatched with its description?
   a. sand/increases porosity of soil
   b. clay/has greatest water-binding surface area
   c. silt/has the smallest particle size and quickly washes into streams
   d. humus/decomposed plant litter that binds water and ions
   e. sand/has the largest particle size

7. Which soil horizon is mismatched with its description?
   a. *O*/plant litter is decomposing
   b. *A*/tends to be darker, humus present
   c. *B*/soil minerals reach here, below the level of most plant roots

d. *E*/this is a highly leached zone below the B zone
e. Trick question alert! All are matched correctly.

8. Which soil is mismatched with its biome?
   a. aridosol/thin soil of the desert
   b. alfisol/highly acidic soil of the deciduous forest
   c. mollisol/loamy soil of the temperate grassland
   d. spodozol/cool, moist soils of the coniferous forest
   e. oxisol/thin, nutrient poor soil of the tropical rainforest

9. John Steinbeck's tome, *The Grapes of Wrath,* considered the plight of farmers that emigrated from which area?
   a. California
   b. Western Oklahoma
   c. West Virginia
   d. Missouri
   e. Illinois

10. Which of the following crops is ideal for restoring nitrogen depleted farm fields?
    a. soybeans
    b. alfalfa
    c. cotton
    d. corn
    e. a and b

11. Which description is accurate concerning the method of no-tillage?
    a. Seeds are inserted into the ground by machines that minimally disturb the soil.
    b. Only contour plowing is done.
    c. The soil is not plowed during the winter.
    d. It increases soil erosion.
    e. Stable surface runoff is kept on the land.

12. The way to check whether your compost heap at home is working is to:
    a. check to see if water is percolating through the surface.
    b. put some of the surface soil in your hand and squeeze it to see if it sticks together.
    c. dig out the center and check if it smells rotten.
    d. stick a thermometer in the center to see if it is hot.
    e. a, b, and c.

13. Alfalfa restores the fertility of nutrient-exhausted fields in which way?
    a. It adds nitrogen to the soil when its roots decompose.
    b. Bacteria living on its roots convert nitrogen gas to ammonia fertilizer.
    c. Alfalfa converts cattle manure to nitrates.
    d. a and b
    e. c only

14. The development of ______ in the late 19th century saved nutrient depleted farms in the Atlantic coastal states.
    a. contour plowing with tractors

b. new strains of alfalfa
c. chemical fertilizers
d. composting
e. plowing pig and cattle manure into fields

15. Which of the following descriptions best explains why compost is aerated?
a. Aeration improves soil porosity so that water can percolate through it.
b. Bacteria and fungi require oxygen for their activities.
c. Nitrogen fixation can occur more easily.
d. Putrefaction/ammonification can turn the animal protein into nitrogen gas.
e. All of the above are correct.

16. Because most U.S. communities have accepted some types of recycling, what can we expect next?
a. using human sewage or residues to fertilize food crops
b. a requirement to build a compost pile at home
c. solid waste recycling for municipal composting
d. all of the above

17. The Soil Conservation Act of 1935 required that farmers use which of the following methods?
a. contour plowing
b. no-tillage
c. strip cropping
d. rice paddy aquaculture/agriculture
e. Trick question alert! Compliance was voluntary.

18. Which of the following methods is best for conserving soil in mountainous areas?
a. strip cropping
b. shelterbelt planting
c. greenbelt planting
d. terracing
e. crop rotation

19. Which method has the Conservation Reserve Program used to get farmers to conserve soil?
a. legal requirements to contour plow
b. paying farmers not to raise crops on highly erodible land
c. requiring strip cropping
d. it requires that erodible land be retired permanently
e. all of the above

20. If you purchase a bag of 10-10-10 fertilizer, what do the numbers represent?
a. percentages of calcium, magnesium and potassium
b. nitrogen as nitrate, phosphorus in phosphate, and potassium
c. iron, humus, and nitrate
d. parts per thousand of ammonia, calcium, and magnesium
e. It cost $10, and you needed three bags.

21. A soil dominated by clay has which set of characteristics?
a. poor aeration and high water retention

b. easy workability and high nutrient retention
c. excellent drainage and high water retention
d. moderate workability, water and nutrient retention
e. is typical for temperate grassland and has high humus content

22. Earthworms, nematodes, ants, moles, and grubs enrich soil by:
    a. eating algae to start a decomposition food chain
    b. moving nutrients from the *A-horizon* to the *O-horizon*
    c. moving nutrients from the *B-horizon* to the *A-horizon*
    d. aerating the soil
    e. b and d

23. From a biological viewpoint, dried sewage treatment plant sludge is not used to fertilize crops for human consumption for this reason:
    a. Harmful organisms are present.
    b. Sewage sludge tends to be high in heavy metal toxins.
    c. It smells too bad!
    d. There are few nutrients left after bacteria eat it.
    e. All of the above are correct.

24. In Georgia, plantation houses were built on soil that is 12 inches thicker than it is today after cotton and other farming for 150 years. Approximately how long would it take to naturally replace that topsoil?
    a. 50 years
    b. 100 years
    c. 10,000 years
    d. 100,000 years
    e. It can never be reformed.

25. In recent times, soil erosion from farming has generally increased under which condition?
    a. prices for their crops were high
    b. demand for their crops was low
    c. crop prices were low
    d. during the Clinton administration
    e. during economic recessions

CYBER SURFIN'

1. Take a look at the U.S. Public Broadcasting System's page, *Surviving the Dust Bowl.* See **www.pbs.org/wgbh/amex/dustbowl/**
2. Professor Gregory S. Okin of the Department of Environmental Sciences, University of Virginia has produced an informative site on desertification. Check out **www.evsc.virginia.edu/~desert/**
3. From the National Sustainable Agriculture Information Service, Appropriate Technology Transfer for Rural Areas, see a primer on soil and fertility: **attra.ncat.org/attra-pub/soil.html**
4. Do your part in controlling soil erosion. See the U.S. Natural Resources Conservation Service website. Join the *Earth Team* at **www.nrcs.usda.gov/feature/volunteers**
5. The USDA has information on organic gardening, see **www.usda.gov/news/garden.htm**

'LIKE' BOOKING IT

1. Steinbeck J, and Demott R (introduction). ***The Grapes of Wrath (20th Century Classics).*** New York: Penguin Putnam, 1992.
2. Stoll S. ***Larding the Lean Earth.*** New York: Hill & Wang Pub., 2002.
3. Thompkins P, and Bird C. ***Secrets of the Soil: New Solutions for Restoring Our Planet.*** Anchorage, AK: Earthpulse, 1998.
4. Kimbrell A (Ed.). ***Fatal Harvest: The Tragedy of Industrial Agriculture.*** Covelo, CA: Island Press, 2002.
5. Howard-Yana S, and Harrisson J. ***Gardening for the Future of the Earth.*** New York: Bantam Doubleday Dell Pub., 2000.

ANSWERS

Seeing the Forest

Area	Soil problems and conservation measures that have been or could be used
1. Green Wall of China	Dust storms blow in from the Gobi Desert carrying and causing soil erosion and damaging dry winds. Tthese effects are mitigated by a wall of 300 million trees.
2. The Everglades of central and south Florida	Agriculture and municipal diversion of freshwater, agri cultureal runoff, municipal sewage and the trans-Florida canal that releases fresh water into the ocean. Limitation of land use for agriculture and cities would help.
3. The high grassland plains of North America	This area includes the location of the American Dust Bowl of the 1930s from the panhandles of north Texas and western Oklahoma to Kansas and Colorado. A severe drought, and poor farming methods in land that is marginally productive combined to produce a great migration of people out of those areas druing the depression era. Greenbelts/shelterbelts and contour plowing will reduce soil erosion.
4. The lower "breadbasket" states along the central Mississippi River and southern Missouri River drainages	Water erosion of soil due to rainstorms and heavy irrigation is particularly severe here. Contour plowing and greenbelts can hold soil.
5. The lower Colorado River Basin	Salt accumulates in the Colorado River water as it flows south due to evaporation and the flow of mineral irrigation water back into the Colorado. The volume of Colorado River water is drastically reduced by the time it reaches Mexico and soils are salinized by its use. Elimination of shallow water irrigation would improve water quality.
6. Sub-Saharan Africa	Overgrazing and the harvesting of wood for fuel have

	reduced the areas that charge underground aquifers and remove the plants that hold soils. Reduction of population and reforestation would help.

Questions to Ponder

1. Describe the agricultural, environmental, and sociological conditions that led to the "American Dust Bowl" described by John Steinbeck in *The Grapes of Wrath*. Go to the U.S. National Public Radio website for more information:
www.npr.org/programs/morning/features/patc/grapesofwrath/ and **www.pbs.org/wgbh/amex/dustbowl/**.

The Dust Bowl conditions in the western high plains of the U.S. were an unfortunate combination of the worldwide "Great Depression," and a prolonged drought. The farms abandoned were in semiarid regions where farming was marginal without artesian water from the Ogallala Aquifer. When the rains did come, the soil dried up. Without vegetation to hold the soil, it was blown around in dust storms. The farmers abandoned many farms and packed all their belongings on 20-year-old model-T trucks and headed westward to California where they could get jobs picking the crops of the Imperial Valley. They were disrespectfully termed "dirty Okies." They moved into government-built shacks in towns called "Hoovervilles," named after President Herbert Hoover, who most people blamed for government inaction that followed the Great Depression. John Steinbeck wrote of the beatings of the migrants, their poor wages and living conditions. For his efforts, Steinbeck was awarded a Pulitzer Prize and ever-lasting infamy in the minds of many California farm owners, who labeled him as a communist.

2. Offer three reasons to plant trees and crops together in arid and semiarid areas.

	Agricultural benefit
1.	Shelterbelts of trees reduce the wind at surface level, reducing blowing dust.
2.	Trees offer cover and habitat for predators of crop-eating insects.
3.	The borders allow for separating diverse crops. Crop diversity reduces insect pests.

3. Historically the discovery of alfalfa by the Persians, a "Darwinian" selection of crop seeds, and the invention of the scouring plow revolutionized medieval farming in Europe. Explain the contribution of each.

	Contribution
Alfalfa	Alfalfa was given to the ancient Greeks by the Persians, who told them it was a good horse feed. Later, it was discovered that fields grown in alfalfa were more fertile as a result of the alfalfa being grown there. In more recent times, the role of nitrogen fixing *rhizobia* bacteria, living on the roots of legumes, was described. Legume crops are highly valued for their protein content. Nitrogen in the forms of ammonia or nitrate allow plants to make protein.
Selection of crop seeds	Ancient farmers were confronted with a problem during the transition from hunter-gatherer societies to farming. Do we eat all the seed or save some for planting next year? When we save the seed, do we save large seeds or small seeds? The ancients knew that large ani-

	mals made large animals and must have concluded the same process worked in plants. Over time, using corn/maize as an example, the seeds became larger as well as the number yielded by each plant. Darwin used domestication of plants and animals as evidence for his proposed mechanism of evolution.
Scouring plow	A scouring plow not only trenches the ground for seeding, but also turns lower levels of the *A-horizon* over, aerating it and bringing nutrients to the surface. The iron scouring plow was first invented by the Chinese in approximately 4th century BC/BCE. Poorer quality iron plows appeared in Europe around 1000 AD/CE and were not used in large numbers until they were improved in the 1600s.

4. Using common materials around the house or available at little expense at a hardware store, list 5 items you would need to start a compost pile. See the site from Virginia Tech's Department of Environmental Horticulture: **www.ext.vt.edu/pubs/envirohort/426-703/426-703.html**

	Needed to construct an inexpensive compost pile
1.	#10 ea. 2–3 foot long wooden stakes.
2.	Porous erosion control cloth, 2–3 feet wide. At least 12 feet are required for the minimum 3×3 square pile. There is a critical mass that must be achieved to allow enough heat to effectively decompose the buried vegetation. Also chicken wire or welded hardware cloth can be used. You will get a large amount of soil leakage with the chicken wire. Better still, chicken wire outside of the erosion cloth works well for aeration.
3.	A large spade shovel.
4.	Coat-hanger or similar wire to make a simple door/gate for the front of the pile.
5.	A long outdoor thermometer, alcohol type, not mercury.

5. Imagine that you are in charge of one of media magnate Ted Turner's "green farms." What practices would you use to make the large farm ecologically sound and productive. Assume a southern Missouri Valley location and that one part of the farm raises chickens for eggs and dairy cattle.

I love these brainstorming questions! Regarding the chicken manure, I would use it and some of the vegetable debris for making biogas on the premises. The biogas would make electricity and steam heat for the operation. The chicken manure is better digested to produce biogas: this controls microorganisms that can infect the chickens from nearby soils. The cattle manure will be mixed into the soil for humus. I would use biological means of fertilizing, such as materials left over from fish processing or algae harvested from canal of organic refuse-holding pond surfaces. No-tillage will be used for winter, or better, a winter-tolerant crop such as canola or winter wheat will hold the soil. Weeds will be scraped mechanically from the soil. The soil will be aerated by the hole punch method, similarly seeds will be planted. The farm would have 15-foot (5-meter) greenbelts separating 10 square-acre fields. The same crop in a given field will be separated by at least one plot of a different crop and two intervening greenbelts. The greenbelts will offer cover for the birds and predatory insects that will feed on the crop pests. I would reintroduce beneficial organisms such as ladybugs, earthworms, green lacewings, and Trichogramma wasps. I would use biological and mechanical means of insect and weed control rather than chemical means. Microirrigation at root level will be supplied from cisterns, gray water, or other sustainable sources. Organic mulch will be applied to control weeds and conserve soil lost from the soil's surface. Nothing would be wasted. Most of these principles are borrowed from organic gardening but applied on a larger scale.

Matching

1. l
2. j
3. f
4. n
5. k
6. b
7. p
8. c
9. m
10. i
11. e
12. h
13. o
14. g
15. q
16. k
17. d

Multiple Choice

1. e
2. a
3. e
4. b
5. e
6. c
7. d
8. b
9. b
10. e
11. c
12. a
13. d
14. c
15. b
16. c
17. e
18. d
19. b
20. b
21. a
22. e
23. b
24. c
25. a

15

Minerals: A Non-Renewable Resource

LEARNING OBJECTIVES

After you have studied this chapter you should be able to:

1. Explain the difference between high-grade ores and low-grade ores, and between metallic and nonmetallic minerals.
2. Describe several processes by which minerals are concentrated in the Earth's crust.
3. Briefly describe how mineral deposits are discovered, extracted, and processed.
4. Relate the environmental impacts of mining and refining minerals and explain how mining lands can be restored.
5. Contrast the consumption of minerals by developing countries and by industrialized nations, such as the U.S. and Canada.
6. Summarize the conservation of minerals by reuse, recycling, and changing our mineral requirements.
7. Explain how sustainable manufacturing and dematerialization can contribute to mineral conservation.
8. Sketch a diagram showing the interrelationships within the industrial ecosystem in Kalundborg, Denmark.

SEEING THE FOREST

It can be said that the economic benefits of mining take place far away from the places where the mining occurs. Give four examples of the negative effects of mining and ore processing on local environments and economies.

	Negative effect on the local area where the mining occurs or the ore is processed
1.	
2.	
3.	
4.	

SEEING THE TREES

VOCAB

To waste, to destroy, our natural resources, to skin and exhaust the land instead of using it so as to increase its usefulness, will result in undermining in the days of our children the very property which we ought by right to hand down to them amplified and developed.

Theodore Roosevelt

1. General mining law of 1872—signed by President U.S. Grant that allowed mining claims for hardrock mineral metals to be staked on federal lands for $2.50 to $5 an acre. There are no environmental protection provisions of this law. If one drives from Ouray, CO, south to Durango, one can see copper-contaminated ponds at the foot of the silver mine spoil. Mine acid not only increases the toxic copper ion (Cu^{2+}) content of the drainage water but kills organisms downstream. President Clinton stopped mining in Yellowstone National Park.
2. Minerals—chemical compounds and elements found in the Earth's crust. Rocks are mixtures of minerals.

Mineral	Commodity extracted or used
Hard minerals—Rock or consolidated material Sandstone, limestone gravel, and other high bulk materials	Construction materials for making and facades.
Unconsolidated materials Silica sand, aragonite (calcium carbonate of marine origin), gypsum; and phosphates; and sulfur	Industrial materials such as, respectively, silicon for computer chips, glass; drywall; fertilizers; sulfur is used to neutralize alkaline soils.

Metallic minerals	Gold, platinum, tin, titanium, and rare earth metals mainly in grains and nuggets in alluvial deposits on surrounding magmatic intrusions.
Metal oxides	Aluminum in bauxite; iron in hematite and magnetite; manganese in ocean nodules and manganite; copper in malachite; uranium in uraninite; zinc in zincite (Ohio). Aluminum is refined by electrolysis, requiring much energy.
Metal sulfides and sulfates	Copper in bournite and chalcocite, lead in galena, zinc, chromium, and gold. Gypsum is calcium sulfate used to make drywall panels for homes.
Metal arsenides	The nickel and iron arsenide minerals include pentlandite and niccolite. The mineral engargite is an ore of copper and arsenic.
Metal silicates	Garnierite, a nickel silicate in Cuba, is the most important single source of nickel ore.
Metal carbonates	Limestone is calcium carbonate and magnesium carbonate, an important fertilizer and neutralizer of acid soils.
Carboniferous rock	Graphite for pencils, lignite, soft and hard coals for fuels.
Organic materials	Peat as a filtration agent and fuel in Ireland

3. Metals—elements that are malleable, shiny, and good conductors of electricity and heat. They typically possess 1–3 valence electrons for chemical bonding with other atoms. Metal-containing minerals may be found as oxides—iron III oxide (Fe_2O_3 hematite); sulfide—iron sulfide (FeS_2 pyrite or fool's gold); and arsenide—nickel arsenide (NiAs or niccolite).

4. Formation of minerals—All the rock of the Earth is thought to have been molten at one time: first as the Earth cooled after its formation, and later as molten rock rose to the crust and descended to melt again. When volcanoes erupt, molten rock or magma appears at the surface. As a continental plate subducts under another plate, solid rock is remelted.

   a. Magmatic concentration—molten rock solidifies as it cools. Lighter minerals rise to the top of the cooling layer, and heavier minerals, such as those with iron and magnesium content, solidify at the bottom of the layer. Chromium oxides (chromites) and platinum are formed this way.
   b. Hydrothermal processes—as heated groundwater moves through layers of rock, soluble minerals (especially if chloride and fluorides are dissolved in the water) are carried along toward the surface. In passage through sulfur-bearing rocks, metal sulfides are formed. Metal sulfides are insoluble: they precipitate (literally fall out) of solution to form distinct deposits of ores, such as iron, copper, and manganese. Gold may form at the edge of volcanic vents in the ocean.
   c. Sedimentation and chemical precipitation—As rock weathers, water carries particles of minerals downhill where they may accumulate in low lying areas, such as lake beds or in oceans. At the edge of the Grand Canyon, one walks on the Coconino limestone formed under oceans. Similarly, when weathered minerals are carried toward the sea, the minerals that dissolve more readily in warm water precipitate in the cold water.
   d. Evaporation—large deposits of sea salt and other salts have been deposited when the bodies of water that they were dissolved in evaporated. The borax deposits in Death Valley, CA formed as a lake evaporated. The lake had been fed by volcanic springs that dissolved the borax salts.

5. Mining—minerals may be extracted by surface mining and subsurface or deep mining.
   a. *Surface mining*—methods include strip mining where an overburden of plants and rock is stripped away by bulldozers or shovels from the surface to expose the minerals below or extract them when they lie at the surface. This may be done in an large *open pit* that looks like an inverted cone. The overburden is pushed into the old trenches or forms piles called *spoil banks*. *Hydraulic mining* is a technique of spraying high-pressure water to remove ores from hillsides. Gold was mined this way 100 years ago, leaving permanent scars in California.
   b. Subsurface or deep mining—underground mining uses vertical or slopes to tunnel underground and remove minerals. Deep mining is less destructive to the surface, generally, but offers one of the most dangerous occupations for residents of the western U.S. and Appalachia.
6. Smelting—the process of heating ores to a very high temperature with a glowing form of coal called coke removes oxygen from the minerals, creating a *slag* of waste minerals that is scraped off the top of the molten iron. Smelting produces air pollution. A copper smelter in Sudbury, Canada, produces large amounts of sulfur dioxides.
7. Mine acid—chemosynthetic bacteria oxidize the sulfur in the sulfur rocks in the mine and in the spoil on the surface, forming sulfuric acid. The acid then dissolves heavy metal toxins from the minerals—lead, cadmium, and mercury. Similarly, the *tailings* of waste minerals from the processing plants contain cyanide, mercury, lead, arsenic, and sulfuric acid.
8. Surface Mining Control and Reclamation Act of 1977—a U.S. law that requires the reclamation of surface-mined land. However, according to research done at Marshall University in West Virginia, the streams are still plagued by mineralization, silt, and mine acid. Moreover, many wells are contaminated. The reclaimed land in mountainous Appalachia looks radically different now than before it was mined. In 1997, less than 1% of land mined in Colorado had actually been "reclaimed."
9. Phytoremediation—some plants, bacteria, and algae not only tolerate certain toxic metals and other elements but also concentrate them in their cells. This bioaccumulation process allows for toxic metal removal from the land. The twist flower, *Streptanthrus polygaloides*, can remove nickel from the soil; the fern *Pteris vittata* can remove arsenic. The plants can be processed to remove the metals for industrial and other use. Nickel, for instance, was used in World War II to plate automobile bumpers, thus conserving the more valuable chrome for military use.
10. Mineral reserves—high-grade ores than can be economically extracted.
11. Mineral resources—low-grade ores that may become economically extractable in the future.
12. World reserve base—total resources are equal to the sum of a certain mineral's reserves and resources.
13. Madrid Protocol—also known as the Environmental Protection Protocol to the Antarctica Treaty, it prevents mineral extraction from Antarctica for 50 years or until 2040 and sets Antarctica apart as a natural reserve.
14. Manganese nodules—deposits of manganese and other metals that lie on the floor of the ocean. These can only be obtained in commercial quantities by destructive dredging. The 1994 U.N. Convention on the Law of the Sea contained provisions preventing dredge mining. The U.S. has not ratified this treaty.
15. Biomining—the *Thiobacillus* bacterium can release copper and gold into an acidic solution, which is then processed to remove the metals.

16. Mineral conservation—this can be done in three general ways:

 a. Recycling—the collection of metals to be used to remake metal products—aluminum cans.
 b. Reuse—when I was a kid, I collected Pepsi and Coke bottles to sell back to the grocery stores for subsequent refilling by bottlers.
 c. Dematerialization—many household appliances and even cars have gotten lighter because of the replacement of metals by plastics.

Table 15.3 **Life Expectancies of Identified Economic World Reserves of Selected Minerals**

Mineral	*1999* Reserves (Thousand Metric Tons)*	*Life Expectancy (Years)†*
Aluminum (bauxite ore)	25,000,000	81
Copper	340,000	22
Iron ore	74,000,000,000	65
Lead	64,000	17
Nickel	46,000	30
Tin	8,000	28
Zinc	190,000	20

* Latest data available.
† Based on a 2% growth rate in primary production.

17. Sustainable manufacturing—many wastes from industrial processes can be recycled. Sustainable manufacturing is the concept of minimizing that waste.

Questions to Ponder

1. Learn about uses of minerals by filling in the table below.

Minerals or material mined	Material extracted	Chemical makeup	Use
Coal			
		Nickel sulfide or silicate	Metal alloys, stainless steel
		Al_2O_3	
Hematite			
	Lead		
		Platinum	
Chalcocite and malachite			
Peat			

2. Research the U.S. and Canadian provinces affected by strip-mining damage and the primary material extracted, by filling in the table below.

Location	Material mined	Damage
West Virginia, Kentucky and Pennsylvania: See **www.penweb.org/issues/mining/** and **www.appvoices.org**		
Arizona: **http://www.wedo.org/ehealth/dineh.htm**		
Georgia: **http://www.gpc.edu/~jaliff/slidesho.htm**		
Idaho—For a positive view: **idahomining.org/environment.html** See the text for an opposing view.		
British Columbia: **www.miningwatch.org/emcbc/publications/amd_water.htm#Waste**		
Ontario: **users.vianet.ca/~dano/histpic.htm** and **www.hort.agri.umn.edu/h5015/99papers/shaw.htm**		

3. Debate the economic benefits of strip mining in Appalachia versus the permanent damage to the landscape. Search the Internet, or see the **Cyber Surfin'** references below.

4. Name four processes of mineral formation and three examples of commercially valuable minerals formed by each.

	Process of mineral formation	Commercially valuable minerals
1.		
2.		
3.		
4.		

5. Describe the industrial ecosystem of Kalundborg, Denmark. Give four examples of industrial wastes that are recycled.

	Recycled material	Use or product
1.		
2.		
3.		
4.		

POP QUIZ

Matching

Match the following terms and definitions

____1. It is always economically feasible to extract metals from a______.	a. Appalachia
____2. A chemical present in copper mine spoil:	b. Antarctica
____3. This metal is associated with magmatic concentration:	c. acidic
____4. This chemical is made by hydrothermal precipitation:	d. derelict/orphan lands
____5. This material is made by evaporation:	e. nickel arsenide
____6. This material is formed by sedimentation:	f. industrial ecosystem
____7. As water flows through sulfur-bearing rocks, it becomes more______.	g. high-grade ore
____8. Contaminants specifically associated with gold extraction are:	h. biomining
____9. Areas mined before the 1977 Surface Mining Control Act are:	i. manganese nodules, scallops
____10. Plants tolerating and concentrating toxic metals or other elements are called:	j. salt mines
____11. Waste steam heats homes in a(n):	k. copper sulfide
____12. One place where mineral extraction is prohibited:	l. borax, limestone
____13. Mountaintop removal occurs here:	m. gold
____14. These are strip mined from the ocean floor:	n. dematerialization
____15. *Thiobacillus* bacteria can help extract copper and gold in the process of:	o. mercury, cyanide
____16. Car engines have been made lighter by the use of aluminum, this is the process of:	p. phytoremediation

Multiple Choice

1. Which statement is incorrect concerning the General Mining Law of 1872?
   a. It allowed mining claims to be made in the west.
   b. It has resulted in huge profits by mining companies.
   c. It has resulted in economic losses to the taxpayers.
   d. It required reclamation of surface mined lands and cleaning of spoil.
   e. Trick question alert! All above are correct.

2. Which of the following is a nonmetallic mineral?
   a. graphite
   b. chalocite
   c. hematite
   d. galena
   e. bauxite

3. Which of the following is a metallic sulfide?
   a. iron pyrite
   b. chalcocite
   c. hematite
   d. a and b only
   e. c only

4. Which pollutant is produced from the smelting of Sudbury, Ont., copper ore?
   a. arsenic
   b. sulfur dioxide
   c. stream siltation
   d. cadmium oxide
   e. nitrogen oxides

5. Which pollutant is found in the solid processing wastes of copper, iron, and nickel ores?
   a. arsenic
   b. copper oxide
   c. sulfur dioxide
   d. tritium-3
   e. fool's gold

6. Which area is incorrectly matching with its resource description?
   a. U.S./coal
   b. Asia/copper
   c. China/tin
   d. U.K./tin
   e. Pennsylvania/coal

7. In magmatic processes of mineral formation, which description is correct?
   a. Molten rock solidifies.
   b. Layering of minerals takes place.
   c. Silicate rise to the top layers.

d. Metals form at the lower layers.
e. All of the above.

8. In hydrothermal processes, which statement an incorrect description?
   a. Hot water flows through rocks carries metallic minerals.
   b. A reaction with sulfur precipitates metal sulfides.
   c. Solid metals precipitate at the edge of geysers.
   d. Metal sulfides are insoluble.
   e. The metal salts dissolve better if they are chlorides.

9. The calcium carbonate and magnesium carbonate of limestone are important as a:
   a. construction material.
   b. fertilizer.
   c. acid scrubber.
   d. acid rain soil buffer.
   e. all of the above.

10. Duckhill, TN, has large gullies and denuded soil due to the smelting of:
    a. iron oxides.
    b. copper sulfides.
    c. limestone to make slaked lime.
    d. gold.
    e. nickel arsenides.

11. To reclaim and restore wetlands and mountainous areas affected by mining requires approximately how much time?
    a. 1 year
    b. 2 years
    c. 10 years
    d. 100 years
    e. 1 million years

12. Which developing country has been damaged most by mining?
    a. Uganda
    b. Argentina
    c. Bolivia
    d. Venezuela
    e. Mexico

13. Which country has the largest reserves of nickel?
    a. Cuba
    b. U.S.
    c. France
    d. Jamaica
    e. Australia

14. Which country has the largest reserves of aluminum?
    a. Cuba
    b. U.S.

c. France
d. Jamaica
e. Australia

15. Which material has been most successfully recycled from both an economic and environmental perspective?
    a. aluminum
    b. glass
    c. iron metal in cans
    d. cardboard
    e. plastics

16. In the past 20 years, plastics, lightweight aluminum and alloys have been most successfully used to reduce the weight of which product?
    a. dishwashers
    b. refrigerators
    c. automobiles
    d. homes
    e. airplanes

17. Which material will be depleted of its known reserves first?
    a. aluminum
    b. copper
    c. lead
    d. nickel
    e. tin

18. The smelting of iron and copper ores must remove ______ bound chemically to the metals.
    a. sulfur
    b. oxygen
    c. hydrogen
    d. mercury
    e. b and c

19. Coke is used in smelting iron ore for which reason?
    a. removal of carbon from the ore
    b. provides heat to oxidize the iron atoms
    c. removes oxygen to create carbon dioxide
    d. reduces the iron atoms by taking electrons away from them
    e. all of the above

20. Of all the products that were dematerialized in the past 50 years, which one was most affected?
    a. automobile
    b. computer
    c. refrigerator
    d. airplane
    e. clothes washer

21. Recycling this material conserves oil:
    a. aluminum
    b. newspaper
    c. glass
    d. plastic
    e. tin cans

22. In Canada, there is a campaign to eliminate this common toxic material in thermostats and thermometers.
    a. arsenic
    b. lead
    c. mercury
    d. tin
    e. cyanide

23. Gold extraction uses this hazardous material that has caused fish kills:
    a. arsenic
    b. lead
    c. mercury
    d. tin
    e. cyanide

CYBER SURFIN'

1. The U.S. Geological Survey has general information on U.S. minerals and their sources by state. Go to **minerals.usgs.gov/minerals**
2. To see the Bush administration's view of surface mining regulation see the U.S. Department of Interior, Office of Surface Mining website: **www.osmre.gov/**
3. To see a concise review of the problems associated with abandoned mines, go to the National Association of Abandoned Mine Land Programs website: **www.onenet.net/~naamlp/hazards.htm,** and the *Charleston (WV) Gazette* online: **wvgazette.com/static/series/mining**
4. The Bush administration's Office of Surface Mining has a view of reclamation that contrasts with the view of environmentalists and local residents on the value of "reclamation." Compare #2, above with the Citizens Coal Council at www.citizenscoalcouncil.org and the Pen (PA) Leadership Team at **www.penweb.org/issues/mining**
5. Miners and mining companies have had major disputes over the years, from the University of Arizona library, see the Bisbee Deportation: digital.library.arizona.edu/bisbee/ and the West Virginia Coal Wars: **http://www.wvculture.org/history/minewars.html**
6. A primer on industrial ecology from Columbia University's Biosphere 2 Center can be seen at **www.eeexchange.org/sustainability/content/d2.html**

'LIKE' BOOKING IT

1. Young JE, Sachs E, and Ayres E. ***The Next Efficiency Revolution: Creating a Sustainable Materials Economy (Worldwatch Paper, 121).*** Washington, DC: Worldwatch Institute, 1994.
2. Schlor J, and Taylor B (Eds.). ***Sustainable Planet: Solutions for the Twenty-first Century.*** Boston: Beacon Press, 2003.
3. Young JW, and Young JE. ***Discarding the Throwaway Society.*** Washington, DC: Worldwatch Institute, 1991.
4. Montrie C. ***Save the Land & People: A History of Opposition to Surface Coal Mining in Appalachia.*** Chapel Hill: University of North Carolina Press, 2003.
5. Giardina D. ***Storming Heaven.*** New York: Norton, 1987. Also go to Bookclub KET: **www.ket.org/bookclub/books/1999_apr/links.htm**
6. Taylor PF. ***Bloody Harlan.*** Lanham, MD: University Press of America, 1989.
7. Manahan SE. ***Industrial Ecology: Environmental Chemistry and Hazardous Waste.*** Boca Raton, FL: Lewis Publishers, 1999.

ANSWERS

Seeing the Forest

It can be said that the economic benefits of mining take place far away from the places where the mining occurs. Give four examples of the negative effects of mining and ore processing on local environments and economies.

	Negative effect on the local area where the mining occurs or the ore is processed
1.	Well water contaminated
2.	Ecotourism eliminated
3.	Wildlife and plant populations displaced and reduced, contributed to an increased loss of songbirds
4.	Sulfuric acid and toxin metal contamination of soil and water

Questions to Ponder

1. Learn about uses of minerals by filling in the table below.

Minerals or material mined	Material extracted	Chemical makeup	Use
anthracite, bituminous, lignite	coal	mostly carbon with some sulfur, silicates and nitrogen	energy source, raw materials for synfuels
garnierite and niccolite	nickel	nickel sulfide or silicate	metal alloys, stainless steel
bauxite	aluminum	aluminum oxide	cars, aircraft frames, cans

hematite	iron	ferric (iron III) oxide	cars, steel beams, stainless steel
galena	lead	lead sulfide	crystal glass, electrical solder, car batteries
platinum	platinum	platinum	important as a catalyst in jewelry, corrosion-resistant electrical contacts and vessels
chalcocite and malachite	copper	copper sulfide and copper arsenide	electrical wire, coinage, in copper sulfate—a water purifier
peat	peat moss	porous organic material	filtration agent, energy source in Ireland

2. Research the U.S. and Canadian provinces affected by strip-mining damage and the primary material extracted, by filling in the table below.

Location	Material mined	Damage
West Virginia, Kentucky and Pennsylvania: See **www.penweb.org/issues/mining/** and **www.appvoices.org**	anthracite and bituminous coal	drinking water contaminated, plumbing failure due to precipitation of minerals in pipes, blast damage to property, displacement and reduction of wildlife, overweight trucks breaking roads at taxpayer expense
Arizona: **http://www.wedo.org/ehealth/dineh.htm**	copper, coal, platinum, gold	displacement of wildlife, etc.
Georgia: **http://www.gpc.edu/~jaliff/slidesho.htm**	kaolin clay, coal	displacement of wildlife, etc.
Idaho—For a positive view: **idahomining.org/environment.html** See the text for an opposing view.	copper, gold	cyanide release and fish kills, acid from sulfur, toxic metals in streams
British Columbia: **www.miningwatch.org/emcbc/publications/amd water.htm#Waste**	copper	cyanide in tailings, acid from sulfur, toxic metals in streams
Ontario: **users.vianet.ca/~dano/histpic.htm** and **www.hort.agri.umn.edu/h5015/99papers/shaw.htm**	copper	cyanide in tailings, acid from sulfur, toxic metals in streams, sulfur dioxide from the smelting plants.

3. Debate the economic benefits of strip mining in Appalachia versus the permanent damage to the landscape. Search the Internet, or see the **Cyber Surfin'** references. *At the local level, economic benefits from coal mining are few. The money in minerals flows out of state. The number of people employed in strip mining is small. There are fewer people living in West Virginia today than there were in 1950.*

4. Name four processes of mineral formation and two examples of commercially valuable minerals formed by each.

Process of mineral formation	Commercially valuable minerals
1. Magmatism	Gold, platinum, palladium. Some mineral deposits containing nickel, chromium (chromite is an oxide), and platinum form by the separation of the metal sulfide or oxides in the molten form before it crystallizes.
2. Hydrothermal	Hot water or hydrothermal solutions form minerals. The *black smokers* are volcanic vents around which copper, gold, and lead form. Along with iron, these also can become chlorides that are dissolved in a hot aqueous solution and precipitated as sulfides after reacting with sulfuric acid.
3. Sedimentation	As calcium ions dissolve in the ocean along with carbon dioxide, carbon dioxide reacts with water to form carbonic acid, the calcium ions react with carbonic acid to form calcium carbonate (limestone, aragonite) that precipitates at the ocean floor. The foraminifera protozoa have calcium carbonate shell that form deposits such as the "white cliffs of Dover."
4. Evaporation	Evaporation of seawater leaves halite/sodium chloride deposits.

5. Describe the industrial ecosystem of Kalundborg, Denmark. Give four examples of industrial wastes that are recycled. *The industrial ecosystem at Kalundborg consists of a coal-fired electrical plant, an international pharmaceutical and chemical company, a drywall/sheetrock (gypsum) producer, and an oil refinery.*

Recycled material and source	Use or product
1. Steam and hot water from the power-generating station that formerly contributed to thermal pollution of the fiord.	Heat for homes and industrial processes.
2. Natural gas formed by the oil refinement process that was vented and flamed in the air.	Gas is recycled to use as heating gas.
3. Sulfur, a by-product of oil refining, is sold to a sulfuric acid producer, limestone used for the sulfur dioxide scrubbers is converted to calcium sulfate in the scrubbing process.	This is sold to the gypsum board, drywall producer.
4. The yeast organisms used to produce drugs and chemicals.	Yeast is given to farmers for use as an organic fertilizer.

Pop Quiz

Matching

1. g
2. e
3. m
4. k
5. j
6. l
7. c
8. o
9. d
10. p
11. f
12. b
13. a
14. i
15. h
16. n

Multiple Choice

1. d
2. a
3. d
4. b
5. a
6. b
7. e
8. c
9. e
10. b
11. d
12. c
13. c
14. d
15. a
16. c
17. c
18. b
19. c
20. b
21. d
22. c
23. e

16

Preserving Earth's Biological Diversity

LEARNING OBJECTIVES

After you have studied this chapter you should be able to:

1. Define biological diversity and distinguish among genetic diversity, species richness, and ecosystem diversity.
2. Discuss five important ecosystem services provided by biological diversity.
3. Contrast threatened, endangered, and extinct species and list four characteristics common to many endangered species.
4. Define biodiversity hotspots and explain where most of the world's biodiversity hotspots are located.
5. Describe four human causes of species endangerment and extinction and tell which cause is the most important.
6. Define biotic pollution and explain how invasive species endanger native species.
7. Define conservation biology and compare in situ conservation and ex situ conservation.
8. Distinguish between conservation biology and wildlife management.
9. Discuss steps that can be taken to slow down or stop the decline in biological diversity.

SEEING THE FOREST

Here in the United States, we have recorded the extinction of the Carolina parakeet, eastern elk, ivory billed woodpecker, and the passenger pigeon. We very nearly lost the buffalo, red wolf, right whale, California condor, and snowy egret. In the last 200 years, what has caused these extinctions or near-extinctions? Do a net search on these species.

	Causes of contemporary extinctions or near extinctions	Example
1.		North Atlantic salmon
2.		ivory-billed woodpecker
3.		snowy egret
4.		American elm and American chestnut
5.		various small Australian marsupials
6.		Atlantic cod and right whale

SEEING THE TREES

VOCAB

In the end we will conserve only what we love; we will love only what we understand; and we will understand only what we have been taught.

Baba Dioum

1. Species—a group of organisms that can interbreed successfully with each other: they are reproductively isolated from other species because they are not compatible by inheritance, structure, and behavior.
2. Biodiversity—the number of species, the richness of species, their genetic diversity, and ecosystem diversity of interactions and habitat.
3. Ecosystem services—these include air and water purification, esthetic value (psychological) and ecotourism, soil and water conservation, novel chemical discovery, food and wood, recycling nutrients in nonliving plant and animal remains or excrement, etc. Trees trap particulate pollutants (pollen, dust, and ash) that can cause respiratory illness. One acre of trees produce enough oxygen for 18 people per day. Annually, they would absorb enough CO_2 over a year's time, to allow the typical car to go 26,000 miles.
4. Evolution and diversity—after catastrophic mass extinctions, the numbers of species increase dramatically, later the rate of evolutionary change slows down. The filling of niches opened by mass extinctions are accompanied by genetic changes in surviving organisms.
5. Extinction—there have been thousands of documented extinctions (and probably millions undocumented) in the fossil record. There are hundreds of species extinctions predicted in the next half-century. Extinction means that the last of a species or subspecies' unique combination of genetic, structural and behavioral traits has disappeared.
   a. *Background extinction*—many extinctions occur due to a lack of genetic variability/diversity that leads to the inability to adapt during an environmental change.

b. *Mass extinction*—the sudden disappearance of large numbers of species after a catastrophic event, such as massive volcanic eruptions, asteroid impact, continental drift, or possibly as a result of the human alteration and consumption of the environment.

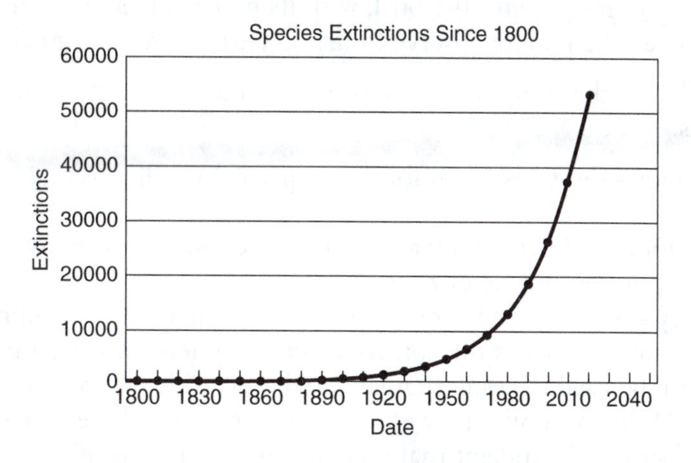

Table 16.1 **Percentages of Imperiled U.S. Species That Are Threatened by Various Human Activities**

Activity	*All Species (1,880)**	*Plants (1,055)*	*Mammals (85)*	*Birds (98)*	*Reptiles (38)*	*Fishes (213)*	*Invertebrates (331)*
Pollution	24	7	19	22	53	66	45
Overexploitation	17	10	45	33	66	13	23
Habitat loss/degradation	85†	81	89	90	97	94	87
Exotic species	49	57	27	69	37	53	27

* Numbers in parentheses are the total number of species evaluated. The "All Species" category represents about 75% of imperiled species in the United States.
† Because many of the species are affected by more than one human activity, the percentages in each column do not add up to 100.

6. Endangered species—the U.S. Endangered Species Act defines such a species as one that is in imminent danger of extinction throughout or in the greater part of its range of occurrence. A threatened species is one in which the population is low and declining. *Endemic species* are found only in one location.
7. Habitat fragmentation—when large areas of forest are broken up into farm fields, pastures, or grassy areas that have isolated areas of forest remaining, habitat fragmentation has occurred—strip-mined areas of Appalachia, the tropical rain forests of Brazil, and the migration of European style farming westward in the United States. This increases *edge effect* that reduces the populations of most wildlife and plants.
8. Adaptive radiation—the process of evolution of organisms into new niches.
9. Biodiversity hotspot—small areas that contain a large number of endemic species—the Polynesian islands.
10. Biotic pollution—the introduction of invasive foreign/exotic species or specific genes into native populations in a specific area—genetically modified corn genes getting into native teosinte corn in Central America.
11. Bellweather or sentinel species—some species are particularly sensitive to environmental change and pollution. The skin-breathing frogs and salamanders are suffering a worldwide col-

lapse of populations. Many factors are involved: tropical rainforest and deciduous forest destruction, acid rain and air pollution, increased U-V radiation caused by ozone depletion, and water pollution. Coal miners 100 years ago would take a canary in a cage down into the mine. If there was poisonous gas present, the bird, with its higher metabolic and breathing rate, would pass out. When the miners observed this, they would get out quickly.

12. Conservation biology—the study of the conservation and preservation of ecosystems, landscapes and species.
    a. *In situ* conservation—*In situ* is Latin for "in its place." Wildlife refuges and nature preserves are examples.
    b. *Ex situ* conservation is illustrated by a zoo or arboretum. The time is coming when there will be very few gorillas outside of zoos.
13. Restoration ecology—arboreta and zoos can reproduce a natural environment that formerly occupied the geographical area as the controlled environment. Artificial insemination and embryo transfer can be used to restore endangered species. If we come up with some good woolly mammoth DNA from Siberia, we may be able to place these genes into an Indian elephant embryo and surgically implant that embryo into the uterus of an extant elephant.
14. Seed banks—seeds and their genetic content can be preserved in a low moisture and temperature environment. The seeds must be germinated periodically as they cannot be stored indefinitely.
15. Endangered Species Act of 1973—a U.S. law that authorizes the U.S. Fish and Wildlife Service to protect endangered and threatened species, construct species recovery programs, and prevent the sale or purchase of any products derived from them. Some construction projects, such as dam projects, have been stopped because endangered species were affected. An amendment in 1982 provided a means to resolve conflicts between development and endangered species on private property.
16. World Conservation Strategy—the World Conservation Union, the World Wildlife Fund, and the United Nations Environment Program devised a plan to protect biological diversity. The 1992 Earth Summit provided that signatory nations inventory their endangered species and plan for the conservation or preservation.

Table 16.2 **U.S. Organisms Listed as Endangered or Threatened, 2002**

Type of Organism	*Number of Endangered Species*	*Number of Threatened Species*
Mammals	65	9
Birds	78	14
Reptiles	14	22
Amphibians	12	9
Fishes	71	44
Snails	21	11
Clams	62	8
Crustaceans	18	3
Insects	35	9
Spiders	12	0
Flowering plants	568	144
Conifers	2	1
Ferns and other plants	24	2

17. Convention on International Trade in Endangered Species of Wild Flora and Fauna—this treaty bans the hunting and collecting of endangered species, and commercial trade in products derived from them. This has slowed down but not stopped the illegal trade of products such as ivory.
18. Wildlife management—the field of conservation biology that applies principles of biology to increase populations of ecosystems and their plants and animals.
19. Commercial extinction—an overexploited food or other commercially valuable species may be harvested to the point that it is no longer profitable to collect—the Atlantic cod and the right whale.
20. Bioaccumulation—many toxic chemicals, such as mercury and DDT, tend to be closeted in fats. Animals that are high on the food chain or have large bodies such as whales tend to have more fat, therefore, more concentrated mercury. Tuna has high levels of mercury.

Questions to Ponder

1. Fill-in the table below by giving two examples for each cause of extinction or near extinction (not including those listed above) and adding a brief explanation of each.

Cause	Examples	Explanation
Climatic change (global warming, ice age)		
Sea level rise or fall		
Natural predation (other than human)		
Disease epidemics		
Overexploitation of food species		
Competition with exotic species		
Habitat destruction or alteration by humans		
Human hunting for sport or ritualistic reasons		
Lack of genetic diversity within a population		
Catastrophic volcanism or astronomical events		
Pollution or toxic chemical buildup		

2. Gene migration is happening as the killer/Africanized bees have hybridized with the European honeybees kept by beekeepers. How might this affect crop pollination in Florida?

3. Match the ecosystem service to the organisms that perform it. Use choices as many times as applicable.

Ecosystem service	Organisms
___1. Shading and temperature reduction in cities, making of oxygen	a. *Penicillium* fungus and *Actinomyces* bacteria
___2. Reduction of carbon dioxide in the atmosphere	b. greenbelts, shelterbelts
___3. Sewage pollution removed	c. freshwater wetlands
___4. Antibiotics made	d. bacteria, fungi, and insect larvae
___5. Recharging groundwater	e. brackish estuary, algae
___6. Decomposition of tree trunks and animal bodies	f. trees
___7. Cover for insect predators on crop pests and soil conservation	

4. List four conditions that have to be met to maintain seeds indefinitely in a seed bank.

	Condition for maintenance of the seeds indefinitely
1.	
2.	
3.	
4.	

5. Check out the U.S. Fish and Wildlife Service site and list six endangered vertebrates and other species in your state at **ecos.fws.gov/servlet/TESSWebpageUsaLists?state=all**

Endangered vertebrates	Other endangered species
1.	
2.	
3.	
4.	
5.	
6.	

6. Name four countries that have a long tradition of hunting whales and usually oppose restrictions or bans on hunting them. Go to **whales.greenpeace.org**

1. ______________________________

2. ______________________________

3. ______________________________

4. ______________________________

POP QUIZ

Matching

Match the following terms and definitions

____1. The biome with the greatest species diversity and richness is:	a. *ex situ*
____2. The American bald eagle and the brown pelican were endangered by the bioaccumulation of:	b. biodiversity hotspots
____3. A group in which the individuals reproduce normal numbers of fertile offspring:	c. Hawaii
____4. This plant produces chemicals that treat cancer:	d. threatened
____5. The dinosaurs disappeared in a ______.	e. tropical rain forest
____6. There is a mass extinction currently underway largely caused by:	f. vinca/periwinkle
____7. If a species is newly identified as having a population decrease, it is ______.	g. amphibians
____8. The fossil record indicates that after a mass extinction, surviving organisms evolve rapidly into open niches. This is termed:	h. DDT
____9. Cats, dogs, and rabbits in Australia are:	i. bush meat
____10. This group make ideal sentinel species:	j. species
____11. 44% of all vascular plants and 20% of the human population lives in:	k. mass extinction
____12. An oceanic island is likely to have a greater number of ______ than a continental area of similar size.	l. adaptive radiation
____13. In Africa, gorillas, chimpanzees, and monkeys are used as:	m. deforestation

____14. A zoo in the U.S. breeding panda bears represents ______ conservation.	n. Endangered Species Act
____15. This act stopped the building of a dam in Tennessee because the snail darter fish inhabited the area.	o. invasive species
____16. This area has the largest number, relative to its size, of extinctions of birds due to invasive species:	p. endemic species

Multiple Choice

1. This material interfered with the female bald eagles and brown pelican estrogen metabolism so that eggs were laid without the calcium carbonate shell:
   a. dioxin
   b. PCBs
   c. sarin
   d. DDT
   e. acid rain

2. Which statement is accurate concerning mass extinctions?
   a. Niches are opened and adaptive radiation takes place.
   b. Evolution rate increases after the events.
   c. They can be caused by volcanism or asteroid impact.
   d. Genetic diversity decreases at the catastrophic event, then it increases.
   e. All of the above are accurate.

3. Which biome has the greatest genetic diversity?
   a. coniferous rainforest
   b. tropical rainforest
   c. deciduous forest
   d. temperate grassland
   e. chaparral

4. Which of the following would not be considered an ecosystem service?
   a. doing photosynthesis
   b. reduction of fecal contamination in water
   c. reduction of carbon dioxide in the atmosphere
   d. recharging groundwater and aquifers
   e. Trick question alert! All of the above are ecosystem services.

5. This animal is threatened because its parts are used to make ritualistic coming-of-age knife handles for Yemeni men:
   a. elephant tusk
   b. black and white rhinoceros horn
   c. South African walrus tusk
   d. lion teeth
   e. swordfish

6. This animal is overpopulated in North America, causing much damage to its nesting grounds (see satellite photos of the Knife River Delta, Canada: **edcwww.cr.usgs.gov/earthshots/slow/Knife/Knife).**
   a. snow geese
   b. whooping cranes
   c. bald eagles
   d. snowy owls
   e. spotted owls

7. Which creature has not become commercially extinct through harvesting? See **darwin.bio.uci.edu/~sustain/bio65/lec03/b65lec03.htm** by Peter Bryant, U. Cal., Irvine.
   a. Atlantic cod
   b. right whales
   c. southern blue whales
   d. tuna
   e. humpback whales

8. This material tends to bioaccumulate in animals with large fat stores—tuna and whales?
   a. DDT
   b. mercury
   c. sulfur dioxide
   d. iron sulfide
   e. a and b

9. This South American country receives payments from drug companies for prospecting in their forests:
   a. Chile
   b. Suriname
   c. Malaysia
   d. Jamaica
   e. Panama

10. Reduction of this organism can cause major losses in crop production, particularly in Florida:
   a. cowbirds
   b. dragon flies
   c. bees
   d. manatees
   e. alligator

11. About 13 thousand years ago, western Europe became much colder, and most of the forests there became tundra. At that time, the Neanderthals retreated to shelter in the valleys and apparently stayed there until they eventually became extinct. Modern humans appear to have migrated seasonally and avoided the cold winters to some extent. This would be an example of a ______extinction.
   a. background
   b. mass
   c. endemic
   d. epizootic
   e. adaptive radiation

12. Which of the following threatened or endangered species have increased their populations under management schemes?
    a. rhinoceros
    b. condors
    c. tigers
    d. chimpanzees
    e. gorillas

13. Which statement is accurate concerning the chief difficulty of maintaining elephant populations in Africa?
    a. Wildlife preserves for elephants do not exist.
    b. Elephants are nomadic.
    c. They raid the crops of people living nearby.
    d. The trade in ivory is so profitable that no African nation has protected them.
    e. Trick question alert! All of the statements are accurate.

14. Which exotic species is mismatched with its effect?
    a. water hyacinths/obstacles to navigation
    b. brown tree snake/caused extinction of 9 species of birds in Guam and power outages
    c. *Caulerpa* algae/causes algal blooms on the surface of the Mediterranean Sea
    d. Japanese beetles/eat grape leaves, green beans in the U.S.
    e. rabbits/eat farm crops in Australia

15. Bear claws and rhinoceros horns are most valuable in Asia for which use?
    a. aphrodisiac properties for males
    b. making jewelry
    c. treatment for cancer
    d. treatment for insomnia
    e. diuretic properties

16. Which animal group makes an ideal sentinel for air and water pollution?
    a. water birds
    b. alligators
    c. frogs and salamanders
    d. insects
    e. bacteria

17. Native Americans and which threatened species depends on the annual salmon migration?
    a. black bears
    b. brown bears
    c. polar bears
    d. walrus
    e. musk oxen

18. The University of Wisconsin has reconstructed an enclosed tall grass prairie. This is an example of:
    a. *in situ* conservation
    b. seed bank

c. *in situ* preservation
d. *ex situ* conservation
e. c and d

19. The Washington zoo has a breeding facility for panda bears. This is an example of:
a. *in situ* conservation
b. seed bank
c. *in situ* preservation
d. *ex situ* conservation
e. c and d

20. This region is a biodiversity hotspot, but has very little area protected:
a. Australia
b. Japan
c. Madagascar
d. Kenya
e. Britain

21. The reintroduction of which species was not successful?
a. gray wolves/Yellowstone
b. buffalo/Yellowstone
c. red wolves/Florida
d. thick-billed parrots/Arizona
e. nene ducks/Kawai and Hawaii

22. Of the following North American threatened and endangered species, which is in the most danger of extinction?
a. sea otter
b. whooping crane
c. Atlantic cod
d. Pacific gray whale
e. mountain lion

23. Of the human activities listed by Table 16-1 above, which ranking of the cause of extinction is correct?
a. 1. habitat degradation, 2. pollution, 3. exotic species, 4. overexploitation
b. 1. pollution, 2. exotic species, 3. habitat degradation, 4. overexploitation
c. 1. pollution, 2. habitat degradation, 3. Exotic species, 4. overexploitation
d. 1. habitat degradation, 2. exotic species, 3. pollution, 4. overexploitation
e. 1. overexploitation, 2. habitat degradation, 3. pollution, 4. exotic species

24. Of vertebrates, which group has the most threatened or endangered species? See Table 16-2.
a. amphibians
b. birds
c. reptiles
d. mammals
e. fishes

CYBER SURFIN'

1. The National Geographic Society has an online article titled *The Sixth Extinction,* after Richard Leakey and Roger Lewin's book of the same name, see **www.nationalgeographic.com/ngm/9902/fngm/index.html**
2. Professor David Ulansky has an excellent list of Internet sites on the current threat of mass extinctions, go to **www.well.com/user/davidu/extinction.html**
3. The *Chattooga Quarterly* online has an excellent article concerning the extinctions of the Carolina parakeet and the passenger pigeon, check out **www.chattoogariver.org/Articles/2000S/Birds.htm**
4. To see a list of extinct vertebrates of the U.S.A., U.S. territories, and Canada since 1492, go to **www.dwave.net/~tony/Mars/extinct2.htm.** Also see the text reference below (Kaufman and Mallory).
5. To learn more about the reasons that species are endangered or extinct, see the Thinkquest.org site: **library.thinkquest.org/25014/english.index.shtml**
6. Go to the American Museum of Natural History's ***Endangered!*** site and learn about the status of rare, threatened and endangered animals: **www.amnh.org/nationalcenter/Endangered**

'LIKE' BOOKING IT

1. Leakey R, and Lewin R. ***The Sixth Extinction: Biodiversity and its Survival.*** New York: Doubleday, 1995.
2. Kaufman I, and Mallory K (Eds.). ***The Last Extinction.*** Cambridge, MA: MIT Press (2/e), 1993.
3. Ehrlich PR, and Ehrlich AH. ***Extinction: The Causes and Consequences of the Disappearance of Species.*** New York: Random House, 1981.
4. Broswimmer FJ. ***Ecocide: A Short History of Mass Extinction of Species.*** London: Pluto Press, 2002.
5. Wilson EO. ***The Future of Life.*** New York: Vintage Books, 2003.
6. Gould SJ (Ed.). ***The Book of Life: An Illustrated History of the Evolution of Life on Earth.*** New York: W. W. Norton, 2001.

ANSWERS

Seeing the Forest

	Causes of contemporary extinctions or near extinctions	Example
1.	Water pollution, salmon and trout require clean, highly oxygenated water; Dam building and over-harvest have also reduced their numbers	North Atlantic salmon
2.	Cutting of old growth/virgin forests in the southeastern U.S. and Cuba. The bird needed old trees for nesting and feeding. There may be a few alive in deep cypress swamps in Louisiana or Cuba.	Ivory billed woodpecker
3.	These white birds had long showy feathers that were popular in ladies hats in the late 19th century and early 20th century.	Snowy egret

4.	Dutch elm fungus/blight carried by bark beetles deforested many midwestern city streets and hit the Appalachian forests hard	American elm and American chestnut
5.	Cats and dogs imported by Australians prey upon the native marsupials (pouched mammals). Imported rabbits out-compete many native herbivores.	Various small Australian marsupials
6.	Over-harvest. The over-fishing of the Georges bank near Newfoundland and over-harvest of the right whale have devastated their populations. The right whale gets its name because it floats after death, making it easier for the old mariners to load on board a boat.	Atlantic cod and right whale

Questions to Ponder (Many of these answers are examples of many possible responses, pleases don't memorize them.)

1. Fill-in the table below by giving two examples for each cause of extinction or near extinction (not including those listed above) and adding a brief explanation of each.

Cause	Examples	Explanation
Climatic change (global warming, ice age)	Woolly mammoth, cave bear, Neandertal, saber-toothed tiger ground sloth	Climatic change can change tundra to coniferous forest (warmer conditions) or vice versa (colder)
Sea level rise or fall	Mollusks, crinoids (sea lilies), and corals	Mass extinctions generally are associated with major reductions in sea level—the Permian extinctions of shellfish and other species living on the continental shelf and similar extinctions at the end of the Cretaceous period. Asteroid impacts may have contributed to both, and volcanism to the latter.
Natural predation (other than human)	Dinosaurs? See exotic species below.	There are many dinosaur extinction hypotheses and theories. One is that egg-eating mammals eliminated dinosaurs. A problem is that the large thick-shelled eggs would be hard for a small mammal to deal with. See exotic species below.
Disease epidemics	Woolly mammoths, ground sloths, saber-toothed tigers? Wolves? American elm and American chestnut trees.	The disappearance of wolves in the U.S. may have been caused by viral diseases acquired from domestic dogs. Parvo virus. Similarly, the extinction of the large mammals of the Pleistocene period may be from viruses acquired from domestic animals of humans. Smallpox from Europeans decreased Native American populations by 85%. In many areas of the eastern U.S. Dutch elm fungus practically eliminated the American elm and chestnut trees.

Overexploitation of food species	European aurochs (long-horned wild cattle), Irish elk, mastodons	On the other hand, the disappearance of mastodons, aurochs (wild cattle), and Irish elk may have occurred due to overhunting in Europe. The eastern elk and the western buffalo were rendered extinct in their natural range by overhunting, some of the buffalo hunting was designed to prevent the Plains indigenous peoples from eating.
Competition with exotic species	Pigs, goats, Indian mongoose, brown vine snakes	Brown vine snakes in Guam eliminated 9 endemic bird species from the island of Guam. The Indian mongoose was imported into Hawaii and Puerto Rico to control rats but has devastated ground-nesting bird populations. Pigs and goats have caused extinctions of plants in Hawaii. If an imported animal is a generalist in feeding or habitat requirements, it has an enormous advantage over endemic species that tend to be specialists. Starlings are aggressive birds that take over eastern bluebird nests and compete with woodpeckers as well.
Habitat destruction or alteration by humans	Lianas, orchids, insects, trees, and amphibians	The conversion of tropical rainforest to grazing and farmland is eliminating hundreds of species annually. Strip mining and changes in coffee growing from forests to fields have also eliminated habitat.
Human hunting for sport or ritualistic reasons	Black rhinoceros (about 300 left) and white rhinoceros (7,000); bear	Regularly in Alaska and Canada, politicians propose reducing wolf populations to increase moose for hunting tourism income. Where wolves are not present, moose occur at population rates of up to four times greater than where wolves are present. Yemeni men traditionally carry a ceremonial knife with a handle made of rhinoceros horn.
Lack of genetic diversity within a population	Cheetah, lowland gorilla	As populations become more isolated and gene migration stops, the gene distribution in a declining population becomes more uniform. A lack of genetic diversity leads to extinction because the organism has no reserve of genes to adapt to a modified environment. The cheetahs and gorillas are examples.

Catastrophic volcanism or astronomical events	Dinosaurs? Large amphibians of the Permian period?	There is evidence that large asteroids hit the Earth at the end of the Cretaceous and Permian periods. These impacts could have reduced sea level and caused volcanic activity that blackened the skies, resulting in a collapse of the food chains that supported these large-bodied creatures.
Pollution or toxic chemical buildup	Bald eagle and brown pelican	Bioaccumulation of DDT caused these birds to lay eggs without hard shells, killing the desiccated embryos.

2. Gene migration is happening as the killer/Africanized bees have hybridized with the European honeybees kept by beekeepers. How might this affect crop pollination in Florida? *The hybridized bees will be hard to handle. Beekeepers who are now paid to take their bees to fields for pollination will go out of business. Many native plants will be affected. Production of certain crops will decrease until another way of pollination is applied.*

3. Match the ecosystem service to the organisms that perform it. Use choices as many times as applicable.

	Ecosystem service	Organisms
f 1.	Shading and temperature reduction in cities, and making of oxygen	a. *Penicillium* fungus and *Actinomyces* bacteria
b,f 2.	Reduction of carbon dioxide in the atmosphere	b. greenbelts, shelterbelts
c,e 3.	Sewage pollution removed	c. freshwater wetlands
a 4.	Antibiotics made	d. bacteria, fungi, and insect larvae
c 5.	Recharging groundwater	e. brackish estuary, algae
d 6.	Decomposition of tree trunks and animal bodies	f. trees
b,f 7.	Cover for insect predators on crop pests and soil conservation	

4. List four conditions that have to be met to maintain seeds indefinitely in a seed bank.

	Condition for maintenance of the seeds indefinitely
1.	Storage at low temperatures
2.	Control of humidity
3.	Periodic germination of seeds to make more seeds
4.	Distribution of seed types to different sites, so that if one is lost, there will be survivors

5. Check out the U.S. Fish and Wildlife Service site and list six endangered vertebrates and other species in your state at **ecos.fws.gov/servlet/TESSWebpageUsaLists?state=all**

	Endangered vertebrates	Other endangered species
1.	indigo snake	green pitcher plant
2.	etowah darter	swamp pink flower
3.	American alligator	Florida torreya (tree)
4.	red cockaded woodpecker	Alabama leather flower
5.	loggerhead sea turtle	smooth coneflower
6.	wood stork	pondberry

Where do I live?

6. Name four countries that have a long tradition of hunting whales and usually oppose restrictions or bans on hunting them. Go to **whales.greenpeace.org** *Japan, Norway, Iceland and Russia.*

1. ___
2. ___
3. ___
4. ___

Pop Quiz

Matching

1. g
2. e
3. m
4. k
5. j
6. l
7. c
8. o
9. d
10. p
11. f
12. b
13. a
14. i
15. h
16. n

Multiple Choice

1. d
2. e
3. b
4. e
5. b
6. a
7. d
8. e
9. b
10. c
11. a
12. b
13. b
14. c
15. a
16. c
17. b
18. a
19. d
20. c
21. d
22. b

17

Land Resources and Conservation

LEARNING OBJECTIVES

After you have studied this chapter you should be able to:

1. Relate at least five ecosystem services provided by natural areas.
2. Summarize current land ownership in the U.S.
3. Describe the following federal lands, stating which government agency administers each and current issues of concern: wilderness areas, national parks, national wildlife refuges, national forests, public rangelands, and national marine sanctuaries.
4. Define deforestation and relate the main causes of tropical deforestation.
5. Define desertification and explain its relationship to overgrazing.
6. Describe the current threats to freshwater and coastal wetlands and explain why the definition of wetlands is controversial.
7. Describe the current trends in U.S. agricultural lands, such as encroachment of suburban sprawl.
8. Contrast the views of the wise-use movement, and the environmental movement regarding the use of federal lands.

SEEING THE FOREST

1. Review (see objective #1): For the 9 ecosystem services provided by the world's various ecosystems, list 9 different biomes or ecosystems represented and 9 different national parks, refuges, or other sanctuaries that protect them.

	Ecosystem service	Biome or aquatic ecosystem	National Park name and location
1.	Air purification		
2.	Water purification		
3.	Ecotourism		
4.	Climate moderation		
5.	Harvest of wood		
6.	Protection of terrestrial biodiversity		
7.	Protection of aquatic biodiversity		
8.	Moderation of climate		
9.	Soil conservation		

To see a list of National Marine Sanctuaries, go to **www.sanctuaries.nos.noaa.gov;** coral reef preserves: **www.nature.nps.gov/wv/coral/Coralnew-01.htm**; U.S. National Parks: **www.cr.nps.gov/history/online_books/glimpses1/index.htm** and **www1.nature.nps.gov/wv**; Central and South American parks: **www.parkswatch.org**; and World Parks by John Uhler: **www.world-national-parks.net**

SEEING THE TREES

VOCAB

The establishment of the National Park Service is justified by considerations of good administration, of the value of natural beauty as a National asset, and of the effectiveness of outdoor life and recreation in the production of good citizenship.

Theodore Roosevelt

1. California Desert Protection Act—a 1994 U.S. law that changed the status of Joshua Tree and Death Valley National Monuments to national parks status.
2. Nonurban or rural lands—sparsely populated grasslands, forests, mountains, wetlands, and deserts that perform many ecosystem services.
3. Ecosystem services—ecosystems directly and indirectly aid humans with clean water, air, and food to consume.

	Ecosystem services
1.	Removal of organic waste from water and its transformation into biomass
2.	Protection of biodiversity
3.	Psychological value as a place for seeing natural beauty and as a place of peaceful retreat
4.	Reduction of carbon dioxide and production of oxygen
5.	Moderation of climate by soaking up solar radiation for photosynthesis
6.	Moderation of climate through transpiration
7.	Decreasing the effects of high winds and waves
8.	Concentration of certain metal toxins (see Ch. 16)
9.	Detoxification of certain wastes and destruction of disease producing microorganisms
10.	Cover and habitat for crop pollinators
11.	Cover and habitat for predators of crop pests
12.	Products for the support of local and distant populations of humans
13.	Soil conservation
14.	Seed dispersal
15.	Recharging of groundwater tables and aquifers

4. Wilderness—areas where humans do not permanently live and where ecosystem disturbance is low.
5. Wilderness Act of 1964—many virgin or old growth forests (uncut) or other areas, and those areas that resemble such areas, have been set aside for protection by U.S. law. These include national parks, national forests, and national wildlife refuges.
6. Wild and Scenic Rivers Act of 1968—a U.S. law that designates the National Park Service to protect 170 areas of river shore (1% of the U.S. total of rivers) from development. Recreational use is allowed.
7. National Parks—the first, Yellowstone National Park, was established by the U.S. Congress in 1872. The National Park Service, created in 1916, protects the natural beauty of scenic wilderness and historic sites.
8. Land and Conservation Fund Act of 1965—a U.S. law that sets aside money to purchase areas to be protected by the National Park Service. Recent additions include the Virgin Islands Reef National Monument and the Minidoke Internment National Monument in Idaho.

Table 17.1 Administration of Federal Lands

Agency	*Land Held*	*Area in Millions of Hectares (Acres)*
Bureau of Land Management (Dept. of Interior)	National resource lands	109 (270)
U.S. Forest Service (Dept. of Agriculture)	National forests	77 (191)
U.S. Fish and Wildlife Service (Dept. of Interior)	National wildlife refuges	37 (92)
National Park Service (Dept. of Interior)	National Park System	34 (84)
Other—includes Department of Defense, Corps of Engineers (Dept. of the Army), and Bureau of Reclamation (Dept. of Interior)	Remaining federal lands	29 (72)

9. Natural regulation—a policy designed to allow "nature to take her course." Natural processes dictate the survival of elk, bears, mountain goats, bighorn sheep, Colorado spruce, quaking aspen, columbine, and Indian paintbrush flowers in a park such as Rocky Mountain National Park. The only exception would be control of invasive species.

10. Transpiration—when water evaporates from a leaf and pulls water through the plant from the root, transpiration occurs. This has a cooling effect on the plant and the surrounding air. You may remember that as water changes from a liquid to a gas, heat makes this change by going into the water. Water gas is added to the air to become clouds. Also remember that the upper tree canopy of the tropical rain forest acts as a cloud-forming area, bringing rain back to the lower levels of the ecosystem. See below.

11. Monocultures—agricultural ecosystems, particularly the agro-business factory farms, specialize in growing one crop over thousands of acres. This actually encourages the pests and diseases that attack the crops. The pests usually reproduce faster than the crops or the natural predators of the pests. The same pests increase their numbers in the remaining national forests.

12. Ecologically sustainable forest management—this aims to sustain biodiversity, yield of wood harvested, soils, and watersheds. Farmers, loggers, the general public, and indigenous peoples cooperate with government to sustain yields. Unfortunately, the biomass extracted from our national forests exceeds its replacement, and the soil loss from clear cutting exceeds that replaced by natural processes. Forestry management techniques include:

    a. Selective cutting—mature trees are selected for cutting in small areas, preserving some old-growth forest for the species that require it. Ecologically sound logging can leave *wildlife corridors* that connect isolated unlogged areas.
    b. Shelterwood cutting—a forest management cycle in which mature and undesirable (?) trees are removed the first year, then a second cycle after a decade, in which some mature trees are harvested, but the saplings continue to be sheltered by a few large trees not harvested.
    c. Seed tree cutting—takes all mature trees, small trees are spared for reseeding.
    d. Clear cutting—almost all trees are cut from an area. For the logging companies, this is the more efficient and profitable way to harvest trees. Soil erosion, mudslides, and habitat fragmentation and loss accompany this practice.

15. Ecologically certified wood—wood that has been obtained from ecologically managed forests is socially responsible and in great demand in the wood market. Certification requires the following:

    a. Sustainability—extraction equals replacement
    b. Socioeconomic benefits to indigenous peoples
    c. Preservation of soil, wildlife and water resources

16. Deforestation—the permanent or temporary loss of forest for agricultural or other reasons.

    a. Tropical rain and dry forests—tropical rain and dry forests are being lost at a rate of more that double the area of the state of North Carolina annually. At this rate, they will disappear in 100 years. See **earthobservatory.nasa.gov/Library/Deforestation/deforestation_2.html.**
    b. Boreal forests—cold or cool coniferous forests of spruce, fir, hemlock, and larch that constitute 11% of the Earth's area—the largest biome. By reason of the low temperature, forest regrowth is slow, and clear cutting is accordingly very destructive with long-lasting effects. Canada is the largest producer of timber; its national policy is to produce wood products sustainably. Similar to mining or mineral rights deeds, the Canadian government grants logging companies *tenures* to cut in national forests.
    c. Deciduous forests—about 50% of U.S. forests are publicly owned. Those areas are under tremendous pressure to develop their land as property taxes increase due to urban encroachment.
    d. Temperate rain forest—use of the Tongass National Forest in Alaska has been the subject of much dispute between logging and environmental interests. The Bush administration has reversed rules made by the Clinton administration to stop road building and to preserve an additional 124,000 acres of old-growth forest.

17. Slash-and-burn agriculture—a subsistence agriculture that the native North, Central, and South Americans used to provide a sustainable agriculture before the Europeans arrived. The indigenous peoples would clear a patch of forest, farm it until the soil was exhausted and abandon it: the surrounding forest area then regenerated forest in the cleared area. This also

increased edge effect of increasing deer and bear to hunt, providing that surrounding forest was substantial in area. When large area of tropical rain forests are cleared, the soils sustain farming and grazing for 10–20 years, then the soil that was thin to begin with is abandoned and *scrub savannah* or banana plantations, subsidized by fertilizer and energy resources, replaces it.

18. Conservation easement—The Forest Legacy Program of the 1990 Farm Bill allows a landowner to sell preservation rights for a forest to the USDA for a designated number of years.

19. Rangeland conservation and management—temperate and tropical grasslands serve as grain food sources and animal feed. There have been wars fought over whether land would be used for farming or grazing, and there are bitter disputes today over public use of government-managed land. Grassland has been subject to degradation and desertification as discussed in Chapter 14. The fibrous root systems of grasses hold soil well as long as the plants are not cropped off too closely by grazing livestock that have exceeded the carrying capacity of the rangeland. When grassland is denuded, soil washes away, and sand dunes appear in the process of grassland desertification.

20. Bureau of Land Management—the agency of the U.S. government that manages the private use of federal lands. It enforces the regulations of three laws. One of these, the Grazing Act of 1934, was enacted in the Dust Bowl era, so that soil could be conserved by reducing the numbers of grazing animals. Another, the Public Rangelands Improvement Act of 1978, has enlisted scientists in the effort to restore sustainable grassland. It also manages the populations of wild horses, burros, and antelopes that graze there.

21. Ecosystem management—scientists, landowners, and governmental agencies cooperate to develop strategies to provide for the ecological and economic health of an area. It is guided by the principles of sustainability and the recognition that the long-term health of humans and the environment are inextricably interrelated.

22. Wetlands management—wetlands not only provide important ecosystem services, such as water purification and groundwater charging, but offer habitat for 1/6 of all endangered or threatened species. The Emergency Wetlands Resources Act of 1986 designated the U.S. Fish and Wildlife Service to inventory wetlands. U.S. administrations from Reagan to Clinton have attempted to preserve and restore wetlands.

23. Coastlines and marine conservation—many coastal estuarine marsh areas have been dredged or filled-in to create *key lagoon* real estate developments (car in the front driveway, boat dock in the back yard), widen waterways, or to control mosquitoes. Where this has been done, commercial fisheries have collapsed. Some areas have been degraded by water pollution, seawall construction, dune erosion, and overfishing. Twelve National Marine Sanctuaries have been established in U.S. waters to allow for sustainable fishing, recreation, education, research, and ecosystem preservation (in the no-take zones). However, today a net loss of wetlands continues. The no-take zones have actually increased in species richness due to the movement of species away from the harvested area—this is called the *spillover effect.*

24. Agricultural land conservation—soil erosion, dam construction, and urban sprawl are destroying agricultural lands that will be needed for the future.

25. Wise-use movement—many economists and citizens believe that the government has too many rules and regulations restricting land use. President Reagan's former Secretary of the Interior and U.S. Chamber of Commerce lawyer James Watt wanted to open up mining and commercial development of wilderness lands, refuges, wetlands and national parks. The preservation of endangered species would not be done without counterbalancing its economic effects.

26. Environmental movement—by contrast this movement assumes that the primary purpose of federally managed lands is to serve as a public trust to protect biological diversity and ecosystem integrity. The extraction of resources from public lands should be paid for by the extractor, and reasonable compensation paid into the public treasury. Those who damage public lands should be held financially accountable.

Questions to Ponder

1. List six types of U.S. federally managed areas, the federal agency responsible, and indicate two areas of concern for each. See objective #3.

Federally managed area	Agency managing	Two concerns
1.		
2.		
3.		
4.		
5.		
6.		

2. Debate the issue of natural regulation of trees in U.S. National Parks and Forests. From the National Geographic Society, hear audio stories at **www.nationalgeographic.com/radiox/yellowstone/**

3. List five negative environmental effects or ecosystem services lost due to tropical rain forest degradation and deforestation.

	Negative environmental effect of tropical rain forest destruction
1.	
2.	
3.	
4.	
5.	

4. List four negative environmental effects of clear cutting in the boreal forest.

	Negative environmental effect of clear cutting
1.	
2.	
3.	
4.	

5. List four reasons why wetlands are threatened.

	Threat to wetlands
1.	
2.	
3.	
4.	

6. List three threats to public grassland ecosystems.

	Action to degrade or cause desertification
1.	
2.	
3.	

7. List three threats to marine ecosystems.

	Action to degrade or cause destruction
1.	
2.	
3.	

8. List three threats to reduce or degrade our current agricultural land.

	Action to reduce or degrade agricultural land
1.	
2.	
3.	

9. List and explain the four criteria for ranking endangered ecosystems and explain why each factor must be considered.

	Criterion for ranking as endangered	Explanation
1.		
2.		
3.		
4.		

POP QUIZ

Matching

Match the following terms and definitions

____1. Death Valley and Joshua Tree National Parks are in this ecosystem:	a. biomass loss = biomass gain
____2. The Tongass National Forest is in this ecosystem:	b. brackish estuarine marsh
____3. The biome of El Yunque National Park in Puerto Rico is ______?	c. monoculture
____4. Area of land that has no permanent human occupation or degradation:	d. carrying capacity is exceeded
____5. This ecosystem offers the services of water purification and groundwater charging:	e. wildlife corridor
____6. This ecosystem forms the basis of a decomposition food chain that nourishes oyster, shrimp and flounder:	f. slash and burn
____7. One danger to the wilderness ecosystem is:	g. poachers and animal collectors
____8. National parks in Africa have to be guarded for this reason:	h. tropical rain forest
____9. Water moves through a plant from the roots to the leaves, where it changes from a liquid to a gas. This is called ______.	i. selective cutting
____10. A farm has 1,000 acres of soybeans. This is called a ______.	j. temperate coniferous rain forest
____11. In logging, this connects unlogged areas so that animals can seek cover and that existing trees can reseed the harvested area:	k. rolling easement
____12. Only mature trees are harvested in ______?	1. wetlands
____13. Ecologically sound logging must have this characteristic:	m. Mohave desert
____14. The most likely long-term effect of the loss of tropical rain forests:	n. off-road vehicles
____15. The type of sustainable agriculture practiced by the first inhabitants of the Western Hemisphere was:	o. transpiration

____16. If grassland is overgrazed, then...	p. wilderness
____17. This local action prevents landowners from building seawalls that run parallel to a beach:	q. global warming

Multiple Choice

1. Which of the following was first designated as a U.S. National Monument and is one of the latest National Parks created?
   a. Yosemite
   b. Yellowstone
   c. Death Valley
   d. Okefenokee
   e. Arcadia

2. Which of the following was the first National Park?
   a. Yosemite
   b. Yellowstone
   c. Death Valley
   d. Okefenokee
   e. Arcadia

3. The percentage of U.S. land held by the federal government is ______?
   a. 3%
   b. 20%
   c. 35%
   d. 42%
   e. 58%

4. The percentage of wilderness in the contiguous U.S. is closest to ______?
   a. 2%
   b. 4%
   c. 7%
   d. 9%

5. The National Park System was created during the administration of which U.S. president?
   a. Grant
   b. Theodore Roosevelt
   c. Taft
   d. Wilson
   e. Franklin Roosevelt

6. Korup National Park of the nation of Cameroon in Africa is in which biome? Hint—it has the largest number of species recorded in an African national park.
   a. tropical savannah
   b. tropical rain forest
   c. semidesert

d. tropical dry forest
e. tropical deciduous forest

7. This animal is overpopulated in Yellowstone National Park, and there are proposals to use contraceptives or culling to control it:
a. grizzly bear
b. raven
c. moose
d. elk
e. black bear

8. Which ecosystem service is provided by the process of transpiration?
a. water purification
b. air purification
c. cooling of air
d. air particulate settling
e. all of the above

9. If satellite surveys indicated a net loss of chlorophyll over time, the result would be which of the following?
a. increase in carbon dioxide
b. desertification
c. decrease in oxygen
d. increase in atmospheric heat
e. all of the above

10. When agro-businesses operate monoculture farms, the effect is which of the following?
a. increase in pest species
b. increase in crop diseases
c. increase in efficiency of producing the crop
d. increased water pollution from herbicides and pesticides
e. all of the above

11. This area connects parts of unlogged forest surrounded by logged areas.
a. shelterbelt
b. wildlife corridor
c. greenbelt
d. rain forest islands

12. Which country has a policy of sustainable wood production?
a. U.S.
b. Canada
c. Russia
d. Cameroon
e. Brazil

13. Which of the following practices produces *ecologically certified wood?*
a. sustainable yields
b. preservation of biodiversity

c. local inhabitants not negatively affected
d. watersheds maintained
e. all of the above

14. The problem with clear cutting an area and then replanting seedlings is which of the following?
a. biomass lost does not equal biomass gain
b. mudslides on steep slopes
c. siltation of stream water
d. floods
e. all of the above

15. The best reason why tropical rain forests do not regenerate easily after they have been farmed and grazed is ______.
a. no seeds are available
b. there is no moisture available
c. the thin soil eroded away
d. intense solar radiation prevents most seed germination
e. Trick question alert! All of these descriptions are equally valid.

16. Slash-and-burn subsistence agriculture is sustainable under which condition?
a. There is much more forest than field.
b. The edge effect continually increases.
c. There is much more field than forest.
d. Fertilizers are withheld.
e. All of the above are correct descriptions.

17. Overgrazing is particularly harmful in this area:
a. temperate grassland
b. tropical rain forest
c. tropical savannah
d. any tropical area bordering a desert
e. Trick question alert! All of these areas are equally damaged.

18. The BLM helped the Dust Bowl problem by doing this:
a. planting shelterbelts
b. reducing grazing on the prairie grasslands
c. replanting crops
d. replanting grass
e. preventing use of water from the Ogallala aquifer

19. An unexpected outcome of the BLM's "Adopt a Horse" program was:
a. increase of the wild horse population
b. decrease of the wild horse population
c. decrease of the burro population
d. horses sent to slaughterhouses

20. Wetlands are preserved under which U.S. act?
a. California Desert Protection Act
b. The Wilderness Act of 1964

c. The Food Security Act
d. The Taylor Grazing Act
e. The 1996 Farm Bill

21. In National Marine Sanctuaries:
 a. No fish can be taken.
 b. Fish can be taken in certain areas.
 c. Fish cannot be taken in specified areas.
 d. Species richness has increased in and around the sanctuaries.
 e. All except a are correct.

22. One cause common to both forest loss and agricultural land loss is which of the following?
 a. increasing property taxes
 b. water pollution
 c. overgrazing
 d. construction on flood plains

23. An agency of the federal government that holds a large amount of land that could be used for nature preserves is which of the following?
 a. USDA
 b. BLM
 c. USFWS
 d. USDOE
 e. USEPA

24. Which of the following is the nation's most popular national park?
 a. Grand Canyon
 b. Yellowstone
 c. Great Smoky Mountains
 d. Yosemite
 e. Acadia

25. The most important farm area threatened by urbanization is which of the following?
 a. South Florida/Miami
 b. North Carolina/Charlotte
 c. Minnesota/Minneapolis
 d. Illinois prairie/Chicago
 e. Virginia's Shenandoah Valley/Washington

26. Which statement accurately reflects the amounts of U.S. land area?
 a. There is more cropland than forest.
 b. There is a nearly equal amount of forests and pastures.
 c. There is less rock, tundra, roads, and urban area than forests.
 d. There is more wetland than pasture.
 e. There are equal amounts of cropland and forests.

CYBER SURFIN'

1. See the following governmental sites on management and protection of federal lands and waters:
   a. U.S. Fish and Wildlife Service' description of wildlife refuges and endangered species protection: **refuges.fws.gov**
   b. U.S. Army Corps of Engineers: **www.usace.army.mil/public.html**
   c. U.S. Bureau of Land Management: **www.blm.gov/nhp/index.htm**
   d. U.S. Park Service: **www.nature.nps.gov and http://www.nps.gov/rivers/**
   e. U.S. Forest Service: www.fs.fed.us/; and the Canadian Forest Service: **www.nrcan-rncan.gc.ca/cfs-scf/index_e.html**
   f. U.S. National Oceanic and Atmospheric Administration (coral reefs): **www.coris.noaa.gov/activities/welcome.html**
2. To learn about the estuarine system of California, see an online paper by Wayne R. Ferren Jr., Peggy L. Fiedler, Robert A. Leidy, and Kevin D. Lafferty, check out **www.mip.berkeley.edu/wetlands/estuarin.html**
3. From the Congressional Research Service library of Congress and the National Council for Science and the Environment, see an overview of federal land management programs at **www.ncseonline.org/NLE/CRSreports/Natural/nrgen-3.cfm?&CFID=7004819&CFTOKEN=28215455**
4. From Lyle Watts, Chief of the U.S. Forest Service, "A program in Range Lands": **www.fao.org/docrep/x5359e/x5359e01.htm**
5. From the U.S. Defenders of Wildlife site, see habitats in danger at **www.defenders.org/habitat/**
6. From Johns Hopkins University, "Population Reports: Winning the Food Race," click on Chapter 3, **www.jhuccp.org/pr/m13edsum.shtml#top**

'LIKE' BOOKING IT

1. Daily G, and Ellison K. ***The New Economy of Nature.*** Covelo, CA: Shearwater Books, 2002.
2. Durning AT, and Douglis C. ***Saving the Forests: What Will It Take?*** (Worldwatch Paper, 117), Washington, DC: Worldwatch Institute, 1993.
3. Williams M. ***Deforesting the Earth: From Prehistory to Global Crisis.*** Chicago: University of Chicago Press, 2002.
4. Gardner GT. ***Shrinking Fields: Cropland Loss in a World of Eight Billion*** (Worldwatch Paper, 131), Washington, DC: Worldwatch Institute, 1996.
5. Davidson A, Wolfe A, Isaac J, and Menchu R. ***Endangered Peoples*** (reprint ed.). San Francisco: Sierra Club Books, 1994.
6. Flores D. ***The Natural West: Environmental History in the Great Plains and Rocky Mountains.*** Norman: University of Oklahoma Press, 2003.

ANSWERS

Seeing the Forest

1. Review (see objective #1): For the 9 ecosystem services provided by the world's various ecosystems, list 9 different biomes or ecosystems represented and 9 different national parks, refuges, or other sanctuaries that protect them.

	Ecosystem service	Biome or aquatic ecosystem	National Park name and location
1.	Air purification	Tropical rain forest	Torup N.P., Cameroon
2.	Water purification	Brackish estuarine	The Waquoit Bay (U.S., MA) National Estuarine Research Reserve
3.	Ecotourism	Desert	Death Valley N.P. (U.S.)
4.	Climate moderation	Eastern deciduous forest	Great Smokies National Park (U.S.)
5.	Harvest of wood	Coniferous rain forest	Tongass National Forest
6.	Protection of terrestrial biodiversity	Tropical rain forest	Laguna del Tigre National Park, Guatemala
7.	Protection of aquatic biodiversity	Coral reef	Virgin Islands N.P., Dry Tortugas N.P. (FL).
8.	Moderation of climate	Coniferous forest, lower montane	Yosemite, Redwoods, and Sequoia N.P., CA (U.S.)
9.	Soil conservation	Temperate grassland	Wind Cave National Park (SD, U.S.) and Grassland National Park (Canada)

To see a list of National Marine Sanctuaries, go to **www.sanctuaries.nos.noaa.gov**; coral reef preserves: **www.nature.nps.gov/wv/coral/Coralnew-01.htm**; U.S. National Parks: **www.cr.nps.gov/history/online_books/glimpses1/index.htm** and **www1.nature.nps.gov/wv**; Central and South American parks: **www.parkswatch.org**; and World Parks by John Uhler: **www.world-national parks.net**

Questions to Ponder

1. List six types of U.S. federally managed areas, the federal agency responsible, and indicate two areas of concern for each. See objective #3.

Federally managed area	Agency managing	Two concerns
1. National parks	U.S. Dept. of the Interior, National Park Service	Too many visitors, air and water pollution, habitat degradation, and commercial/political efforts to extract natural resources
2. Rangelands	Bureau of Land Management	Overgrazing, loss of biodiversity, commercial/political desires to extract natural resources

3. Scenic and wild rivers	National Park Service	Pollution from cities flowing downstream, encroachment of urban areas
4. Forests	U.S. Forest Service	Overharvesting of trees leading to soil erosion, water pollution, mudslides, loss of diversity
5. National wildlife refuges	U.S. Fish and Wildlife Service	Water and air pollution, encroachment of urban areas
6. Rivers	U.S. Corps of Engineers	The Corps maintains more than 12,000 miles (19,200 km) of inland waterways and operates 235 locks. Concerns lie with dam building: flood plain land loss and loss of water by evaporation.

2. Debate the issue of natural regulation of trees in U.S. National Parks and Forests. From the National Geographic Society, hear audio stories at **www.nationalgeographic.com/radiox/yellowstone/** *Fires occurring in very dry conditions and where there has been a buildup of dead wood under the trees can be very destructive—Yellowstone National Park in 1988 and New Mexico in 2000. The National Parks Service and U.S. Forest Service have modified policy to allow for some controlled burning under conditions of low temperature and humidity. On the other hand, these ecosystems will regenerate and the nutrients released by burning will be cycled into new growth. However, it takes a longer period of time to regenerate according to increasing altitude or increasing latitude. President Bush just announced a program to clear federal forests of trees to reduce the occurrence of forest fires and save logging jobs (May 20, 2003).*

3. List five negative environmental effects or ecosystem services lost due to tropical rain forest degradation and deforestation.

	Negative environmental effect of tropical rain forest destruction
1.	Increase in carbon dioxide
2.	Decrease in oxygen
3.	Increase in global temperatures
4.	Loss of biodiversity, drug resources
5.	Desertification

4. List four negative environmental effects of clear cutting in the boreal forest.

	Negative environmental effect of clear cutting
1.	Soil erosion
2.	Water pollution due to siltation and mineral contamination

3.	Loss of biodiversity
4.	Mudslides

5. List four reasons why wetlands are threatened.

	Threat to wetlands
1.	Filling-in of salt marshes and fresh water marshes for real estate development
2.	Toxins in water pollution
3.	Property tax considerations
4.	Withdrawal of water by urban areas (Everglades) and farming

6. List three threats to public grassland ecosystems.

	Action to degrade or cause desertification
1.	Overgrazing
2.	Water extraction
3.	Encroachment of urban and farming areas

7. Describe three threats to marine ecosystems.

	Action to degrade or cause destruction
1.	Water pollution—toxic and sewage
2.	Ozone layer degradation causing UV radiation damage (killing the zooanthellae of corals)
3.	Overfishing and fish "strip mining."

8. Describe three threats to reduce or degrade our current agricultural land.

	Action to reduce or degrade agricultural land
1.	Property tax pressures to sell land for development
2.	Soil erosion
3.	Toxic residues of insecticides and herbicides

9. List and explain the four criteria for ranking endangered ecosystems and explain why each factor must be considered.

Criterion for ranking as endangered	Explanation
1. Area lost or degraded since the European colonization of North America	The Native American "Earth as Mother" philosophy was replaced by the pioneer philosophy that assumed that land was infinite for development. Forests were destroyed for wood, farming, and minerals.
2. The number of existing examples of a particular ecosystem, or the total area	Many ecosystems, their fauna and flora, are shrinking, notably the tropical rain forest and the Atlantic deciduous forest of Brazil.
3. An estimate of the likelihood that a significant area will be degraded in the next 10 years	Strip mining is presently destroying a significant area of eastern U.S. deciduous forest.
4. The number of threatened and endangered species living in that area.	Extinction is forever. Biodiversity loss of endemic species may be permanent.

Pop Quiz

Matching

1. m
2. j
3. h
4. p
5. l
6. b
7. n
8. g
9. o
10. c
11. e
12. i
13. a
14. q
15. f
16. d
17. k

Multiple Choice

1. c
2. b
3. c
4. a
5. d
6. b
7. d

8. c
9. e
10. e
11. b
12. b
13. e
14. e
15. c
16. a
17. d
18. b
19. d
20. c
21. e
22. a
23. d
24. c
25. a
26. b

18

Food Resources: A Challenge for Agriculture

LEARNING OBJECTIVES

After you have studied this chapter you should be able to:

1. Identify the main components of human nutritional requirements and differentiate among undernourished, malnourished, and overnourished conditions.
2. Define world carryover stocks and explain how they are a measure of world food security.
3. Describe the beneficial and harmful effects of domestication on crop plants and livestock.
4. Define deforestation and relate the main causes of tropical deforestation.
5. Relate the benefits and problems associated with the green revolution.
6. Describe current food safety issues.
7. Describe the environmental impacts of industrialized agriculture.
8. Define sustainable agriculture and contrast sustainable agriculture with industrialized agriculture.
9. Identify the potential benefits and problems with genetic engineering.
10. Contrast fishing and aquaculture and relate the environmental challenges of each activity.

SEEING THE FOREST

Describe the major food problems listed below:

Food problem	Food problem description
Undernourishment of calories	
Undernourishment of protein	
Inadequate distribution of food	
Monoculture agriculture increasing	
Toxic, mutagenic, and hormonal constituents of food. Presence of antibiotics in food and their overuse.	
Soil degradation	
Economic inability to subsidize crops with chemical fertilizers and energy for planting and harvest, or purchase farm equipment.	
Lack of water	
Overnourishment	

SEEING THE TREES

VOCAB

As we watch the sun go down, evening after evening, through the smog across the poisoned waters of our native earth, we must ask ourselves seriously whether we really wish some future universal historian on another planet to say about us: "With all their genius and with all their skill, they ran out of foresight and air and food and water and ideas."

U Thant, former secretary general of the United Nations

1. Organic Food Production Act—a U.S. law enacted in 1990 that defined *organic foods* as those grown on land that has been free of inorganic fertilizers, pesticides, and synthetic herbicides for 3 years. If the land has been officially inspected, it is termed *certified organic;* if it has been raised in open pastures or fields, it is termed *free range* or *naturally raised.* Organic seed must be used to grow organic crops. Standards were added in 2002 that prevented the use of genetically modified crops and domestic animals.

2. Food nutrients—food contains nutrients—chemicals with caloric value, minerals and vitamins used in metabolism.

Carbohydrates	The basic food of cells is glucose or "blood sugar." Cells extract energy from glucose to use in metabolism. The ideal dietary intake of carbohydrates is about 55% of 1,600–2,400 total calories. Daily caloric intakes have been revised downward from the 2,200–3,000 calories formerly recommended.
Proteins	Proteins are broken down into amino acids, which are then reassembled into the body's proteins. Proteins generally serve as structural materials for cells, enzymes, or hormones. Amino acids or short chains of amino acids serve as neurotransmitters. Certain *essential amino acids* must be obtained in the diet because the body cannot make them. The ideal dietary intake of proteins is 15%.
Lipids	Fats store more calories per gram than do any other food group. They are broken down into fatty acids that are used to make hormones, and the fatty acids are broken down into acetates (glucose residues) that make much ATP energy. Cholesterol is a lipid that the body uses to make steroid hormones. An ideal diet for an active person has 30% in calories from fats. Today's total is over 40%.
Vitamins	Vitamins generally serve as parts of enzymes necessary for cell metabolism.
Minerals	Calcium, sodium, and potassium are needed for neural and muscular function. Also calcium is needed to make bone. Zinc and magnesium form parts of enzymes.
(Water and oxygen)	(Water and oxygen are not nutrients because they do not have caloric, mineral, or vitamin functions, nevertheless they are required to support metabolism.)

3. Nutrient deprivation or excess terms are as follows:

   a. Undernourishment—an undersupply of calories, minerals, and vitamins that results in poor growth of children.
   b. Malnourishment—A person may be receiving an adequate supply of calories, but an undersupply of protein, essential amino acids, vitamins, or minerals.
   c. Marasmus—children who are undernourished of calories and other nutrients waste away. Their limbs are stick-like, and their ribs show through the skin prominently.
   d. Kwashiorkor—protein malnutrition. The abdomen of children swells prominently due to fluid accumulation, and growth is slowed. Protein malnutrition in infancy and early childhood can result in the slowing of mental development.
   e. Overnourishment—the oversupply of calories leads to morbid obesity. The oversupply of specific vitamins leads to hypervitaminosis—effects include loss of function in the liver and kidney, and diarrhea. Excessive calcium in the diet contributes to heart disease, arteriosclerosis, and kidney stones and selenium is toxic in overdose.

4. World grain carryover stock—the amount of rice, wheat, corn, rye, etc. reserves held from a previous year's harvests are a measure of world food security. Since 1987, world grain carryover stocks have decreased as population has increased.

5. Principal types of agriculture are as follows:

   a. Industrialized agriculture—Industrialized or high-input agriculture is subsidized (E.P. Odum's term) by large inputs of capital, water, fossil fuels energy, mechanization, and agrochemicals (fertilizers, synthetic herbicides, and pesticides) to produce large yields. Monocultures are used extensively.
   b. Subsistence agriculture—agricultural methods that require heavy labor by the farmers and

their domestic animals and a large amount of land. Subsistence agriculture produces food to feed the producer's family. Droughts, wind, and hail may damage crops and cause farm failure, or on a wider scale, famine.

c. Shifting cultivation—a form of subsistence agriculture that allows fields to become fallow or planted in alfalfa and allowed to recover. Native Americans used slash-and-burn agriculture in the eastern deciduous forest this way—they cut out a section, planted until the soil was exhausted, then abandoned the field to succeed ecologically back to forest. They "shifted" the location of their corn fields in locations other than flood plains where they were maintained.
d. Nomadic herding—nomadic shepherds move freely over rangelands, searching for good grazing.
e. Intercropping—subsistence or intensive sustainable agriculture that involves growing several crops in nearby plots—the opposite of monoculture.
f. Polyculture—A type of intercropping that uses plants that mature at different times. The term is also used in fisheries to indicate that several species of fish are raised together.

6. Domestication of animals—humans have by domestication or artificial selection, in Darwinian terms, propagated those farm animals with certain traits of body size, fat content, hide texture, and passive behavior. A loss in genetic diversity accompanies domestication. Remember that genetic diversity provides for the evolutionary survival of species.
7. Vegetarianism—some cultures, religions, or other philosophies either limit or prohibit the intake of meat or other animal product foods. Ecologically speaking, feeding an increasingly large population necessarily means that more people will have to get protein from sources that waste less energy in transfer among trophic levels in a food chain (see Chapter 4). Some compromised by eating only fowl and/or eggs; this approach takes advantage of the chicken's relatively high efficiency in turning chicken feed to meat. *Ovovegetarians* only allow eggs. *Lactovegetarians* eat milk and milk products. Vegetarian diets require that essential amino acids and vitamins, normally provided by meat, milk and eggs, are ingested.
8. Germplasm preservation—seed plants and cryopreservation of animal tissues preserves genetic information in cells. The International Plant Genetics Resources Institute in Rome, Italy, and the U.S. National Plant Germplasm System in Colorado hold major reserves.
9. International Treaty on Plant Genetic Resources for Food and Agriculture—ratified by 53 countries, not including the U.S., the treaty affirms the rights of farmers to save, use and distribute native germplasms in seeds and prevents agricultural companies from patenting native germplasms.
10. Green revolution—both native germplasm/genes can be interbred into plants or genetically engineered genes can be put into strains of crop plants to improve yields and reduce fertilizer, fungicide, or insecticide use. Norman Borlaug won the Nobel Prize in agriculture in 1970 for developing high-yield varieties of wheat. The use of chemicals, equipment, water, and energy subsidies presents major problems for local economies and cultures.
11. Monocultures—agricultural ecosystems, particularly the industrialized farms, specialize in growing one crop over thousands of acres. This actually encourages the pests and diseases that attack the crops. The pests usually reproduce faster than the crops or the natural predators of the pests. The same pests increase their numbers in the remaining national forests.
12. Overuse of antibiotics—to do a final fattening up/weight gain before slaughter, cattle and other livestock are crowded together in feces-covered feeding lots. That increases the chances that wounds will become infected, and their weight gain will be affected by infection, so antibiotics

are added to their feed there and also back on the farm for the same reason. Bacteria evolve resistance to bacteria quickly, after all, they have been on this Earth longer than any other type of organism. In that time, they have been adapting to changing conditions constantly, acquiring genes to evolve into forms that can survive most environmental conditions and chemicals that nature produces.

13. Food additives—as occurs in nature, bacteria, fungi, and maggots will begin to decompose our food after harvest. Food additives are added to processed food to enhance the taste, color (MSG), texture of the food, and to prevent its natural spoilage. Some additives like MSG and sulfites cause allergic reactions, others, such as nitrites used in meat preservation, are mutagenic and related to childhood leukemia. Many bacteria produce toxins that are secreted or released when the bacteria die. Preventing their growth is a good idea. See some notes about common food additives below.

Table 14-4.

Food additive	Food	Effects
Alginates	Ice cream	A thickener. No harmful effects.
Alpha tocopherol	Vegetable oils and margarine	A form of Vit. E antioxidant that replaces BHA and BHT. Prevents fats from becoming rancid.
Ascorbic acid	Fruit drinks, cereals	Color stabilizer in fruit drinks, it detoxifies nitrosamines formed from nitrates and nitrites. Makes collagen. Also is an antioxidant.
Beta carotene	Margarine coloring	Makes Vit. A in the body. In large doses, enhances tumor growth in smoker's lungs.
BHA and BHT	Oils, potato chips, gum	Antioxidants that mimic Vit. E. May be carcinogenic. Prevents fats from becoming rancid due to bacterial decomposition.
Calcium or sodium propionate	Bread	Antifungal preservation. Contributes calcium. No harmful effects known.
Calcium or sodium lactylate	Bread dough, whipped creams	Stiffens dough and whipped creams. No harmful effects known.
Casein	Milk and milk products	Whitening agent. Some people are allergic to it. Otherwise it is a nutritious protein.
Citric Acid	Jellies, candy	Adds tart flavor. A nutrient in cell metabolism.
Gelatin	Yogurt, jello, breath mints	Made from cattle hides. Because there is fear of mad cow disease, some people did not want to eat British breath mints containing gelatin. The processing removes infective agents, however.
Monsodium glutamate (MSG)	Chinese vegetables, stir fry, soup	Flavor and color enhancer. Some people are sensitive to it.

Table 14-4 *(continued)*

Food additive	Food	Effects
Nitrates, nitrites	Preserved meats: sausage, hot dogs, bologna, salami	Mutagenic/carcinogenic. One of the first meat preservatives used.
Potassium bromate	Flour	Used to produce batter-type breads. Most breaks down in cooking, but the remainder may be carcinogenic.
Red dyes #3, 8, 9, 19, citrus red dye #2	Maraschino cherries, fruit cocktail, candy	May be carcinogenic.
Sulfites	Dried fruit, canned and frozen vegetables, bread, etc.	Some are allergic to these preservatives.

14. Feedlot pollution—"factory" feedlots increase transmission of pathogenic organisms like *E. coli* O157:H7, *Salmonella*, *Listeria*, and *Campylobacter*. As the contaminated carcass is processed, bacteria are spread to the meat. If the meat is ground up, the pathogens are mixed in. Feedlots are a local nuisance because of the smell and water pollution.
15. Sustainable agriculture—sustainable agriculture is also termed low input or alternative agriculture. Indigenous peoples naturally used a subsistence agriculture that was sustainable. Updating that approach, sustainability applies ecologically sound principles to farming—maintain crop and farm animal growth without antiobiotics, using polyculture instead of monoculture, using soil conservation practice, using water conservation practice, encouraging the available of predators of crop pests by using greenbelts, using animal wastes as fertilizer, and rotating legumes with other crops.
16. Integrated pest management—this method uses small amounts of pesticides or fungicides, precisely applied in terms of location and time, the use of crop rotation, disease resistant varieties, and biological insect controls—trichogammid wasps to attack caterpillar pests.
17. Genetic engineering—GE or GM crops are those that have been genetically modified by the modern techniques of gene slicing, recombination, and transfer into cells, eggs, or ovules. The transfer of genes may allow the crops to be resistant to a specific pest or disease by producing an antibiotic that kills or repels it, a gene to prevent roots from absorbing aluminum, a gene to cause rhizobia nitrogen-fixing bacteria to grow on roots, to produce higher yields, or a gene to produce a certain vitamin.
18. Biosafety protocol—fearing the transfer of GM genes to native plants and the possibility that food allergies would be increased by exposure to GM gene products, in 1992 the United Nations established a treaty (the U.S. has not agreed to the treaty) establishing procedures for handling GM crops.
19. Destructive factory fishing practices—*longlines* of baited hooks, stripping of sea floor by drag nets, and *trawlbags*, and catching and killing fish or mammals that are not food items *(bycatch)*, are practices that not only threaten biodiversity and ecosystems but also the future food supplies from the ocean.

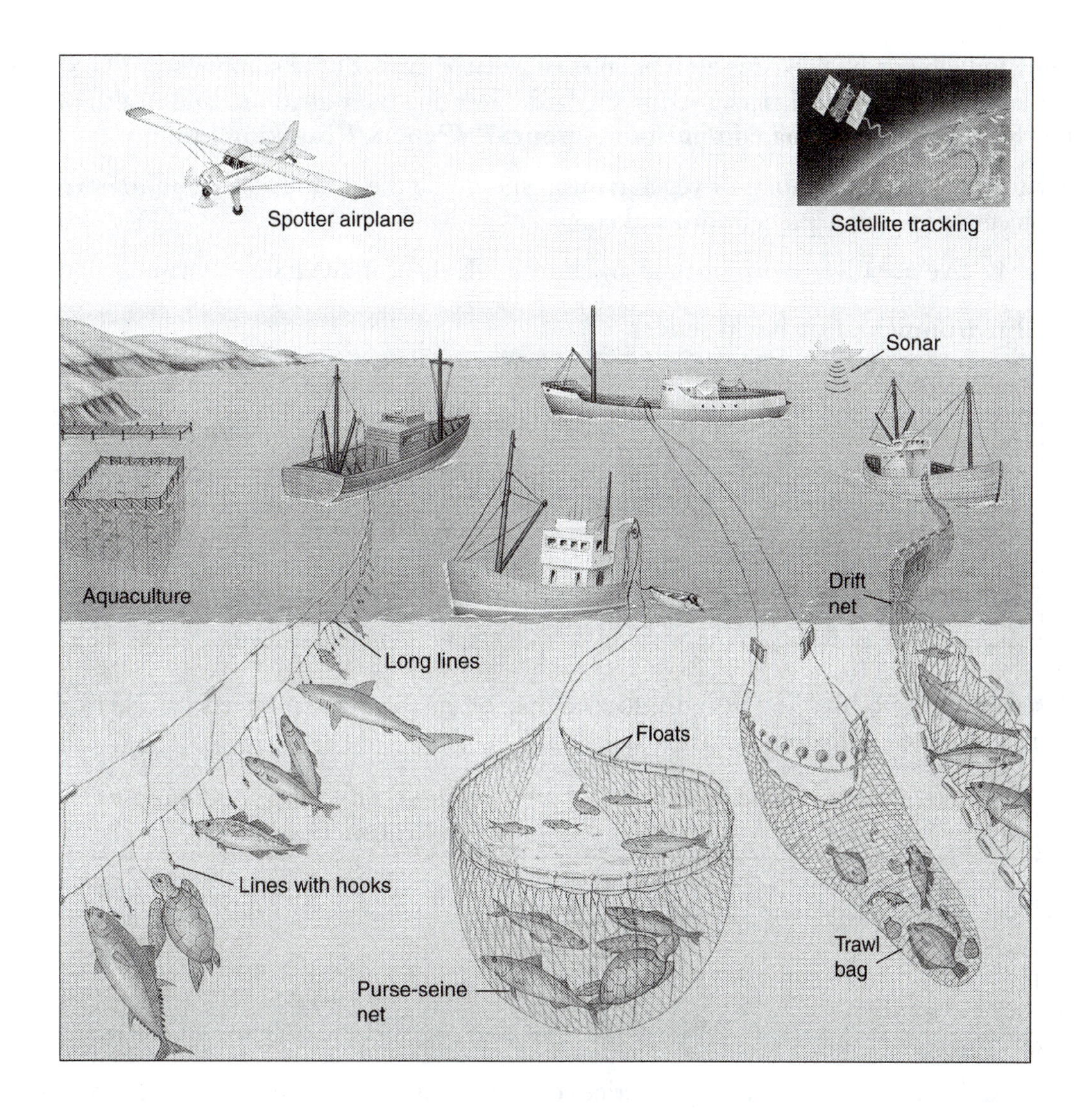

20. Magnuson Fishery Recovery Act of 1977 and the Magnuson-Stevens Fishery Conservation and Management Act of 1996—develops conservation management plans, habitat protection for 600 species of fish, reduction of overfishing, restoration of fish populations, and the minimization of *bycatch*.
21. Marine conservation—twelve National Marine Sanctuaries have been established in U.S. waters to allow for sustainable fishing, recreation, education, research, and ecosystem preservation (in the no-take zones).
22. Aquiculture—fish farming. There is extensive fresh water aquiculture of channel catfish in the southern U.S. and of carp in the rice paddies of the Far East. Mariculture of oysters, clams, crabs, shrimp, salmon, and other species is common worldwide.

Questions to Ponder

1. Explain how carryover stocks of grain are related to world food security. Why is this an issue now?

2. Many indigenous Native American peoples of isolated areas of the southwestern U.S. live on a diet largely of corn meal tacos fried in pig lard. Describe the nutritional and health problems they have. See **www.as.ua.edu/ant/bindon/ant570/Papers/King/king.htm**

3. Debate whether meat eating or vegetarianism should be the diet of today and tomorrow. Try to be objective in your argument pro and con.

4. Describe five negative environmental and health effects of industrialized farming.

	Environmental or health effect
1.	
2.	
3.	
4.	
5.	

5. Describe three positive and negative food supply or environmental effects of genetically engineered/modified crops or animals.

	Positive effect on food supply or environment	**Negative effect on food supply or environment**
1.		
2.		
3.		

6. List and describe four destructive practices of factory fishing on the environment or food supply.

	Destructive fishing technique	**Description**
1.		
2.		
3.		
4.		
5.		

POP QUIZ

Matching

Match the following terms and definitions

____1. This crop and land have been inspected to see that no pesticides and inorganic fertilizers were applied:	a. essential amino acids
____2. This material may not be used to raise certified organic food:	b. low genetic diversity
____3. This food material is needed for proper growth and mental development and to make metabolic enzymes:	c. overfishing
____4. The food materials that store the most calories and supply the materials to make estrogen and testosterone are ______?	d. germplasm
____5. This is the reason why the right kinds of protein must be in a vegetarian diet:	e. industrialized
____6. Swelling of the abdomen, slow or absent growth in children describes ______?	f. certified organic
____7. Stick-like arms and legs and protruding ribs indicate ______?	g. inorganic fertilizers
____8. This type of agriculture produces little surplus to be traded in markets:	h. lipids
____9. This type of monoculture farming is heavily subsidized by expenditures of energy, water, and agrochemical:	i. protein
____10. The problem with domestication of farm animals is that it produces ______?	j. "Green Revolution"
____11. The opposite of monoculture farming is ______?	k. kwashiorkor
____12. The International Treaty on Plant Resources for Food and Agriculture prohibits companies from patenting native ______?	l. slaughterhouse by-products
____13. Interbreeding and genetic engineering of wheat and rice created the ______?	m. bycatch
____14. The overuse of antibiotics enhances ___ in bacteria.	n. traditional subsistence

____15. Mad cow disease was spread in Britain because ___ was added to cattle feed.	o. intercropping
____16. Porpoises caught while commercial tuna fishing is taking place is termed ______?	p. marasmus
____17. The Magnuson-Stevens Act prohibits ______ in U.S. waters.	q. evolution of resistance

Multiple Choice

1. Which of the following was not required by the Organic Food Production Act?
   a. no inorganic fertilizers used
   b. chickens raised in open pastures
   c. biological control of insects
   d. no pesticides used
   e. no radiation sterilization can be used

2. The reason explaining the higher cost of organically raised crops and animals is which of the following?
   a. high labor costs
   b. small yields
   c. supply exceeds demand
   d. food contaminated with animal waste
   e. a and b

3. Ideal nutrient intake in percentages of calories per food group should be:
   a. 10% protein, 58% carbohydrates, and 30% fats as it was in 1916
   b. 12% protein, 58% carbohydrates, and 40% fats as it is today
   c. go on the Atkins diet of 20% carbohydrates, 40% fat, and 40% protein.
   d. 15% protein, 55% carbohydrates, and 30% fat
   e. No sugars or starches! Only 100% protein is allowed.

4. Which statement is correct about current world food supply and consumption?
   a. World food reserves have risen faster than population due to the Green Revolution.
   b. World food consumption exceeds annual grain production.
   c. The highest amount of grain produced per capita was in 1950 when grain carryover capacity was also greatest.
   d. Population increase and grain production are keeping pace.

5. Which nation has not suffered famine at least once since 1960?
   a. China did during the Cultural Revolution in the 1960s
   b. North Korea
   c. Sudan
   d. Somalia
   e. Trick question alert! All of these have suffered a famine.

6. If a cow consumed 3,000 calories of plant material, which amount ends up being stored in the cow's body?
   a. 10
   b. 130
   c. 300
   d. 420
   e. 1,000

7. Of the animals below, which is unusually efficient in transforming food into biomass?
   a. beef cattle
   b. chicken
   c. goat
   d. lamb
   e. pig

8. Vegetarian diets must be planned so that they provide an adequate intake of which of the following?
   a. protein
   b. essential amino acids
   c. calories
   d. calcium and B_{12}
   e. all of the above

9. Which of the following is not a characteristic of industrialized agriculture?
   a. sustainability
   b. monocultures
   c. high use of energy
   d. high use of inorganic fertilizers
   e. high use of pesticides and synthetic herbicides

10. Which of the following practices increases genetic diversity?
    a. monoculture
    b. polyculture
    c. intercropping
    d. seed banks
    e. domesticating the alpaca

11. Which of the following does not happen when antibiotics are overused?
    a. antibiotic resistant bacteria increase their populations
    b. genes are selected by an environment that contains antibiotics
    c. evolution
    d. extinction of pathogenic bacteria
    e. disease-causing pathogen infections increase

12. Which of the following food additives has been linked to childhood leukemia?
    a. Vitamin C
    b. sodium nitrite
    c. sugar

d. sodium propionate
e. MSG

13. Factory feedlots are problematic. Which is an incorrect description of them?
    a. *E. coli* O157:H7 moves quickly from cow to cow.
    b. Water released from manure lagoons has killed fish in streams.
    c. Because of the poor environmental condition, cattle weight decreases.
    d. The smell can be whiffed a mile away.
    e. Not a blade of grass can be seen.

14. The principal reason explaining the drop of water in the Ogallala Aquifer is:
    a. factory farming of pigs.
    b. soil erosion.
    c. a prolonged drought.
    d. wheat production.
    e. excessive grazing.

15. In the last 5 years, which country has increased in irrigated land area?
    a. U.S.
    b. China
    c. Egypt
    d. France
    e. U.K.

16. Which technique increased crop productivity and soil fertility?
    a. application of manure and crop rotation with corn and soybeans
    b. crop rotation between corn and soybeans without fertilizers added
    c. fertilizers used to raise corn alone
    d. large subsidies of energy and farm equipment applied
    e. crop rotation and no fertilizers

17. The country that has banned the import of genetically modified food is which of the following?
    a. France
    b. United Kingdom
    c. Germany
    d. Norway
    e. Trick question alert! All of these countries have banned GM food.

18. A beneficial application of genetic engineering would be to prevent the absorption of which mineral in acidic soils?
    a. aluminum
    b. phosphate
    c. calcium
    d. potassium
    e. magnesium

19. Which animal is losing population because it is in the *bycatch* of factory fishing operations?
    a. dolphins/porpoises
    b. sea turtles

c. tuna
d. Atlantic cod
e. a and b

20. The destruction of which coastal area has great effects on the fish and shellfish populations nearby?
a. brackish estuaries
b. mangrove swamps
c. tidal marshes
d. all of the above

21. The type of fishing technique that "strip mines" the ocean bottom is which of the following?
a. purse seine net
b. trawlbag
c. drift net
d. longline
e. all of the above

22. Which country has the highest consumption of poultry in kg/year? Refer to the figure below.
a. India
b. U.S.
c. Italy
d. China

Table 18.1 **Annual Per-Capita Consumption of Meat in Selected Countries**

Country	*Beef (kg)*	*Pork (kg)*	*Poultry (kg)*	*Mutton (kg)*
India	1	0.4	1	1
China	4	30	6	2
Italy	26	33	19	2
United States	45	31	46	1

23. Which country is most likely to have protein malnutrition?
a. India
b. U.S.
c. Italy
d. China

24. Which is the world's largest grain crop?
a. wheat
b. corn
c. rice
d. barley
e. dried beans and peas

25. Which is the world's largest ground crop?
    a. sweet potato roots
    b. cassava root
    c. peanuts
    d. Irish potatoes
    e. sugar cane

26. Which crop had the largest drop in diversity in the last 100 years?
    a. bean
    b. beet
    c. pea
    d. carrot
    e. onion

CYBER SURFIN'

1. For links and information on organic food and alternative farming, see the U.S.D.A. site: **www.nal.usda.gov/afsic/ofp/**
2. This site is a good source of general information about GM crops, check out "Transgenic Crops: An Introduction and Resource Guide" from Colorado State University: **www.colostate.edu/programs/lifesciences/TransgenicCrops/**
3. To a read positive views on the role of GM crops in developing nations, see the testimony given by Roger Beachy of the Danforth Plant Science Center (from the AgBio World site): **www.agbioworld.com/biotech_info/topics/agbiotech/beachy.html**
4. To see a compilation of negative opinions of GM foods, from Ron Epstein of San Francisco State University, "Genetic Engineering and its Dangers": **online.sfsu.edu/~rone/GEessays/gedanger.htm;** and Global Issues.org: **www.globalissues.org/EnvIssues/GEFood/Hunger.asp**
5. Take a look at the World Hunger Year site on "Innovation Solutions Top Hunger and Poverty" (through self-reliance, food security and economic justice): **www.worldhungeryear.org/**
6. Learn about the estuarine system of California, see an online paper by Wayne R. Ferren Jr., Peggy L. Fiedler, Robert A. Leidy, and Kevin D. Lafferty, check out: **www.mip.berkeley.edu/wetlands/estuarin.html**
7. To read about the biology of "good bugs" and *integrated pest management* go to the BioResources site in Australia: **www.bioresources.com.au/**
8. Greenpeace has extensive reports on the depletion of world fisheries:
    a. Ecologically Responsible Fisheries: **archive.greenpeace.org/~comms/fish/part11.html**
    b. Amazing facts: **archive.greenpeace.org/~comms/fish/amaze.html**
    c. Canadian fisheries collapse (Atlantic cod): **archive.greenpeace.org/~comms/cbio/cancod.html**
9. To see the U.S. government position on fisheries conservation see the National Marine Fisheries Service of NOAA: **www.nmfs.noaa.gov**; and a rebuttal of one Greenpeace report on Alaska fisheries: www.atsea.org/nmfssumm.html (the link comes from a fisheries industry organization, At Sea Processors).
10. Native peoples have a vested interest in maintaining sustainable fisheries—food! Check out **www.indians.org/library/fish.html**

'LIKE' BOOKING IT

1. Postel S. ***Dividing the Waters: Food, Security, Ecosystem, Health & the New Politics (Worldwatch Paper, 132).*** Washington, DC, 1996.
2. Moore FM, Collins J, Rosset P, and Esparza L. ***World Hunger: 12 Myths*** (2nd ed.) New York, New York: Grove/Atlantic Press, 1998.
3. Lappe I, and Bailey B (Eds.). ***Engineering the Farm: The Social and Ethical Aspects of Agricultural Biotechnology.*** Covelo, CA: Island Press, 2002.
4. Lambrecht B. ***Dinner at the New Gene Cafe: How.*** New York: St. Martin's Press, 2002.
5. Vandana S. ***Biopiracy: The Plunder of Nature and Knowledge.*** Boston: South End Press, 1997.
6. Sloan S. ***Ocean Bankruptcy: World Fisheries on the Brink of Disaster.*** Guilford, CT: Lyons Press, 2003.
7. Cone J. ***A Common Fate: Endangered Salmon and the People of the Pacific Northwest.*** Corvallis: Oregon State University Press, 1996.
8. McGinn PM. ***Rocking the Boat: Conserving Fisheries and Protecting Jobs (Worldwatch Paper 142).*** Washington, DC: Worldwatch Institute, 1998.

ANSWERS

Seeing the Forest

1. Describe the major food problems listed below:

Food problem	Food problem description
Undernourishment of calories	In recent times, famines are usually caused by a population exceeding its environmental carrying capacity after droughts, internal political upheavals, or wars. A total lack of calories or *marasmus* is characterized by stick-like limbs, weakness, and dry, brittle hair. The skin is thin with loose folds, and the ribs are prominent. Growth and brain weight are slowed. In their weakened condition and low reserves of fluids, people die most often due to shock caused by diarrheic diseases transmitted easily in refugee camps.
Undernourishment of protein	In children aged 1–5, early symptoms of *kwashiorkor* include fatigue, irritability, and lethargy. As protein deprivation continues, one sees growth failure, loss of muscle mass, abdominal swelling (edema), and impaired immunity. It is differentiated from marasmus by the abdominal edema and vitiligo (increased skin pigmentation). Similar to *marasmus*, death is usually caused by hypovolemic shock due to diarrheic diseases. There can be permanent loss of mental function and physical disabilities. Approximately 50% of nursing home residents suffer from protein malnutrition.
Inadequate distribution of food	Developing countries suffer from a lack of refrigeration and transportation for food distribution.

Monoculture agriculture increasing	There is a catastrophic loss in the genetic diversity of crops that will contribute to massive crop failures when pests adapt to the new crops. We may be facing the same type of problems with microorganism pests that we now face with antibiotic resistance in pathogenic bacteria.
Toxic, mutagenic, and hormonal constituents of food. Presence of antibiotics in food and their overuse.	Some food additives are carcinogenic; pesticide and herbicide residues reside in soils and crops. Mercury is in tuna. Animals are fed steroid hormones to increase growth. Some insecticides are estrogen mimics or antagonists (DDT). The overuse of antibiotics has led to the evolution of antibiotic resistance. Meanwhile, overall cancer rates increase and sperm counts go down.
Soil degradation	Toxic residues of fertilizers, pesticides, and herbicides reside in soil. Soil erosion continues in most areas of the world.
Economic inability to subsidize crops with chemical fertilizers and energy for planting and harvest, or purchase farm equipment.	Developing countries have few economic resources and capital to purchase the equipment and supplies needed to operate an industrialized farm.
Lack of water	Many areas in the world do not have enough water for raising crops and supporting an increasing population.
Overnourishment	While children in the Sudan, Somalia, Eritrea, and Ethiopia are suffering from marasmus and kwashiorkor, the U.S. has an epidemic of obesity due to overnourishment and lack of physical activity. Specific overnourishment of vitamins is called hypervitaminosis. Excessive doses of vitamin A, D or E can damage the liver.

QUESTIONS TO PONDER

1. Explain how carryover stocks of grain are related to world food security. Why is this an issue now? *In the event of a catastrophic collapse of grain production (like WW III), the amount of grain harvested in a previous year that is not consumed in that year represents a cushion against starvation. In the 1990s, the increase in world population kept pace with the increase in carryover stocks of grain. Since 2000, the carryover stocks increases have lagged behind population increases.*

2. Many indigenous Native American peoples of isolated areas of the southwestern U.S. live on a diet largely of corn meal tacos fried in pig lard. Describe the nutritional and health problems they have. See ***www.as.ua.edu/ant/bindon/ant570/Papers/King/king.htm.*** *Historically, the indigenous people did hunting and gathering, some animal herding, some raising of corn and growing of nuts and fruits. Many were nomadic: all had high levels of physical activity. As work and leisure became more sedentary Native American adopted (by choice or coercion) European foods. Eating a high carbohydrate and milk meal (lactose) causes first hyperinsulinemia (excessive secretion of insulin) followed by insulin resistance as the insulin receptors of cells downregulate (Type II sugar diabetes). Eating a low pro-*

tein, high-fat carbohydrate diet can result in a form of protein malnutrition where the person is obese but has edema, impaired kidney function, reduced cognition, and low immune function. When this diet is combined with alcohol, cirrhosis of the liver occurs more readily.

3. Debate whether meat eating or vegetarianism should be the diet of today and tomorrow. Try to be objective in your argument pro and con. *The needs for protein worldwide will stimulate research in increasing protein content of plant foods. Taking into account the energy losses that occur in transfer through more than two trophic levels of a food chain, inefficient animals like beef cattle will be decreasingly used. Threatened commercial and practical animal and plant extinctions will occur, then most of us will be forced to become vegetarians.*
4. Describe five negative environmental and health effects of industrialized farming.

	Environmental or health effect
1.	Increase of pest and crop disease microorganisms due to monoculture: concentrating one crop also concentrates their pests, and with no natural predators available because there are no greenbelts, their populations increase. This causes even more application of pesticides and herbicides.
2.	Famine due to a catastrophic failure of one crop of wheat, or corn, or rice, or soybeans. With only a few "super crops" being grown, the effects of failure of one will cause starvation of an overpopulated earth.
3.	Genetic diversity will be lost due to the selection of the "best crops" for sugar, fat, protein, or vitamin yield.
4.	There will be increased development of new genes. GM/GE genes will move into native plant population with unknown effects. Consider what happened when the African bees hybridized with domestic honeybees.
5.	Soil erosion will continue to increase, as well as fertilizer leaching into groundwater and streams. The Louisiana Dead Zone and others, where major rivers meet the ocean, will get larger, having disastrous effects on recreation and commercial fishing.

5. Describe three positive and negative food supply or environmental effects of genetically engineered/modified crops or animals.

	Positive effect on food supply or environment	**Negative effect on food supply or environment**
1.	Crops can become more resistant to pests.	Loss of genetic diversity as local seed varieties disappear.
2.	Crops can produce larger yields.	Gene escape resulting in food allergies or other as yet unforeseen consequences.
3.	Crops can be modified to live in acidic or other poor soils.	Starvation if one of the few super crops fails.

6. List and describe five destructive practices of factory fishing on the environment or food supply.

	Destructive fishing technique	Description
1.	Trawlbag	This is the most destructive practice that literally "strip mines" the ocean bottom, collecting shellfish, cod, flounder, and other fish. Many other types of organisms are killed haphazardly: sea turtles, sponges, corals, etc.
2.	Longline	Baited hooks are strung in long lines and left to catch whatever bites. Again, many species that are not commercially valuable are caught and killed.
3.	Drift net	This is used to catch salmon and tuna in the open sea. *Bycatch* includes dolphins/porpoises and sea turtles. Nets may drown dolphins and turtles injure the creatures or make them susceptible to attack by sharks.
4.	Purse seine	Purse seines are used to catch anchovies (an important animal feed protein supplement), herring, and mackerel.
5.	Sonar and satellite tracking	Unregulated, "pirate" overfishing is done by shadow fishing companies in Taiwan, Japan, South Korea, China, Spain, U.S., and Norway. The factory ships use satellite and sonar tracking of fish so that their huge nets can scoop up large numbers of fish. Various estimates (by their nature imprecise—who knows what gets thrown away?) range between 17–47 million tons/year, or approximately 27% of the total catch.

Pop Quiz

Matching

1. f
2. g
3. i
4. h
5. a
6. k
7. p
8. n
9. e
10. b
11. o
12. d
13. j
14. q

15. l
16. m
17. c

Multiple Choice

1. c
2. e
3. d
4. b
5. e
6. d
7. b
8. e
9. a
10. d
11. d
12. b
13. c
14. d
15. b
16. a
17. e
18. a
19. e
20. d
21. b
22. c
23. a
24. b
25. d
26. a

19

Air Pollution

LEARNING OBJECTIVES

After you have studied this chapter you should be able to:

1. List the seven major classes of air pollutants and describe their characteristics and effects.
2. Relate the adverse health effects of specific air pollutants and explain why children are particularly susceptible to air pollution.
3. Describe industrial smog, photochemical smog temperature inversions, urban heat islands, and dust domes.
4. Summarize the effects of the Clean Air Act on U.S. air pollution.
5. Contrast air pollution in highly developed and developing countries.
6. Describe the global distillation effect and tell where it commonly occurs.
7. Summarize the sick building syndrome.
8. Describe the physiological effects of noise pollution on the human body.

SEEING THE FOREST

List ten outdoor or indoor hazardous air pollutants, their source, and the environmental or health effects of each. Consult Tables 19-1 and 19-2, and the chemistry appendix.

	Air pollutant	Source	Environmental or health effect
1.			
2.			
3.			
4.			
5.			
6.			
7.			
8.			
9.			
10.			

Table 19.2 **Health Effects of Several Major Air Pollutants**

Pollutant	*Source*	*Effects*
Particulate matter	Industries, motor vehicles	Aggravates respiratory illnesses; long-term exposure may cause increased incidence of chronic conditions such as bronchitis
Sulfur oxides	Electric power plants and other industries	Irritate respiratory tract; same effects as particulates
Nitrogen oxides	Motor vehicles, industries, heavily fertilized farmland	Irritate respiratory tract; aggravate respiratory conditions such as asthma and chronic bronchitis
Carbon monoxide	Motor vehicles, industries	Reduces blood's ability to transport oxygen; headache and fatigue at lower levels; mental impairment or death at high levels
Ozone	Formed in atmosphere (secondary air pollutant)	Irritates eyes; irritates respiratory tract; produces chest discomfort; aggravates respiratory conditions such as asthma and chronic bronchitis

SEEING THE TREES

VOCAB

In 2 hours, an operating jet ski engine produces as much air pollution as a 1998 car driven for 130,000 miles.

California Air Resources Board

1. Air pollution—gases, liquids, or particulate solids in the atmosphere that can harm human, organisms, or structural materials. Air pollution fatalities exceed traffic accidents by 3:1.

Selected megacities in the world for air pollution and disease risk, a rating of risk factors for sulfur dioxide, suspended particulate matter, lead, carbon dioxide, carbon monoxide and nitrogen oxides derived from U.N. and the World Health Organization data for 1992:

City	Derived rank in risk factors	Rank in population	Notes
Mexico City	1	1	All categories heavy to serious
Jakarta, Indonesia	Tie for 2–3	8	Particulates serious, carbon monoxide and carbon dioxide heavy
Seoul, South Korea	Tie for 2–3	11	Sulfur dioxide and particulates serious
Beijing, China	Tie for 4–5	16	Particulates and sulfur dioxide serious
Los Angeles, CA	Tie for 4–5	17	Heavy in particulates and carbon, serious in ozone
Bangkok, Thailand	6	19	Serious in particulates, heavy in lead
New York City	Tie for 7	4	Heavy in carbon monoxide and ozone
Cairo, Egypt	Tie for 7	13	Serious in particulates and lead
Moscow, Russia	Tie for 9–10	20	Heavy in particulates, carbon monoxide and nitrogen oxides
London, U.K..	Tie for 9–10	18	Heavy in carbon monoxide

2. Primary air pollutants—harmful materials that affect the atmosphere directly and have an effect on the health of humans and the environment or damage structural materials.
3. Secondary air pollutants—primary air pollutants react with other air pollutants or chemicals naturally present in the air to form hazardous products.

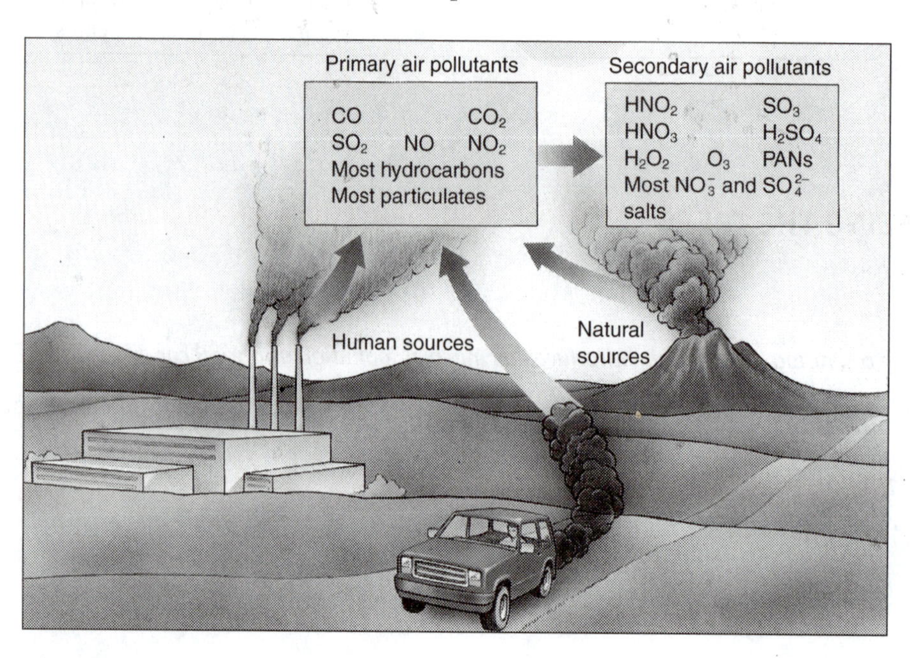

4. Particulate matter—solids or liquid droplets that coalesce around solid particles that are suspended in the air including dust from farming, dust storms, and volcanic ash; smoke from fires; and aerosols.

5. Nitrogen oxide—nitrous oxide, N_2O; nitric oxide, NO; and nitrogen dioxide (orange in color), NO_2 are formed by oxidation of atmosphere nitrogen, N_2, particularly the high temperatures and pressures of fossil fuel combustion. Before the explosion of the first atomic bombs, physicists were worried that the heat of the explosion might ignite the whole atmosphere—it does so immediately in the fireball. Nitrogen oxides contribute to the secondary formation of street level ozone in smog, and dissolved in water, nitric acid, HNO_3, contributes to acid rain.

6. Sulfur oxides—sulfur in coal, gasoline, and other fuels oxidizes to form sulfur dioxide, SO_2 (yellow in color), and sulfur trioxide, SO_3. Respectively, these gases dissolve in water to form sulfurous acid, H_2SO_3, and sulfuric acid, H_2SO_4. The acids dissolve the limestone of monuments and buildings and dissolve the calcium and magnesium out of conifer needles and soils in general, causing plant roots to soak up toxic aluminum, in other words, acid rain.

7. Hydrocarbons—compounds of carbon that contain some combination of carbon and hydrogen atoms. Methane, propane, octane, and acetylene are hydrocarbon fuels. Some hydrocarbon fuels have oxygen—ethyl and methyl alcohol. Trees emit hydrocarbon turpenes and isoprenes that contribute to smog.

8. Carbon oxides—complete combustion of carbon produces carbon dioxide, CO_2. Incomplete combustion produces soot (elemental carbon), smoke (unburnt carbon particles and aerosols), and carbon monoxide, CO. Carbon dioxide traps infrared radiation, therefore contributing most of the global warming effect and forms carbonic acid, H_2CO_3 that contributes to acid rain.

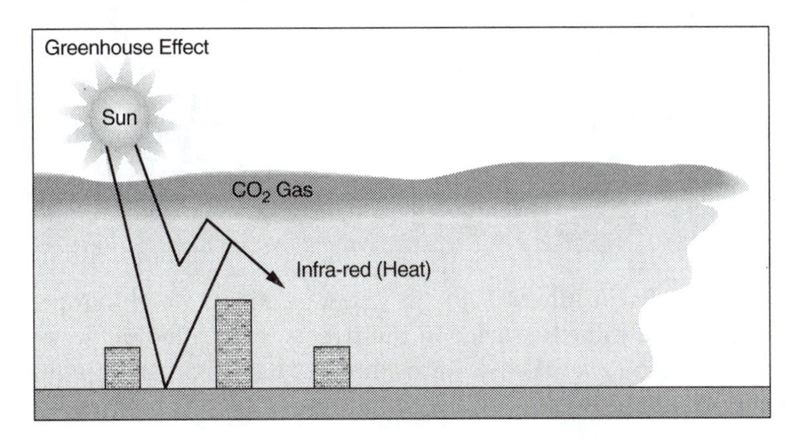

9. Ozone—oxygen gas that consists of three oxygen atoms, O_3, produced by the ultraviolet radiation energized splitting of O_2 and combination of O_1 with O_2. The U-V photons are transformed to heat or infrared photons in the process of ozone formation and breakdown. In the troposphere, ozone is a secondary air pollutant produced by photochemical reactions that produce smog. Ozone causes lung diseases, particularly in those who have compromised lung health by smoking.

10. Smog—the term is a combination of the words smoke and fog. There are two types:

    a. Industrial smog—a fog due to a combination of sulfur oxides, nitrogen oxides, and smoke from industrial and municipal sources of combustion of fossil fuels. London, U.K., has had the single worst incident of industrial smog in 1952 and other incidents going back to the Middle Ages, caused by the burning of sea coal and peat.

b. Photochemical smog—every winter a brownish orange haze lies in the "Valley of the Sun," where Phoenix, AZ, is located, due to smoke of wood fireplaces and automobile pollution. One measure taken was the requirement to use oxygenated fuels in the winter. In Los Angeles, geography also plays a role with mountains to the north, east, and southeast; and Mexico City, which lies in a bowl surrounded by mountains. Sunlight energy, U-V, drives reactions between hydrocarbons and nitrogen oxides to produce ozone and an eye-burning chemical peroxyacetyl nitrate (PAN).

11. Temperature inversion—The normal atmospheric circulation pattern is for air to be heated at the earth's surface and then rise as convection currents. Hot air, being less dense, rises; cold air, being more dense, sinks. However, if there is a layer of warmer air over a layer of colder air (see below) the layers will not mix and air pollution will fill the lower colder air layer. People with lung problems will suffer. This is a common condition in Denver, CO (due to the mountain range that creates an inversion), Phoenix in the winter, Los Angeles, Mexico City, and London. A cold front or otherwise strong winds can blow the pollution away, and the air will clear for a while.

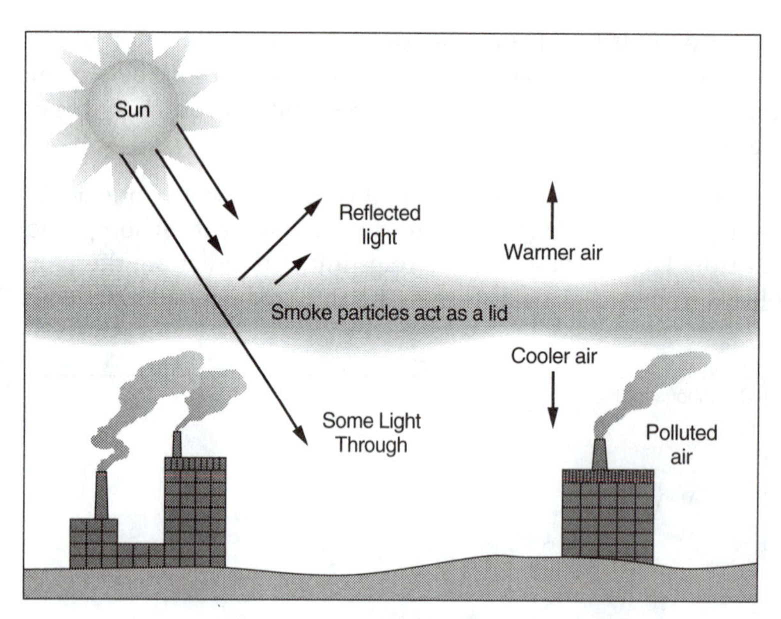

12. Urban heat island—one day in July of 1988, in Phoenix, AZ, the high temperature was 110°F and the low was 100°F. Dust devils swirled in the streets. One hundred years before, the night low would have been much lower. Why? Autos, homes, businesses, and industries (in cities) emit heat from combustion processes, and air conditioner or heat pump compressors also produce heat.

13. Dust domes—the heated air of the city rises in convection currents taking dust with it. Winds may then blow the dust to other areas.

14. The Clean Air Act and amendments of 1970, 1977, 1990, and 1997—U.S. law has focused on reducing primary and secondary air pollutants. The law eliminated leaded gas. Lead compounds were used to increase the "octane rating" (resistance to preignite or "knock") of poor quality gasoline, and as valve lubricants. The law has successfully reduced the amounts of hazardous air pollutants in U.S. air, but the problem continues in major urban areas. The law spec-

ifies lower limits on hazardous pollutants over time, leading to charges that the EPA sets rules for cleaner air, even as the air has improved somewhat over time. Based on these new standards, Atlanta may be reclassified as moving from a serious classification to severe, even as the air has slightly improved.

15. Global distillation effect—persistent air pollutants may be carried from developing countries nearer the equator to colder countries by the general atmospheric circulation pattern. Volatile hydrocarbons (such as the carcinogenic gasoline additive benzene, DDT, PCBs and solid particulates containing lead) are deposited in the Arctic Hadley zone or they may "leap frog" from the equatorial Hadley zone to the next one toward the pole.

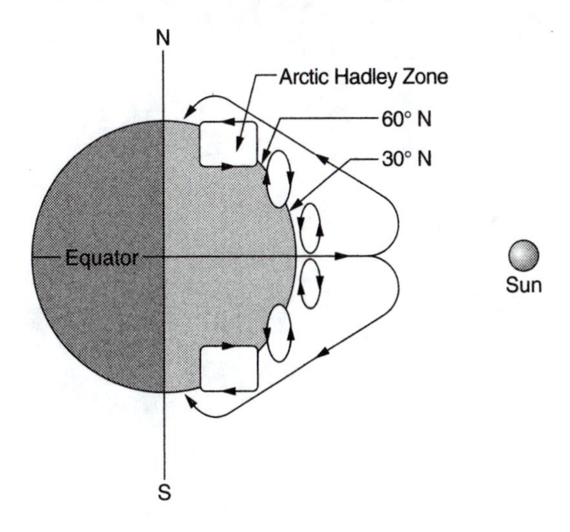

Wind circulation from the equator to the poles and in cells at 30 degrees of latitude from the equator.

16. Sick building syndrome—also called *indoor smog*, causes include the following: butyric acid from ceiling panels; formaldehyde from particleboard; mold spores and odor; benzene, xylene and toluene in ink; trichloroethane in photocopiers; dust; dust mites; and various allergies to latex and other materials. Illness resulting is likely to be multifactorial in cause.

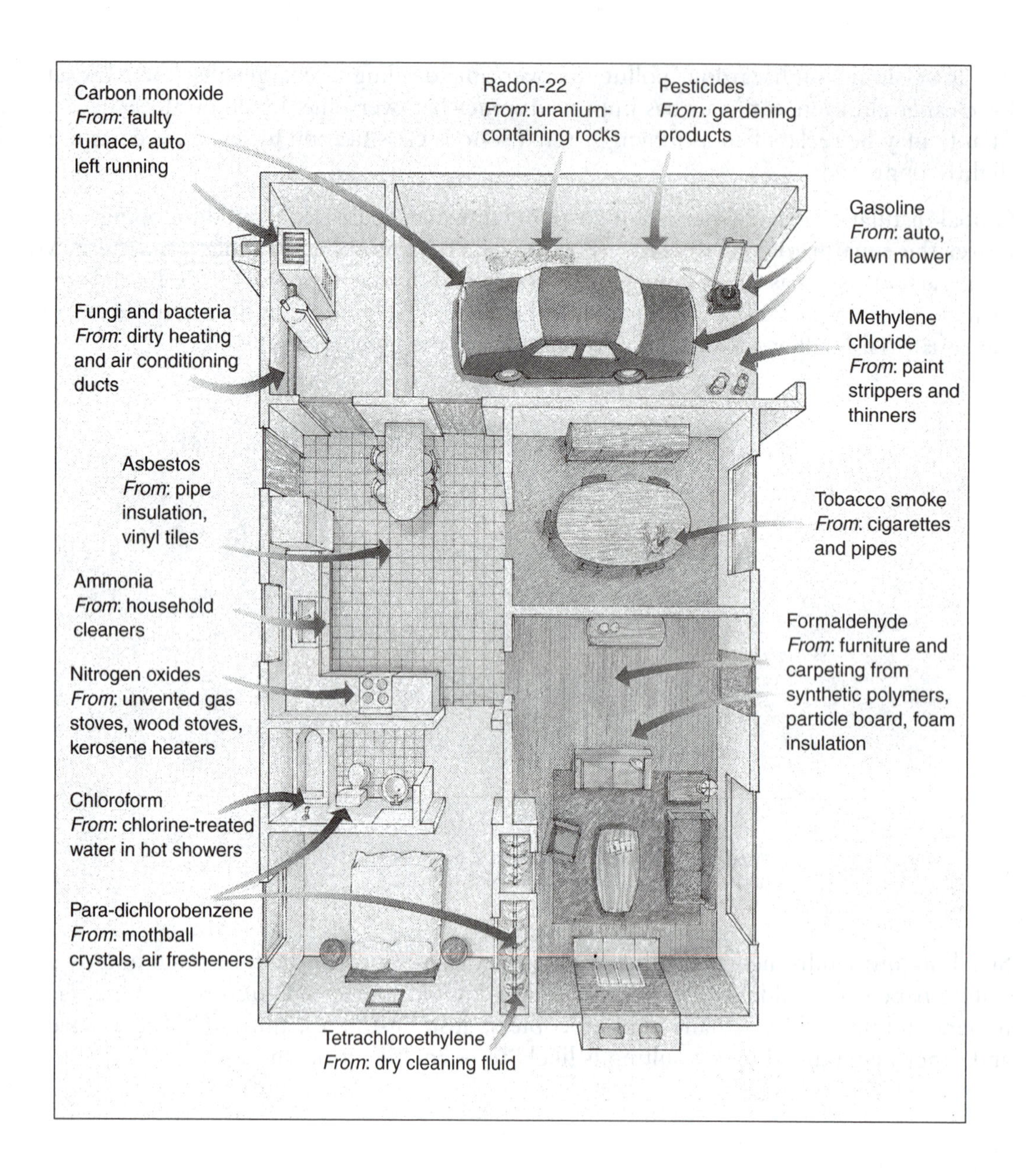

17. Radon—a serious pollutant of slab houses and basements of houses built on layers of uranium-containing sheared rocks (Pennsylvania, Maryland, New Jersey, and New York [the *Reading Prong* area]); or glacial till formed by ice age glaciers grinding bedrock in Iowa (71% of homes with radon), North Dakota, South Dakota, Minnesota, Wisconsin, Illinois, and Ohio. When Uranium-238 breaks down to form radium, the radium further decays to radon-222, a radioisotope that quickly decays to form radioisotopes of lead and polonium. The alpha radiation emitted is harmless to skin, but on the cells lining the lungs, they are mutagenic. In addition to the more serious problem of cigarette smoking, radon is an important cause of lung cancer in these geographical areas.

18. Cigarette smoking as air pollution—cigarette smoking worsens the effects of other air pollution, such as radon, sulfur dioxide, and ozone. Cigarettes, burning leaves and diesel smoke contain the potent aerosol carcinogen group, the hydrocarbon benzopyrenes, a tumor growth stimulator; nicotine, hydrogen cyanide, and carbon monoxide damage the cell lining of the lungs

and blood vessels throughout the body. Emphysema results when air sac walls are destroyed, and the larger sacs remaining stiffen from scar tissue (collagen) formation. Lung cancer results when the cells lining the bronchioles of lungs undergo a series of mutations.

19. Mesothelioma—asbestos, an inert fibrous mineral used to insulate steam pipes and to make floor and ceiling tiles at one time, causes cancer of the lining of the lungs the mesothelium. Apparently, the fibers (when inhaled) are carried out of the air sacs by immune cells; the immune cells die and leave the fibers behind because they cannot be broken down. The fibers then work themselves like splinters toward the outer covering of the lungs where they cause so much irritation and inflammation, they speed up cell division, and, as mistakes in DNA replication accumulate, cancer cells appear and spread.

20. Noise pollution—sound is caused by vibrations or waves of energy in the air that cause alternative compacting and thinning of air molecules. Sound has frequency or pitch, and amplitude or loudness. The frequency of vibrations sensed by the human ear are approximately 16 cycles per second to 20,0000 cps. Both low-frequency *infrasound* and high-frequency *ultrasound* can be damaging to the organs of hearing. The decibel scale measures loudness. The hair cell processes are like little fingers sticking up; they make contact with a tectorial membrane and deflect during the sensing of sound. These processes are rubbed off by very loud/high amplitude sound.

21. Electromagnetism—radio waves, microwaves, and electromagnetic waves (generated by high-tension electrical wires or electrical equipment or appliances) are in the same electromagnetic spectrum with gamma, X, and U-V radiation. See below. Cell biologists have demonstrated changes in cells when exposed to this radiation. A few studies have shown a weak correlation, others have shown no correlation with disease.

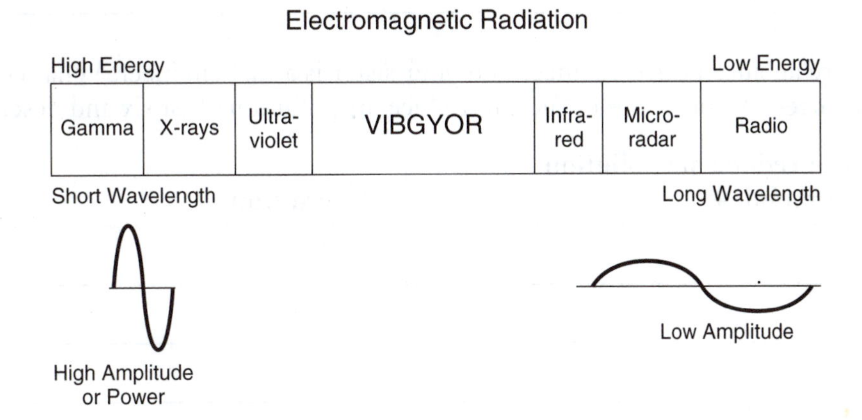

Questions to Ponder

1. Chattanooga, TN, lies in a valley created by the Tennessee and Cumberland rivers, surrounded by mountains that were the site of fierce battles in the Civil War. The U.S. government regarded it the worst city for air pollution before the Clean Air Act. List six actions that the city took to improve their air.

	Air pollution control action
1.	
2.	
3.	
4.	
5.	
6.	

2. The state of California continues to have the most stringent air pollution standards in the U.S. List four ways California has reduced air pollution. Go to **www.arb.ca.gov/html/brochure/9arb.pdf.** Also see video of their successes and challenges: **www.arb.ca.gov/videos/clskies.htm**

	Action taken to reduce air pollution by California
1.	
2.	
3.	
4.	

3. Imagine you are an automotive engineer and your sister is a fuel engineer. What changes can you make to current automobiles or fuels to reduce air pollution? List six and describe them.

	Method to reduce air pollution from automobiles	Description
1.		
2.		
3.		
4.		
5.		
6.		

4. Is it necessarily true that hydrogen fuel combustion is cleaner than other fuels? Debate this pro and con.

5. Why are children highly susceptible to adverse health effects from air pollution? See the website developed by Michael Kleinman of U. Cal. Irvine: **www.aqmd.gov/forstudents/Kleinman_article.htm** (an excellent primer on the health effects of air pollution).

6. How do activities at home, directly or indirectly, affect air pollution? See **209.208.153.222/air/air_indoor_redux1.html**; the Mayo Clinic: **www.mayoclinic.com/invoke.cfm?id=FL00012**, and Fig. 19-16.

7. Summarize the successes and failures of the Clean Air Act and amendments in reducing the amounts of the ambient and emission pollutants listed below. See **www.epa.gov/airtrends/sixpoll.html** and Fig. 19-11.

Pollutant	**Indicate *Success!* or *Failure!***
Carbon dioxide. See **www.eia.doe.gov/oiaf/1605/gg97/rpt/chap2.html**	
Nitrogen oxides	
Sulfur oxides	
Ozone	
Carbon monoxide	
Smog	
Particulates	
Lead	

8. Compare and contrast the effects of thinning stratospheric ozone with increasing ozone at street level.

9. Compare and contrast the greenhouse effect and the temperature inversion effect of particulates.

10. List four sources or types of noise pollution and describe the effect on health of each. See **www.nonoise.org/library/levels/levels.htm** and **www.nonoise.org/library/fctsheet/wildlife.htm**

	Source/type	**Health effect**
1.		
2.		
3.		
4.		

POP QUIZ

Matching

Match the following terms and definitions

____1. One measure to reduce air pollution from automobiles has been laws requiring the reduction of ______ in fossil fuels.	a. global distillation
____2. An ecosystem service of the atmosphere is:	b. benzene
____3. Hydrogen fuel cars will not make this acid rain and global warming gas when their engines are running:	c. provisions of Clean Air Act
____4. A high-compression, high-temperature automobile engine will make lots of this acid rain gas that also contributes to smog formation:	d. secondary air pollutants
____5. The photochemical reaction that makes smog also produces ______	e. ozone
____6. If your car is not tuned up and the exhaust smokes, this hazardous pollutant gas ______ is present and the car may fail the emission test:	f. temperature inversion
____7. Ozone and sulfur trioxide are:	g. U-V rays in sunlight
____8. This carcinogenic liquid was used in place of gasoline by the Germans in WW II and today is used as an octane enhancer in U.S. gasolines:	h. nitrogen oxides
____9. Ozone presence is wonderful here:	i. U-V rays blocked
____10. A layer of warmer air lies over a layer of cooler air describes a ______.	j. mesothelioma
____11. This explains how Atlanta could be reclassified as moving from the serious to severe air pollution classification even as some aspects of their air quality improve:	k. benzopyrenes
____12. A potent carcinogen in diesel smoke and cigarette smoke is (are) ______.	l. sulfur
____13. Aerosol pollutants and particulates are carried from developing countries near the equator to Arctic and Antarctic regions. This is called ______.	m. carbon monoxide

____14. The energy to make smog that is highest in concentration at noon is ______.	n. stratosphere
____15. Other than cigarette smoking, this is a major cause of lung cancer:	o. hair cell processes
____16. Asbestos fibers in the outer covering of the lungs cause ______?	p. carbon dioxide
____17. Extremely loud noise causes ______ to permanently disappear:	q. radon-222

Multiple Choice

1. Which statement is incorrect about Chattanooga, TN, efforts to clean their air?
   a. Electric buses decreased street level pollution.
   b. Electrical generating plants increased air pollution to charge the buses.
   c. Sulfur concentration in fuels was reduced.
   d. Outdoor burning with permits was banned.
   e. Trick question alert! All of these are correct.

2. Which statement is correct about ozone?
   a. Ozone is O_4.
   b. U-V rays are transformed into infrared/heat by stratospheric ozone.
   c. Ozone is harmful 10 miles up.
   d. U-V radiation is blocked in the troposphere where ozone is formed.
   e. Ozone is harmless to lungs.

3. The smoke that named the Smoky Mountains comes mostly from:
   a. Atlanta automobiles
   b. coal-burning power plants in Georgia and Alabama
   c. hydrocarbons emitted by trees
   d. industrial smog
   e. steam from nuclear power plant cooling towers

4. A 0.3% mixture of carbon monoxide and air will kill if breathed long enough. The reason why is:
   a. Carbon monoxide paralyzes nerve cells.
   b. Carbon monoxide binds to hemoglobin better that oxygen.
   c. Carbon monoxide arrests breathing in the medulla of the brain.
   d. Carbon monoxide reacts with oxygen to make carbon dioxide in the blood.
   e. All of the above are correct.

5. A secondary air pollutant in smog is which of the following?
   a. nitrogen dioxide
   b. carbon dioxide
   c. peroxyacetyl nitrate (PAN)
   d. ozone
   e. c and d

6. Radon is formed when rocks containing ______ are sheared or ground up.
   a. U-238
   b. radium
   c. radon-222
   d. tritium
   e. strontium-90

7. The most likely explanation for the *sick building syndrome* is a synergism between gaseous or volatile air pollutants and ______.
   a. formaldehyde in particle board
   b. color printer paper
   c. ink solvents
   d. allergens like dust and mold
   e. all of the above

8. Diesel engines are by design, high-compression engines that ignite their oily fuel by the heat of compression instead of spark plugs. Therefore, without catalytic converters they are particularly high polluters of______.
   a. nitrogen oxides
   b. particulate pollution
   c. benzopyrenes in black exhaust
   d. a only
   e. a, b, and c

9. Catalytic converters reduce the amounts of which gas in automobile and truck exhaust?
   a. carbon monoxide
   b. nitrogen oxides
   c. ozone
   d. carbon dioxide
   e. a and b

10. Relative to their number and time of use, which device produces the most air pollution in terms of carbon monoxide and particulates?
    a. automobile
    b. truck
    c. lawn mower
    d. gas furnace
    e. modern diesel car

11. The Clean Air Act of 1970 was most successful in reducing the amount of ______ in the air.
    a. ozone
    b. lead
    c. carbon monoxide
    d. nitrogen oxides
    e. sulfur dioxide

12. The presence of this material in gasoline exhaust not only causes exhaust systems and catalytic converters to fail but also is a serious hazard for lung health.
    a. nitrogen oxides
    b. ozone

c. carbon monoxide
d. sulfur dioxide
e. particulates of unburnt gasoline and oil

13. This type of automobile greatly increased air pollution because it was not required to have high-efficiency gasoline engines or all of the pollution controls of most cars:
a. V-8s
b. V-12s
c. SUVs
d. convertibles

14. The worst megacity in the world for air pollution is which of the following?
a. Bangkok
b. Los Angeles
c. London
d. Mexico City
e. Moscow

15. Which city historically had the highest number of deaths during a single smog?
a. Bangkok
b. Los Angeles
c. London
d. Mexico City
e. Moscow

16. Which city in the U.S. has the most air pollution?
a. Denver
b. Detroit
c. Chicago
d. Los Angeles
e. New York

17. The "cleanest" automobile fuel is ______.
a. natural gas
b. gasohol
c. MTBE (an oxygenating additive) and gasoline
d. hydrogen
e. diesel fuel

18. The cleanest hydrocarbon fuel is ______.
a. natural gas
b. benzene
c. MTBE and gasoline
d. ethyl alcohol and gasoline
e. diesel fuel

19. Which statement is not true concerning electric or hybrid electric cars?
a. The batteries charge when the gasoline engine runs.
b. The gasoline engine and the electric engine run alternately.

c. The electric car produces no pollution directly or indirectly.
d. Electric cars produce more emissions than combustion engine cars if coal-burning power plants supply the electricity.
e. Trick question alert! All statements above are correct.

20. In developing countries, which fuel is used that produces a large amount of air pollution?
    a. animal dung
    b. wood
    c. coal
    d. leaded gasoline
    e. All of the above are used

21. Which compound is not transported by the atmosphere from tropical areas to the Arctic?
    a. DDT
    b. PCBs
    c. Carbon monoxide
    d. Dioxins
    e. Furans

22. Which state has the greatest problem with radon emissions from the ground?
    a. Maryland
    b. Minnesota
    c. Iowa
    d. Pennsylvania
    e. New York

23. Which of the following materials was not historically produced with asbestos?
    a. pipe insulation
    b. furnace or boiler insulation
    c. brake pad and shoe linings
    d. fiberglass insulation
    e. high-tension electrical insulation

24. Which of the following devices produce EMFs?
    a. cell phones
    b. electrical transformers
    c. high-tension electrical wires
    d. microwave ovens
    e. all of the above

25. Refer to the figure on the following page. Which of the following ranking (in descending order of air pollutants) is correct?
    a. carbon monoxide, nitrogen oxides, sulfur oxides
    b. carbon dioxide, nitrogen oxides, particulates
    c. ammonia, lead
    d. carbon dioxide, particulates, nitrogen oxides
    e. nitrogen oxides, sulfur oxides, volatile organics

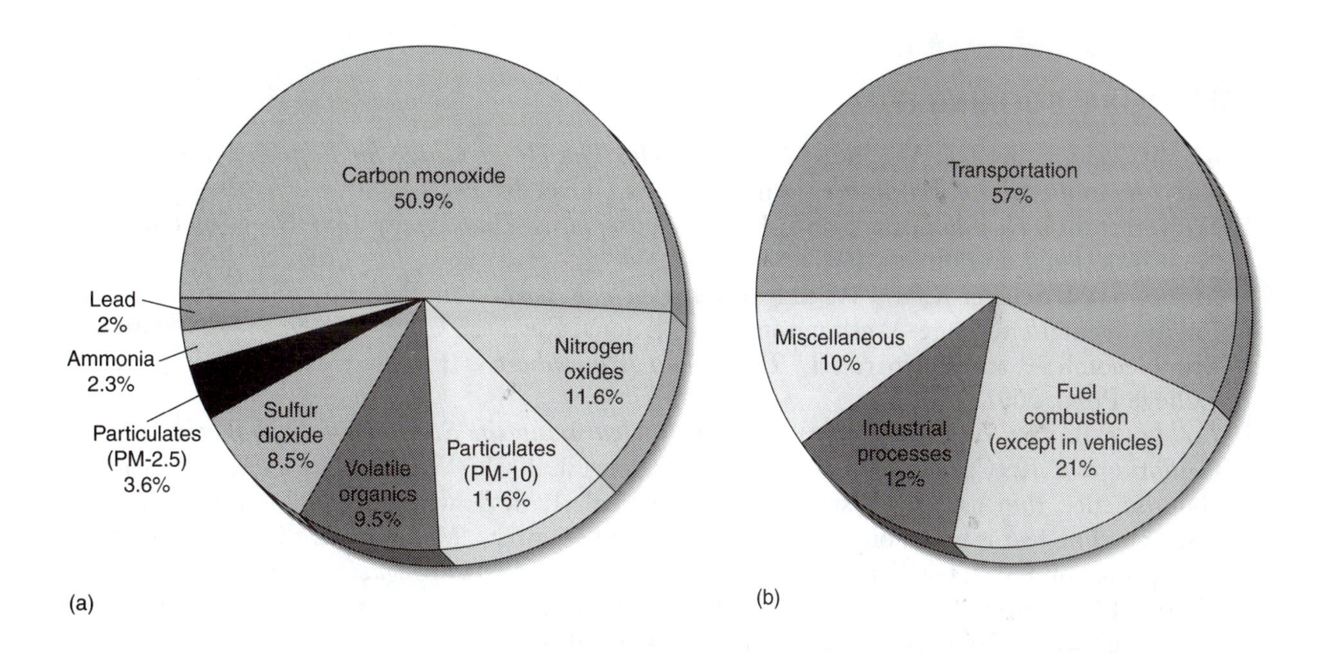

26. Which of the sources below contributes the most air pollution?
    a. industry
    b. transportation
    c. power generation
    d. lawn and landscape care

CYBER SURFIN'

1. From the US EPA, "The Plain English Guide to the Clean Air Act," see **www.epa.gov/oar/oaqps/peg_caa/pegcaa03.html** and The Health Effects of Air Pollution, **www.epa.gov/air/concerns/**
2. To get information on air pollution and health from the American Lung Association go to **209.208.153.222/air.** Click on indoor and outdoor air quality.
3. To learn about radon, its sources and dangers, go to the USGS site: **energy.cr.usgs.gov/radon/radonhome.html**
4. To see an informative slide show on temperature inversion, from the California State Department of Pesticide Regulation: **www.cdpr.ca.gov/docs/drftinit/confs/2001/ramsey.ppt**
5. For the Bush administration's predictions on how health effects of air pollution will decrease under their Clear Skies initiative, see **www.epa.gov/air/clearskies/benefits.html**
6. For all you want to learn about noise, go to The Noise Clearinghouse at **www.nonoise.org/library.htm**
7. For a primer on what is known and what is not known about the health effects of electromagnetic fields radiation exposure, from the World Health Organization: **www.who.int/inffs/en/fact205.html, www.who.int/inf-fs/en/fact182.html** and **www.niehs.nih.gov/emfrapid/** from the U.S. National Institute of Environmental Health Sciences.

'LIKE' BOOKING IT

1. May JC, and Samet JM. ***My House Is Killing Me! The Home Guide for Families With Allergies and Asthma.*** Baltimore: Johns Hopkins University Press, 2001.
2. Wolverton BC. ***How to Grow Fresh Air: 50 Houseplants That Purify Your Home or Office.*** New York: Penguin Putnam, 1997.
3. Malaspina M, Schafer K, and Wiles R. ***What Works Report – Air Pollution Solutions.*** The Environmental Exchange, Washington, DC: National Research Council, 1992.
4. Brimblecombe P, and Maynard RL. ***The Urban Atmosphere & Its Effects.*** London: Imperial College Press, 2001.
5. Lomborg, B. ***The Skeptical Environmentalist: Measuring the State of the Real World,*** Cambridge, U.K.: Cambridge University Press, reprint, 2001. (When scientists want to learn about science, they usually consult scientists, particularly those with outstanding records in research. The text is based on the work of respected scientists. Mr. Lomborg is a political science professor and statistician. Please go to the websites of *Grist Magazine, Scientific American,* Union of Concerned Scientists, and the World Resources Institute sites for scientific opinions of Lomborg's work at **www.wri.org/wri/press/mk_lomborg.html.**)
6. Brown LR. ***Eco-Economy: Building an Economy for the Earth.*** New York: W.W. Norton, 2001. (This is in opposition to the reference above.)
7. Walker J, and Flindell I. ***Noise Pollution (Modules in Environmental Science).*** New York: John Wiley & Sons, 2003.

ANSWERS

Seeing the Forest

1. List 10 outdoor or indoor hazardous air pollutants, their source, and the environmental or health effects of each. Consult Tables 19-1 and 19-2, and the chemistry appendix.

	Air pollutant	Source	Environmental or health effect
1.	Nitrous oxide (N_2O)	Fossil fuel combustion, agricultural soils with manure as fertilizer, inorganic fertilizer breakdown, nitrogen fixing crops, workers exposed to anesthesia	Greenhouse gas, birth defects in animal models according to NIOSH
2.	Nitric oxide (NO), nitrogen dioxide	Fossil fuel combustion, high-compression diesel and gasoline engines, fertilizers	Combines with water and oxygen to make nitric acid in acid rain, a reactant in the process of making smog including PAN. Damaging to structures and lungs
3.	Sulfur dioxide and trioxide	Fossil fuel combustion, mostly from coal, fuel oil, and gasoline	Combines with water and oxygen in air to made sulfuric acid in acid rain. Damaging to structures and lungs

4.	Carbon monoxide	Mostly automobiles, industries	Suffocation, deprivation of oxygen to brain, damage to blood vessels (perhaps make the initial lesions that become fatty plaques)
5.	Ozone	Secondary pollutant that is a by-product of smog formation—gasoline and diesel fuel combustion, and industries	Damaging to lungs, worsens asthma
6.	Particulates solid (dust, lead, carbon soot, coal dust, cotton dust)	PM-10 dust from farmlands, deserts, and volcanic eruptions. Lead from combustion of leaded gasoline in developing countries. Carbon soot from any incomplete combustion that produces smoke—diesel smoke, barbecue smoke. Coal dust and silica dust from mining, and cotton dust from cotton processing in developing countries and historically in the U.S.	Dust particles serve as condensation nuclei for raindrops. Irritates lungs. Coal dust and cigarettes produce black lung disease when the lung tissues fill up with carbon particles. Brown lung results when cotton particles are involved, silicosis when rock dust is involved.
7.	Aerosols of volatile compounds including hydrocarbons	Industries, gasoline vaporization, insecticide vaporization, PCBs, benzene	Lung irritants, many are carcinogenic
8.	Formaldehyde	Used in glues that hold particleboard together (most houses have this material under the siding and as a base for floors and roofs). Preservative used in biology and medicine	Carcinogenic, damages lungs
9.	Ethylene chloride	Paint and varnish stripper, industries	Carcinogenic, damages lungs
10.	Chlorine	Industries, train and barge wrecks	Reacts with water to make hydrochloric acid, very hazardous to breathe

Questions to Ponder

1. Chattanooga, TN, lies in a valley created by the Tennessee and Cumberland rivers, surrounded by mountains that were the site of fierce battles in the Civil War. The U.S. government regarded it the worst city for air pollution before the Clean Air Act. List six actions that the city took to improve their air.

	Air pollution control action
1.	Open burning by permit only
2.	Limits placed on industrial odors
3.	Particular emissions limits established
4.	Prohibited smoking car and truck exhaust
5.	Reduced sulfur content of fuels
6.	Business installed electrostatic precipitators for smoke stacks

2. The state of California continues to have the most stringent air pollution standards in the U.S. List four ways California has reduced air pollution. Go to **www.arb.ca.gov/html/brochure/9arb.pdf.** Also see video of their successes and challenges: **www.arb.ca.gov/videos/clskies.htm.**

	Action taken to reduce air pollution by California
1.	Power plants use natural gas instead of coal to reduce nitrogen and sulfur oxides. Outdoor burning restricted
2.	First to require crankcase emission control in autos
3.	Compression ratios reduced in gasoline engines. California was the first to limit particles and carbon monoxide tailpipe emissions.
4.	First state to have ambient air quality standards to limit total suspended particulates, photochemical oxidants, sulfur dioxide, nitrogen dioxide, and carbon monoxide

Historically, California has stricter air pollution controls than other states.

3. Imagine you are an automotive engineer and your sister is a fuel engineer. What changes can you make to current automobiles or fuels to reduce air pollution? List six and describe them.

	Method to reduce air pollution from automobiles	Description
1.	Hydrogen fuel	But only if the electricity to breakdown the water to make hydrogen is produced by solar and wind power, perhaps natural gas, definitely not coal burning
2.	Ban high-friction wide and low aspect tires.	This reduces emissions by making cars more fuel efficient. (Forgive me for not being cool.)
3.	Further reduce the weight of cars by replacing aluminum and iron with plastics.	Aluminum production requires large amounts of electricity generated by coal-burning power plants.

4.	Eliminate high-octane gasoline	Most of it is wasted in engines that don't require it. It does not produce any more power than the gasoline that the engine was designed for. Carcinogenic benzenes are added to high-octane gasoline.
5.	Eliminate high-octane-requiring engines and reduce the horsepower of engines	Ditto. SUVs and high-performance cars have brought about an engine size and horsepower war by manufactures.
6.	Use diesel-electric and gasoline-electric hybrid engine	Also use gasohol instead of gasoline/MTBE mixtures. Alcohol is naturally oxygenated, and the supply is renewable. MTBE is carcinogenic and adds to the hydrocarbon load because of its high volatility (vapor pressure).

4. Is it necessarily true that hydrogen fuel combustion is cleaner than other fuels? Debate this pro and con. *Hydrogen fuel produces heat and water as the main by-products of combustion. There will be some incidental formation of nitrogen oxides proportional to combustion temperate and pressures created by compression of the hydrogen/oxygen mixture. If the electricity to breakdown the water to make hydrogen is produced by solar and wind power, that is best. Natural gas power generation comes in second place, but it makes carbon dioxide and incidental nitrogen oxides from the combining of atmospheric oxygen and nitrogen at high temperature. Definitely not coal burning—the clean coal technology advertised on T.V. is not available or economic to use!*
5. Why are children highly susceptible to adverse health effects from air pollution? See the website developed by Michael Kleinman of U. Cal. Irvine, **www.aqmd.gov/forstudents/Kleinman_article.htm** (an excellent primer on the health effects of air pollution). *Children breathe 20% to 50% more air per unit body weight than an adult in a given level of exercise. They also may breathe more polluted air before they feel discomfort.*
6. How do activities at home, directly or indirectly, affect air pollution? See **209.208.153.222/air/air_indoor_redux1.html.**
7. Summarize the successes and failures of the Clean Air Act and amendments in reducing the amounts of the ambient and emission pollutants listed below. See **www.epa.gov/airtrends/sixpoll.html**

Pollutant	**Indicate *Success!* or *Failure!***
Carbon dioxide. See **www.eia.doe.gov/oiaf/1605/gg97rpt/chap2.html**	Failure in controlling emissions and ambient air levels!
Nitrogen oxides	Failure in emissions, success in ambient air?
Sulfur oxides	Success!
Ozone	Success!
Carbon monoxide	Success!

Smog	Success, generally equivalent to ozone trends!
Particulates	Apparent success in reducing PM-10—there is no long-term data more than a decade old. No data for PM 2.5.
Lead	Success!

8. Compare and contrast the effects of thinning stratospheric ozone with increasing ozone at street level. *The thinning of stratospheric ozone increases U-V B and A exposure at the surface. Exposure to both forms of radiation are linked to increasing occurrence of skin cancer. Increased tropospheric or street level ozone increases lung disease.*
9. Compare and contrast the greenhouse effect and the temperature inversion effect of particulates. *The two processes are opposite in effect on atmospheric temperature change. When heat is trapped by heavy gases (carbon dioxide, nitrous oxide, and nitrogen dioxide, sulfur dioxide, methane) in the lower levels of the troposphere, the effect is global warming. Check out the planet Venus for extreme global warming. When Krakatoa erupted, an extremely cold year in the northern hemisphere occurred because of the screening of sunlight by dust. Similarly, dust from farms or particulates in smoke screen the sun. When the sun is reflected from the top of the layer of particles, the air is warmed immediately above it, creating a layer of warmer air over a layer of cooler air—a temperature inversion.*
10. List four sources or types of noise pollution and describe the effect on health of each. See **www.nonoise.org/library/levels/levels.htm** and **www.nonoise.org/library/fctsheet/wildlife.htm.**

	Source/type	Health effects
1.	Aircraft—jet and propeller	Disturbance to colonial nesting species—abandonment of nesting sites. Bighorn sheep stampede
2.	Trains	Reductions in hemoglobin—anemia, circulating = blood sugar. Changes to hypothalamus
3.	Heavy traffic	Disturbance of sense of balance and shifting of fields of vision. Interference with communication. Psychological effects—loss of temper
4.	Chain saws, grinders, rock crushers, jack hammers, air conditioners, etc.	Hearing loss may be frequency or amplitude specific

Pop Quiz

Matching

1. l
2. i
3. p
4. h
5. e
6. m

7. d
8. b
9. n
10. f
11. c
12. k
13. a
14. g
15. q
16. j
17. o

Multiple Choice

1. e
2. b
3. c
4. b
5. e
6. a
7. d
8. e
9. e
10. c
11. b
12. d
13. c
14. d
15. c
16. d
17. d
18. a
19. c
20. e
21. c
22. c
23. d
24. e
25. d
26. b

20

Regional and Global Atmospheric Changes

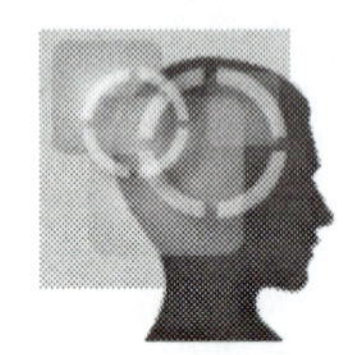

LEARNING OBJECTIVES

After you have studied this chapter you should be able to:

1. Describe the greenhouse effect and list the five main greenhouse gases.
2. Discuss the ramifications of some of the potential effects of global warming, including rising sea level, changes in precipitation patterns, effects on organisms, effects on human health, and effects on agriculture.
3. Give examples of several ways to mitigate and adapt to global warming.
4. Describe the importance of the stratospheric ozone layer and distinguish between tropospheric and stratospheric ozone.
5. Explain how ozone depletion takes place and relate some of the harmful effects of ozone depletion.
6. Relate how the international community is working to protect the ozone layer.
7. Explain how acid deposition develops and relate some of the effects of acid deposition.
8. Describe how North American lakes are affected by interactions among global warming, ozone depletion, and acid deposition.

SEEING THE FOREST

1. Global warming, ozone depletion, and acid deposition may change the environment of the future. Assuming the worst case scenarios for the three, what changes in the health, numbers, types and habitats of organisms are in store?

	Health or environmental effects
1.	
2.	
3.	
4.	
5.	
6.	
7.	
8.	
9.	
10.	
11.	
12.	
13.	
14.	
15.	

Table 20.1 Health Effects of Several Major Air Pollutants

Pollutant	*Source*	*Effects*
Particulate matter	Industries, motor vehicles	Aggravates respiratory illnesses; long-term exposure may cause increased incidence of chronic conditions such as bronchitis
Sulfur oxides	Electric power plants and other industries	Irritate respiratory tract; same effects as particulates
Nitrogen oxides	Motor vehicles, industries, heavily fertilized farmland	Irritate respiratory tract; aggravate respiratory conditions such as asthma and chronic bronchitis
Carbon monoxide	Motor vehicles, industries	Reduces blood's ability to transport oxygen; headache and fatigue at lower levels; mental impairment or death at high levels
Ozone	Formed in atmosphere (secondary air pollutant)	Irritates eyes; irritates respiratory tract; produces chest discomfort; aggravates respiratory conditions such as asthma and chronic bronchitis

Sign posts along the way to global climate change

1. Reductions and extinctions in populations of ecological specialists and increased populations of generalists including the following: pollinators (e.g., bats, bees, beetles, birds, and butterflies) and the plants supported by them and animal and plant populations isolated on mountain ranges or those in polar regions.
2. Algal blooms along ocean coasts. Coral reef reduction
3. Emerging infectious diseases and expansion of the ranges of insect vectors.
4. Increasing numbers of skin cancer and melanoma.
5. Increasing severity of storms, mid-latitude droughts, and semiarid grassland and tropical rainforest desertification.

SEEING THE TREES

VOCAB

Only when the last tree has died and the last river been poisoned and the last fish been caught will we realize we cannot eat money.

Cree Indian Proverb

1. Agroforestry—a combination of agricultural and forestry techniques to improve degraded soils. Often crops are planted along tree seedlings. Electrical power-generating companies attempt to offset the carbon dioxide they produce by planting trees.
2. Greenhouse effect—like glass, the troposphere is transparent to visible sunlight, but when that sunlight strikes the Earth, the photons of violet-indigo-blue-yellow-orange-red are transformed into infrared photons or heat. Heavy tropospheric gases such as nitrogen oxides, carbon dioxide, methane, and water reflect these infrared photons back toward the surface, creating the greenhouse effect. The planet Venus has a *runaway greenhouse effect* that has produced temperatures high enough to evaporate all the planet's liquid water to form a thick cloud layer of water vapor and sulfuric acid 100 miles above the surface. The surface temperature of Venus is approximately 460°C, and the atmospheric pressure is 100 times more than Earth. Most climatologists predict that the Earth will experience an enhanced greenhouse effect after 2050.
3. Aerosols—with particulates, aerosols in the troposphere, can reflect sunlight back to higher altitudes, increasing the temperature inversion effect of surface cooling, and increasing the luminosity of the Earth. Aerosol cooling effects are typically seen over and in the downwind plume of industrialized areas. Volcanoes produce dust particles and sulfur aerosol hazes that have cooling effects—the explosion of Karatoa and recently Mount Pinitubo.
4. Climate models—the two great unknowns in greenhouse climate modeling is whether increased heat resulting in increased evaporation will cause low clouds to form that decrease surface temperatures (negative feedback) or high clouds that serve to reflect heat back to the surface (e.g., Venus; positive feedback). Some models suggest a weakening of the *ocean conveyor belt* circula-

tion current studied in Chapter 6. If water flowing to the North Atlantic gets too warm, it will not sink. When this current sinks, it carries carbon/dissolved carbon dioxide and carbonic acid into deep water to form limestone at the ocean bottom. A positive feedback mechanism will cause even more heating.

5. International efforts to control greenhouse emissions include:

    a. UN Framework Convention on Climate Change—In 1992, 174 countries agreed to stabilize greenhouse emissions.
    b. UN Climate Change Convention—In 1994 nations agreed to set timetables for reducing greenhouse gases.
    c. Kyoto Protocol—In 1997, the U.S. agreed to operational rules to reduce emissions of carbon dioxide, methane, and nitrous oxide. In 2001, the Bush administration withdrew from this agreement, instead favoring voluntary compliance. Even with the Kyoto Protocol agreements in force in developed countries, developing countries will increase carbon dioxide emissions above present levels.
6. Carbon dioxide mitigation strategies—planting trees will soak up 10–15% of the extra greenhouse gas in the atmosphere. Some have suggested pumping carbon dioxide underground, fertilizing the ocean with iron II to increase photosynthetic algae and carbon management.

7. Carbon management—separating carbon from other emissions and disposing of it in geological formations.

8. Ultraviolet radiation—an electromagnetic radiation of a longer wavelength and less power than X-rays, but a shorter wavelength and greater power than visible light. Wavelength and power are inversely proportional. It occurs in three ranges of wavelength, from shorter to longer as UV-C, UV-B, and UV-A. UV-C is intercepted in the Earth's magnetic belts, UV-B is screened out by ozone in the stratosphere, UV-A warms and tans skins on the Earth's surface. UV-B is mutagenic, it changes the chemical structure of DNA and the genes made by DNA. Typically, the genes that control cell division are inactivated by the chemical changes in DNA caused by UV-B.

9. Chlorofluorocarbons (CFCs)—compounds of carbon, fluorine, and chlorine used as refrigerant gases in refrigerators, air conditioners, and heat pumps, as propellants in spray cans, as a foam plastic expander. Freon, also known as CFC-12 or R-12 was used in U.S. automobiles until 1993. The production of R-12 was banned in the U.S. in 1995. See **www.epa.nsw.gov.au/soe/97/ch1/2_4t1.htm**

10. Stratospheric ozone depletion—ozone is attacked by organic compounds that release chlorine or bromine (halide elements) there. These include CFCs halons, methyl bromide, chloroform, and carbon tetrachloride (a solvent). One chlorine or bromine atom can break down thousands of ozone molecules, O_3 to oxygen, O_2.

11. The ozone hole—near the South Pole and over Antarctica, the ozone layer is the thinnest, creating the *ozone hole*. There are areas of thinning over the industrial northeast in the U.S. and Canada, and similarly from England over northern Europe in France and Germany. Thinning of the ozone layer over New Zealand has caused increased rates of melanoma and other skin cancers. The *ozone hole* occurs annually over Antarctica between September and November, when the Antarctic spring when sunlight returns. See **www.noaa.gov/ozone.html.** The circumpolar vortex of cold air forms ice crystals to which chlorine and bromine adhere. When the circumpolar vortex breaks up the chlorine, brome are dispersed northward to New Zealand and South Australia.

12. Melanoma—the most aggressive and fatal form of skin cancers is caused by a series of mutations to melanocytes that make the pigment melanin at the base of the upper cellular layer cells of the skin (epidermis). Once the cancer cells have penetrated into the dermis of the skin where blood vessels and lymphatic vessels are located, it quickly moves to lymph nodes, and to other tissues and organs. Squamous cell carcinoma and basal cell carcinoma are also caused by UV-B exposure, the latter are usually not fatal.
13. Montreal Protocol—a 1987 agreement to reduce the world's production of CFCs. Industries quickly developed HFCs (R-134a), hydrofluorocarbons, and HCFCs (hydrochlorofluorocarbons to use as refrigerants). The latter one still attacks ozone. Both are greenhouse gases. The existence of many older freon compressor systems has resulted in CFC smuggling from China, India, and Mexico where they are still manufactured.
14. Acid deposition—rain or dust fall brings *wet deposition* (acid rain or mists) and *dry deposition* when acids adhere to dust or other particulates.
15. Acid rain—carbonic, nitric, and sulfuric acids in rain can dissolve limestone on buildings and monuments. The basements of the Jefferson and Lincoln Memorials in Washington have stalagtites of dissolved limestone. Acid rain causes the *forest decline.* Calcium and magnesium are leached by acids from the leaves or needles of the plant and from the soils. Roots then absorb more aluminum that is toxic. The weakened plants are then attacked by insects or microorganisms.
16. pH scale—is an inverse logarithmic scale of H^+ ion concentration in moles per liter of solution. Mathematically, pH = 1/log H^+ ion concentration in moles/liter. Because the formula is 1 over the concentration of H^+ ion, as H^+ ion concentration increases, pH decreases and vice versa. The pH scale runs from 0 to 14. A pH of 0 to just below 7 means the solution is acidic and has more H^+ ions than OH^- ions. A pH above 7.00 means a basic solution where OH^- is more concentrated than H^+. *Beware: the scale, at first look, appears backward or if the pH number is high,* H^+ *concentration is low; if the pH number is low,* H^+ *concentration is high!*

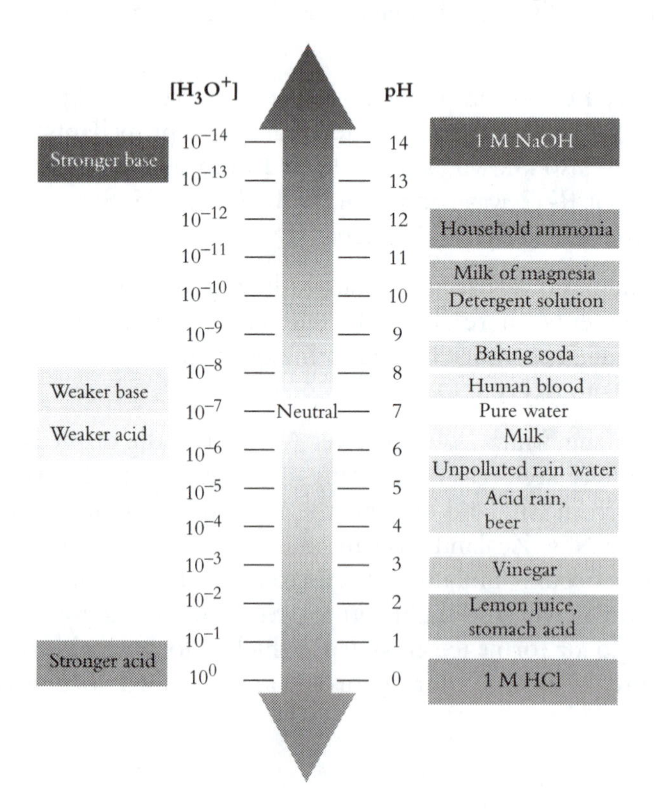

17. Mitigation of sulfur dioxide emissions—over most of the U.S., the only measures taken by power companies to reduce sulfur dioxide emissions were to use *low sulfur coal* and *electrostatic precipitators* of fly ash. The result is continuing large emissions of sulfur dioxide. Clean coal technology is possible but currently not available for economic reasons.

18. Nitrogen oxides mitigation—nitrogen oxides can be formed when nitrogen gas reacts with oxygen used in high temperature and pressure combustion reactions. Generally lower temperature and pressure combustion produces nitrous oxide, N_2O, higher temperatures and pressures produce nitric oxide, NO, and the very high temperatures and pressures of coal-fired power plants make nitrogen dioxide, NO_2. Nitrous oxide and nitric oxide can be converted into ammonia by catalytic converters, and the ammonia collected for use as a fertilizer or heat transfer gas in ocean thermal energy conversion. Nitrogen dioxide can be easily converted to nitrates for fertilizers. (Of course, many say we are using too much fertilizer now and much of that is leaching into our water. Remember the *dead zone* in the Gulf of Mexico? Delivering fertilizer in root-level irrigation would be better than broadcast methods.)

Questions to Ponder

1. List five greenhouse gases and their principal sources.

Greenhouse gas	Sources
1.	
2.	
3.	
4.	
5.	

2. Describe how nature homeostatically regulates carbon dioxide. How might humans intervene to assist nature in reducing carbon dioxide?

3. Explain the relationship between stratospheric ozone level, skin cancer, plant health, and coral reef health.

4. Describe the ozone hole, thinning ozone, and the process that causes ozone concentration at the poles. See **http://www.cpc.noaa.gov/products/stratosphere/sbuv2to/**, an ozone hole animation at **http://www.cmdl.noaa.gov/info/ozone_anim.html**, and the ozone index at **http://www.cpc.ncep.noaa.gov./products/stratosphere/uv_index/uv_current_map.html.**

5. List four actions that the international community have taken and four actions that can be used to prevent global warming, ozone depletion, and acid rain.

International actions taken	Prospective international actions
1.	
2.	
3.	
4.	

6. Refer to the figure below and explain the annual fluctuations in tropospheric carbon dioxide.

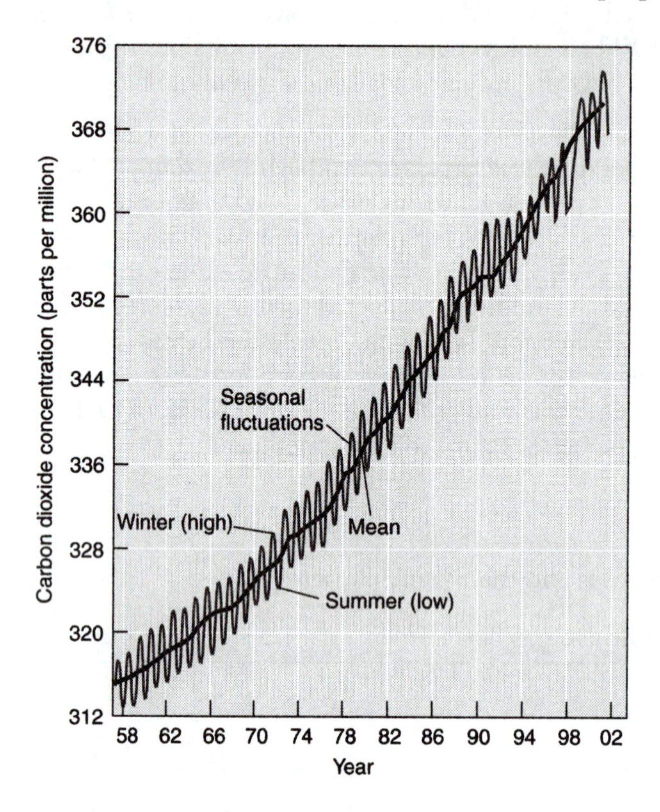

7. For the acid deposition chemicals listed below, describe the air pollutant gas, source, and process of formation of each. See **www.gpc.edu/~jaliff/wileychem.htm** and **royal.okanagan.bc.ca/mpidwirn/atmosphereandclimate/acidprecip.html**

Acid	Pollutant gas	Source	Process of formation
H_2SO_3—sulfurous			
H_2SO_4—sulfuric			
HNO_2—nitrous (a mutagen)			
HNO_3—nitric			
H_2CO_3—carbonic			

POP QUIZ

Matching

Match the following terms and definitions

____1. The effect of sulfur aerosols is tropospheric ______.	a. negative feedback
____2. When high compression engines react atmospheric gases incidentally, ______ is(are) produced.	b. current conveyor belt
____3. The most important global warming gas is ______.	c. UV-B
____4. In 1994, this refrigerant gas was no longer being using in the U.S. for automotive air conditioning:	d. coniferous forest
____5. A refrigerant gas containing bromine and chlorine is ______.	e. cooling
____6. Carbon dioxide traps ______ radiation.	f. temperature inversions
____7. The atoms in freon that destroy ozone are ______.	g. freon
____8. When ozone is cyclically formed and broken down naturally in the stratosphere, ______ radiation is degraded into heat.	h. nitrogen oxides
____9. Atmospheric carbon dioxide has been increasing rapidly since the beginning of the ______.	i. melanoma
____10. Smoke particulates create lids of warm air over cooler, stagnant air in cities called:	j. limestone and dolomite
____11. The most dangerous form of skin cancer is ______.	k. halon
____12. If water evaporates and forms high clouds instead of low clouds, the effect will be ______.	l. chlorine
____13. Most natural homeostatic processes would cool the Earth in reaction to an increase in carbon dioxide. This type of homeostatic regulation is called:	m. carbon dioxide
____14. Global warming may affect the Arctic National Wildlife Refuge by changing tundra into ______.	n. mosquitoes
____15. The advance of global warming toward the north pole will cause the spread of diseases carried by ______.	o. Industrial Revolution

____16. Global warming may weaken the ______ that sequesters carbon in the ocean.	p. infrared
____17. This geologic material mitigates the effects of acid rain in soils:	q. runaway global warming

Multiple Choice

1. The combustion of pure coal fuel always produces this product.
   a. sulfur dioxide
   b. nitrogen oxides
   c. carbon dioxide
   d. carbon monoxide
   e. water

2. The production of which substance is common to the combustion of pure methane and pure hydrogen?
   a. sulfur dioxide
   b. nitrogen oxides
   c. carbon dioxide
   d. carbon monoxide
   e. water

3. Which of the following materials can create a temperature inversion?
   a. volcanic dust
   b. farm dust
   c. smoke aerosols and soot particles
   d. sulfur aerosols
   e. all of the above

4. Which of the following environmental effects can be predicted for global warming?
   a. melting of permafrost
   b. melting of glaciers
   c. rising of sea levels
   d. flooding of coastal cities
   e. All of the above are expected.

5. Which of the following effects is not predicted for global warming?
   a. more tornadoes
   b. increased populations of penguins and whales
   c. movement of tropical species northward
   d. movement of the coniferous forest southward
   e. extinction of the deciduous Atlantic coast forest of Brazil

6. Which of the following nations still produces CFCs?
   a. Canada
   b. U.S.

c. U.K.
d. China
e. France

7. Although water screens out U-V radiation, in shallow waters enough leaks through to directly reduce populations of ______.
   a. coral zooanthellae
   b. krill
   c. whales
   d. c only
   e. a and b

8. If global warming is enhanced, this effect will occur in the middle latitudes:
   a. increased droughts
   b. advancement of coniferous forest southward
   c. increased hurricanes
   d. melting of permafrost
   e. all of the above

9. Chlorine and bromine atoms are held in ______ in the ozone hole and are dispersed northward in the Arctic summer.
   a. CFCs
   b. volcanic dust
   c. ice crystals
   d. halons
   e. smoke particles

10. Flooding caused by global warming will have the greatest impact on world food resources in which area?
    a. Mississippi River delta
    b. Nile River delta
    c. Chesapeake Bay estuary
    d. Everglades
    e. Ogallala Aquifer

11. U-V radiation and global warming effects that would reduce Antarctic krill populations would adversely affect the populations of which animal?
    a. blue whale
    b. Adele penguin
    c. polar bear
    d. musk ox
    e. a and b

12. When the ocean conveyor belt cools and sinks in the North Atlantic, this ecosystem service occurs:
    a. reduction of carbon dioxide
    b. feeding of North Atlantic krill
    c. El Niño results in more anchovies in Peru
    d. La Niña results in more anchovies in Peru
    e. all of the above

13. Which ecosystem *will not* be adversely affected by global warming?
    a. coral reefs
    b. tundra
    c. polar oceans
    d. coastal wetlands
    e. Trick question alert! All of these will be adversely affected.

14. Which disease will appear in mid-Canada when global warming is enhanced?
    a. viral encephalitis
    b. bacterial meningitis
    c. melanoma
    d. pneumonia
    e. all of the above

15. Global warming will increase water evaporation. Which statement is correct concerning the possible effects?
    a. Low clouds of water droplets will cause enhanced warming of the surface.
    b. High clouds of water droplets will result in enhanced warming.
    c. The result will always be a negative feedback mechanism.
    d. High clouds enhance negative feedback in temperature regulation.
    e. All of the above are correct.

16. The global warming-caused increases in sea level will produce the most deaths of inhabitants by starvation and water borne disease in which region?
    a. Bangladesh
    c. Egypt
    d. Florida
    b. California
    e. New York

17. Which of the following countries will not regulate carbon dioxide emissions by law?
    a. Australia
    b. Canada
    c. Russia
    d. U.S.
    e. all of the above

18. Which country signed the Kyoto protocol and then backed out?
    a. France
    b. U.K.
    c. U.S.
    d. Japan
    e. all of the above

19. This material is a significant limiting factor for algal growth in the oceans:
    a. carbon dioxide
    b. calcium
    c. iron

d. copper
e. oxygen

20. Hydrogen can be produced in this way that does not involve the combustion of carbon or hydrocarbons:
    a. solar electrolysis
    b. algae deprived of oxygen make hydrogen as a production of glucose metabolism
    c. hydrogen can be produced by combusting methane
    d. c only
    e. a and b

21. Which compound is *not* an acid deposition chemical?
    a. nitric acid
    b. sulfuric acid
    c. dust particulates with acids adhering
    d. CFCs
    e. Trick question alert! All of these are acid deposition chemicals.

22. Which form of radiation is less powerful than UV-B?
    a. X
    b. gamma
    c. UV-C
    d. UV-A
    e. Trick question alert! All of these are less powerful.

23. Of the greenhouse gases listed in the table below, which has increased the least in percentage and which the most from the beginning of the industrial age to 2003?
    a. CFC-12 (R-12) has increased the least and carbon dioxide the most.
    b. R-12 has increased the most and nitrous oxide the least.
    c. Methane has increased the most and R-12 the least.
    d. Carbon dioxide has increased the most and CFC-11 the least.
    e. Nitrous oxide has increased the least and carbon dioxide the most.

Table 20.2 **Increase in Selected Atmospheric Greenhouse Gases Preindustrial to Present**

Gas	*Estimated Preindustrial Concentration*	*Present Concentration*
Carbon dioxide	288 ppm*	370.9 ppm†
Methane	848 ppb**	1,783 ppb‡
Nitrous oxide	285 ppb	315 ppb‡
Chlorofluorocarbon-12	0 ppt***	541 ppt‡
Chlorofluorocarbon-11	0 ppt	262 ppt‡

* ppm = parts per million.
** ppb = parts per billion.
*** ppt = parts per trillion.
† 2001 annual average.
‡ 1999 values.

24. Since 1971, global temperature has increased by approximately _____ degrees?
    a. 0.1 degrees C
    b. 0.5 degrees C
    c. 1.0 degrees C
    d. 10 degrees F

25. The developed nation with the lowest per capita emissions of carbon dioxide is ______?
    a. Canada
    b. U.S.
    c. Japan
    d. Germany

26. The aquatic animal most susceptible to acid deposition is ______?
    a. salamander
    b. mayfly
    c. mussels
    d. trout
    e. wading birds

CYBER SURFIN'

1. From the US EPA, "Global Warming," see **yosemite.epa.gov/oar/globalwarming.nsf/content/index.html** US NOAA **www.ncdc.noaa.gov/ol/climate/globalwarming.html**
2. There is an animation of greenhouse effect courtesy of MSNBC at **www.msnbc.com/news/106332.asp**
3. This site is a great source of information and graphs on trends of global warming gases, including ice core data going back 400,000 years: **cdiac.esd.ornl.gov/trends/co2/contents.htm**
4. How will global warming affect animal plant and animal populations? See **yosemite.epa.gov/oar/globalwarming.nsf/content/Impacts.html** and click on the items in the left blue menu: for birds, see **www.co2science.org/journal/1999/v2n12c3.htm**; for butterflies: **www.co2science.org/journal/1999/v2n12c4.htm**
5. How will human health be affected? See Scientific American.com: **www.sciam.com/article.cfm?articleID=0008C7B2-E060-1C73-9B81809EC588EF21** and **www.pbs.org/journeytoplanetearth/johnshopkins/** from US PBS and Johns Hopkins University.
6. To see a basic description of stratospheric ozone from US NOAA, see **www.ogp.noaa.gov/library/rtnf92.htm**
7. From the the U.S. National Cancer Center, see a paper of melanoma risks at **seer.cancer.gov/publications/raterisk/risks163.html**
8. Take a visual tour of the ozone hole courtesy of University of Cambridge, U.K., **www.atm.ch.cam.ac.uk/tour/index.html**
9. Check out a paper by Rod Jenkins that includes human, plant, and animal risks to increasing ultraviolet radiation exposures at **www.digitaldeli.net/ozone/browserpp/healthrpt/healthrpt.html,** and from Lee Anne Thompson of the University of Florida on human effects at **//ess.geology.ufl.edu/ess/labs/termpapersfall9900/thompson/ozone%20depletion%20human%20%23ab6.html**

'LIKE' BOOKING IT

1. Willis H. ***Earth's Future Climate.*** Coral Springs, FL: Lumina Press, 2002.
2. Christianson GE. ***Greenhouse: The 200-Year Story of Global Warming.*** New York: Penguin (reissue), 2000.
3. Brown P. ***Global Warming: Can Civilization Survive?*** London: Blandford Press, 1997.
4. Read a book free online at National Academy Press, ***Ozone Depletion, Greenhouse Gases, and Climate Change*** (1989), Commission on Physical Sciences, Mathematics, and Applications (CPSMA): **www.nap.edu/books/0309039452/html/**

ANSWERS

Seeing the Forest

Global warming, ozone depletion, and acid deposition may change the environment of the future. Assuming the worst case scenarios for the three, what changes in the health, numbers, types and habitats of organisms are in store?

	Health or environmental effects
1.	Decrease in coral reefs
2.	Increase in heat-related deaths
3.	Mid-latitude droughts and desertification
4.	Stronger hurricanes with wave-related damage to reef and shoreline communities
5.	Further decrease in phytoplankton of the California Ocean current with concomitant losses of sea life.
6.	Submersion of some islands and coastal areas
7.	Thawing of permafrost, reduction of lichen food for reindeer (Santa can't get south!)
8.	Reduction in western U.S. snowfall and stream flow.
9.	A mistiming of the flowering and emergence patterns of plants and their pollinators, with a decline in populations of both.
10.	An increase in insect vector populations and the diseases they transmit.
11.	Decrease in buffalo grass in the prairies with an increase of invasive species.
12.	U-V damage to photosynthetic chemicals' concurrent reduction of the ability of plants to mitigate carbon dioxide increase.
13.	Increased skin cancers and leaf damage to plants.
14.	Further decreases seen in mountaintop trees and in those northeastern areas of North America and Europe due to acid rain.
15.	Further increases in the death of amphibians, insect larvae, and fishes in the lakes affected by acid rain.

Questions to Ponder

1. List five greenhouse gases and their principal sources.

	Greenhouse gas	Sources
1.	Carbon dioxide	Combustion of carbon compounds, volcanoes, and cellular respiration of organisms
2.	Nitrous oxide, nitric oxide, and nitrogen dioxide	Coincidental to combustion at high temperature and pressure, joining of atmospheric nitrogen and oxygen
3.	Methane	Produced by bacterial decomposition of organic materials in swamps, soils, feces, and plant material in the intestines of humans and herbivores
4.	CFCs	The refrigerant gases R-11 and R-12 released by leakage
5.	Sulfur dioxide	Sulfur contamination in fossil fuels—coal, gasoline, and heating oil. Also volcanoes.

2. Describe how nature homeostatically regulates carbon dioxide. How might humans intervene to assist nature in reducing carbon dioxide? *Nature can dispose of carbon dioxide in 2 ways: 1) photosynthesis incorporates carbon into the bodies of plants and that carbon is passed to animals. When the plants are buried, the carbon may be stored geologically as coal or oil. (The question for the contrarians to answer is this: "Where does the carbon dioxide go if we are destroying plants? Will plants be increasingly replaced by algae?) When carbon dioxide dissolves in the ocean, it forms carbon acid that then reacts with calcium ions to precipitate limestone. Also certain protozoa and most mollusks make shells that contain carbonates. The most important action humans can take in reducing carbon dioxide is to reduce or stop burning fossil fuels and allow plant populations to increase.*

3. Explain the relationship between stratospheric ozone level, skin cancer, plant health, and coral reef health. *Ultraviolet radiation, UV-B, and to a lesser extent, UV-A, cause direct damage to cells by overheating, destruction of chlorophyll and related photosynthetic chemicals, and mutations to DNA.*

4. Describe the ozone hole, thinning ozone, and the process that causes ozone concentration at the poles. See **www.cpc.noaa.gov/products/stratosphere/sbuv2to/**, an ozone hole animation at **www.cmdl.noaa.gov/info/ozone_anim.html**, and the ozone index at **www.cpc.ncep.noaa.gov./products/stratosphere/uv_index/uv_current_map.html**. *The ozone thinning is caused by organic compounds that release chlorine or bromine (halide elements) in the stratosphere:* $Cl + O_3 \rightarrow ClO + O_2$; $ClO + O \rightarrow Cl + O_2$; the net result is: $O_3 + O \rightarrow 2O_2$. The ozone hole represents an extreme thinning of zone annually over Antarctica between September and November, the Antarctic spring when sunlight returns. The circumpolar vortex of cold air forms ice crystals to which chlorine and bromine adhere. When the circumpolar vortex breaks up the chlorine and brome are dispersed northward to New Zealand and South Australia, it causes thinning there and accompanying increases in melanoma and other skin cancers.

5. List 4 actions that the international community has taken and 6 actions that can be taken in the future to prevent global warming, ozone depletion, and acid rain.

	International actions taken	Prospective international actions	
1.	UN Framework Convention on Climate Change—In 1992, 174 countries agreed to stabilize greenhouse emissions.	Rules dictating the replanting of forests and other ecosystems subjected to overharvest or other damage	
2.	UN Climate Change Convention—In 1994, nations agreed to set timetables for reducing greenhouse gases.	Rules to regulate the extraction of fossil fuels for combustion	
3.	Kyoto Protocol—many nations agree to operational rules to reduce emissions of carbon dioxide, methane, and nitrous oxide.	Policies and economic incentives to encourage alternative technologies for transportation and energy generation	
4.	Montreal Protocol—a 1987 agreement to reduce the world's production of CFCs.	Policies and economic incentives to discourage overconsumption of natural resources with its accompanying production of carbon dioxide.	A total ban on CFCs and related compounds

6. Refer to the figure on page 356 and explain the annual fluctuations in tropospheric carbon dioxide. *There is an increase of carbon dioxide in the winter in the northern hemisphere and a decrease in the summer. Why? Photosynthesis!*

7. For the acid deposition chemicals listed below, describe the air pollutant gas, source, and process of formation of each. See **www.gpc.edu/~jaliff/wileychem.htm** and **royal.okanagan.bc.ca/mpidwirn/atmosphereandclimate/acidprecip.html**

Acid	Pollutant gas	Source	Process of formation
H_2SO_3—sulfurous (onion acid)	Sulfur dioxide, SO_2	Burning of fossil fuels contaminated with sulfur	$SO_2 + H_2O \rightarrow H_2SO_3$
H_2SO_3—sulfuric	Sulfur dioxide, SO_2 and sulfur trioxide, SO_3	Burning of fossil fuels contaminated with sulfur	$SO_2 + H_2O \rightarrow H_2SO_3 + O_2 \rightarrow H_2SO_4$
HNO_2—nitrous (a mutagen)	NO	Coincidental to combustion	$NO + H_2O \rightarrow HNO_2$
HNO_3—nitric	NO_2	Coincidental to combustion and nitrogen contamination of coal	$NO_2 + H_2O \rightarrow HNO_3$
H_2CO_3—carbonic	CO_2	Always produced in the combustion of carbon or organic compounds	$CO_2 + H_2O \rightarrow H_2CO_3$

Pop Quiz

Matching

1. e
2. h
3. m
4. g
5. k
6. p
7. l
8. c
9. o
10. f
11. i
12. q
13. a
14. d
15. n
16. b
17. j

Multiple Choice

1. c
2. e
3. e
4. e
5. d
6. d
7. e
8. a
9. c
10. b
11. e
12. a
13. e
14. a
15. b
16. a
17. e
18. c
19. c
20. e
21. d
22. d
23. b
24. b
25. c
26. c

21

Water and Soil Pollution

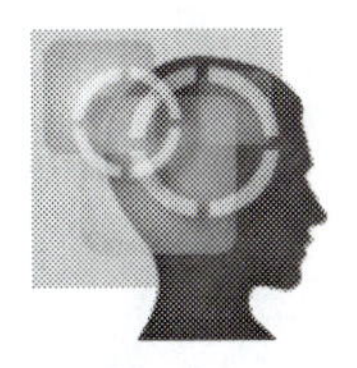

LEARNING OBJECTIVES

1. List and briefly define eight categories of water pollutants.
2. Discuss how sewage is related to eutrophication, biochemical oxygen demand (BOD), and dissolved oxygen.
3. Distinguish between oligotrophic and eutrophic lakes and explain how humans induce artificial eutrophication.
4. Contrast point source pollution and nonpoint source pollution.
5. Describe how most drinking water is purified in the U.S. and discuss the chlorine dilemma.
6. Distinguish among primary, secondary, and tertiary treatments for wastewater.
7. Compare the goals of the Safe Drinking Water Act and the Clean Water Act.
8. Define soil pollution and briefly discuss the specific problem of salinization.

SEEING THE FOREST

List eight categories of water pollutants, including principal sources and method of remediation for each.

	Water pollutant	Principal source	Method of remediation
1.			
2.			
3.			
4.			
5.			
6.			
7.			
8.			

SEEING THE TREES

VOCAB

Water has become a highly precious resource. There are some places where a barrel of water costs more than a barrel of oil.

Lloyd Axworthy, foreign minister of Canada

1. Water pollution—any physical or chemical change in water that adversely affects humans and organisms.
2. Sustainable water use—if as much water is used that is cleaned, the water use is sustainable.
3. Biochemical (biological) oxygen demand (BOD)—a measure of the amount of oxygen needed by bacteria and fungi to decompose organic material. Sewage decomposition produces more odor at high BODs (low dissolved oxygen) than at low BODs (high dissolved oxygen). BOD is an index of sewage pollution.
4. Disease-causing agents in water—organisms that cause disease acquired from water include these examples: (See Table 21-1)
    a. Protozoa—amoebic dysentery, *Cryptosporidium*, and *Giardia*.
    b. Bacteria—cholera and typhoid (these probably killed approximately 30% of all migrants on wagon trains to the west between 1840–1880): *E. coli* O157:H7; *Shigella*, *Clostridium perfringens*, and *Salmonella*.
    c. Viruses—hepatitis A and E, polio, rotaviral diarrhea.
    d. Parasitic worm eggs and larvae—schistosomiasis, liver and lung flukes (flat worms); Guinea worms, hookworms, *Strongyloides*, and pinworms (round worms).
5. Sewage pollution—sewage includes large amounts of undigested fecal wastes and bacteria. Bacteria can make up to 50% of the weight of feces. Feces contain organic materials that are high in the nutrients nitrate, phosphate, and potassium, therefore water *nutrient enrichment* occurs when sewage is present in large quantities.

6. Fecal coliform bacteria count—the common fecal coliform bacterium is *Escherichia coli*, better known as *E. coli*. The vast majority of these are harmless mutualists in the colon where they decompose waste material in the gut and make vitamins for us—vitamin K for blood clotting. See Chapter 5. The strain *E. coli* O157:H7 causes severe diarrhea and kidney failure. Raven-Berg 4e describes its transmission in water supplies in Walkerton, Canada. It is also transmitted in poorly chlorinated swimming pools and inadequately cooked hamburger. The number of fecal coliform bacteria per unit volume of water is an index of sewage pollution. *Listeria* is a flagellated bacterium that can cause birth defects and abortions, septicemia, and meningitis. It is found in water, soil, fish, ground meat, and contaminated milk products. There was a large outbreak of *Listeria* in California caused by contaminated cheese imported from Mexico.
7. Sediment pollution—soil particles are washed into streams where they cloud the water, interfering with photosynthesis, and cover the nesting grounds and their eggs with silt. Sediment or silt is a natural process of erosion, but it is greatly enhanced by soil disruption. Farming (see Chapter 14), construction, clear cutting of forests, and strip mining are major sources of sediment pollution.
8. Inorganic plant and algal nutrients

Nutrient	Ionic or atomic form absorbed	Function in plants
Nitrogen or nitrate (first number on a bag of general purpose fertilizer) and ammonia	NO_3^- (nitrate) and NH_3 (ammonia)	Very important for stem growth. Needed to make plant proteins, DNA (genetic chemical), RNA (gene copies), and ATP (energy chemical)
Phosphorus or phosphate (2nd number on the bag of general purpose fertilizer)	PO_3^-	Encourages root growth. Needed for making DNA, RNA, and ATP
Potassium (3rd number)	K^+, K_2O (potash)	Important for retaining water inside cells. Ionic balance
Calcium	Ca^{2+} (in limestone)	Part of cell walls. Stimulates leaf and root development. Inhibits the absorption of aluminum ions by roots
Magnesium	Mg^{++} (in dolomite)	Needed to make chlorophyll, metabolic enzymes
Iron	Fe^{2+} (iron II)	In photosynthesis and cellular respiration chemicals. Many shrubs need extra amounts
Sulfur (Sulphur)	SO_4^- (sulfate)	In proteins and vitamins
Molybdenum	MoO_4^{2-}	Needed for rhizobia bacteria to do nitrogen fixation. See Chapter 5.

Copper, manganese, zinc	Cu^+ or Cu^{2+}, Mn^{2+} and Zn^{2+}	In photosynthesis and cellular respiration enzymes
Chloride	Cl^-	Needed to make oxygen in photosynthesis. Balances cations in cells

9. Algal bloom—generally a combination of high nutrient loads, and higher temperature contribute to explosions in the populations of algae. One type, a red dinoflagellate, causes the red tide. These organisms produce a neurotoxin, saxotoxin, that can make humans or animals sick. The endangered monk seal population in Africa lost one third their population and manatees in Florida have been killed by red tide poisoning. During red tides, oyster and clam harvesting is prohibited. These mollusks are filter feeders and have large numbers of eaten red tide organisms. Another dinoflagellate that feeds on hog farm sewage, *Pfiesteria,* causes fish kills in North Carolina estuaries, coastal areas, and aquaculture operations. Its toxins cause an Alzheimer's-like neurological disease there among local fisheries workers.
10. Dead zones—when algae grow in thick mats on the water surface, the lower, dark side of the mat begins to rot. As bacteria consume the dead algae, they use up oxygen in the water, producing low oxygen or *hypoxia* (high BOD) and fish kills. Recent reports shows that mild hypoxia interferes with reproduction of carp.
11. Organic compounds—excreted drugs and hormones from humans and animals, pesticides, herbicides, solvents, industrial wastes, oils, antifreeze, and plastics find their way into water supplies.
12. Inorganic chemicals in water—these include the following major pollutants:
    a. Lead—the principal sources of lead historically are leached lead from old paints (lead prevents the growth of microorganisms, mostly algal, on paint), lead from the tetraethyl lead additive for gasoline, incinerator ash, lead-acid batteries, lead shot or bullets, and lead sinkers for fishing lines. Lead has neurological effects that include learning disabilities and attention deficit disorder, it causes high blood pressure (secondary hypertension) and miscarriages.
    b. Mercury—mercury is a liquid metal used in thermostats, thermometers, electrical switches, old paints, and dental fillings. Most mercury pollutant comes from burning coal and municipal wastes. Methyl mercury is passed in marine food chains to bioaccumulate in tuna, sharks, and swordfish. Mercury causes depression and aggressive behavior, kidney failure, and heart failure.
    c. Radioactive isotopes—radioisotopes of iodine, strontium, cesium, and radon. Radon comes from well water in those areas with uranium-238 sheared or ground rock described in Chapter 15.
13. Thermal pollution—water may be heated in industrial processes or steam-generated electrical power stations. Fishes like the warm water in the winter, but in the summer it causes major problems with aquatic ecosystems. Due to the fact that warmer water can hold less oxygen than cooler water, hypoxia may result.
14. Eutrophication—the process of nutrient enrichment of lakes, estuaries, wetlands, or streams is a natural one. Beginning as an oligotrophic lake (one with little nutrients and sediments), it will naturally fill. Then it will be invaded by plants such as cattails, eventually ecological succession will transform the filled-in area into a typical climax community of the biome. *Cultural or artificial eutrophication* occurs due to human-source overloading of nutrients—algal blooms result.

15. Point source pollution—water pollution discharged from specific sites through pipes, sewers, ditches, sewage treatment plants, and injection wells.

16. Nonland pollutants that are present in surface runoff and aquifer discharge over a large area. This includes fertilizer runoff, pesticides, herbicides, soil erosion, mine acids, sewer overflow in rainstorms, septic tank leakage, etc.

17. Municipal water pollution—wastewater from cities may be industrial sources, organic sewage, or storm water. A combined sewer system carries all three and is likely to overflow during rainstorms. The city of Atlanta, GA has many consecutive years of EPA fines for such releases into the Chattahoochee River. Combined sewage system also complicate the cleaning of the water, tending to concentrate heavy metals in the treated water and the solid materials produced by sewage treatment facilities.

18. Groundwater pollution—inorganic fertilizers (particularly nitrate), pesticides, herbicides, gasoline, and MTBE from leaking underground tanks, etc., can leach into groundwater tables and aquifers. In the late 1980s in Phoenix, AZ, some industries were injecting waste solvents into wells. Phoenix gets most of it water from groundwater.

19. Water purification—municipal water from reservoirs or streams is filtered to remove sediment and large organic particles and then treated with chlorine, ozone, or ultraviolet light to kill infective organisms. High chlorine concentrations are required for public swimming pools for that reason and to control the growth of algae. Chlorine combines with small hydrocarbon molecules to produce carcinogenic chemicals that cause colorectal and bladder cancer.

20. Fluoridation—Fluoride ions (F^-) strengthen the calcium phosphate/apatite of tooth enamel, allowing teeth to resist the attack of acids that are made by bacteria living on them. Many communities have naturally adequate fluoride levels, a few have too much and blackening of the teeth can result. This program naturally has resulted in a 50–60% reduction in tooth decay of children.

21. Sewage treatment—wastewater may be filtered, sewage organic solids digested by bacteria, or treated with chemicals to remove nitrogen and phosphate.

    a. *Primary treatment*—pools' wastewater containing sand or silt and gravitational settling removes large particles, creating *primary sludge.*
    b. *Secondary treatment*—uses trickling filters and bacteria to degrade the organic waste. In one type of secondary treatment, *activated sludge processing,* oxygenated water is circulated through bacterial filters. *Secondary sludge* is a slimy mixture of bacteria-laden solids. Dried sludge can be used as fertilizer for flower gardens but not human food. It also can be digested to make biogas.
    c. *Tertiary treatment*—water is treated with chemicals to remove nitrogen and phosphorus. Water can be used after tertiary treatment.

22. Septic systems—typically used in rural communities and lakeside homes, septic systems are individual sewage treatments works that use bacterial decomposition and soil filtration to clean water as it percolates into groundwater. The water is high in nitrates so there is usually a thick stand of green grass near the drain fields. The tank must be cleaned periodically to remove sludge. Septic systems are notoriously ineffective, depending on the soil and groundwater drainage pattern.

23. U.S. water quality laws—these include the following:

    a. Refuse Act of 1899—a law that prohibited the release of solid wastes into navigable rivers.
    b. 1974 Safe Drinking Water Act—authorized the EPA to set standards for drinking water, determining the maximum contaminant levels for pollutants. Twenty-five percent of U.S.

municipal water systems had substandard water in 1998. Amendments in 1986 and 1996 required cities to tell consumers which contaminants are present in drinking water.

c. Clean Water Acts of 1972, 1977, 1981, and 2002—many municipalities acquired funds to build secondary sewage treatment facilities as a result of the original law. Nonpoint sources are not effectively regulated under this law, although it has helped to limit sediment in streams from construction.

d. Solid Waste Recovery Act of 1965 and the Resource, Conservation and Recovery Act (RCRA) amendments of 1970 and 1976—laws enacted to improve solid waste disposal methods and the management of hazardous and nonhazardous wastes. RCRA also promotes resource recovery and reduction of hazardous waste generation from point sources.

e. Ocean Dumping Ban Act—This 1988 U.S. act prohibited dumping of sludge, industrial waste, and garbage in the ocean.

24. Great Lakes Toxic Substance Control Agreement—this treaty was agreed to by the U.S. states and the two Canadian provinces bordering the Great Lakes. The Great Lakes have been affected by PCB pollution so that the state of Michigan advises no human ingestion of Lake Michigan fish during pregnancy and limited ingestion otherwise by all. Also drinking water has an unacceptable level of PCBs in many lakes' communities. Zebra mussels and sea lampreys invaded the Great Lakes, moving from the Saint Lawrence River through the locks constructed to allow navigation from one lake to another. Sea lampreys are jawless vertebrate parasites of fishes—yellow perch. Commercial fishing was a big business in the early 1900s in all the Great Lakes west of Niagara Falls, which acted as a natural barrier. After the 1940s, the exploding lamprey population destroyed commercial harvesting of large fishes.

25. Soil pollution—soil contaminants include pesticides, herbicides, selenium, cadmium, mercury, lead, salts from salinization, etc. Selenium is a major pollutant from denuded soils in the west and also Copper Basin, Tennessee. See Chapter 15 Salinization of soils occurs when shallow irrigation is applied in drier climates. The salts are leached to root level where they are toxic to standard crops.

26. Soil remediation—many soils from Super Fund sites are literally incinerated to remove organic contaminants like tars. Deep-level irrigation uses the process of *dilution* to reduce contaminating salts. *Vapor extraction* uses forced air to vaporize volatile compounds like benzene. Chapter 15 introduced you to *bioremediation* or *phytoremediation* using plants that not only tolerate certain toxic chemicals, but concentrate them in their cells.

Questions to Ponder

1. Cattle and chicken feedlots and pig farms have major problems with water pollution. List four effects of manure releases or runoff into streams. See **www.bmts.com/~jkaminski/kincard/Spills.htm**, and **cfpub.epa.gov/npdes/faqs.cfm?program_id=7** and **www.aeconline.ws/pollution.htm** (Alabama Environmental Council).

	Effects of manure releases or runoff into streams
1.	
2.	
3.	
4.	

2. List and briefly describe four point and nonpoint sources of water pollution. See **www.epa.gov/owow/nps/qa.html** and the UN Food and Agricultural Organization: **www.fao.org/docrep/W2598E/w2598e04.htm**

	Point source of water pollutants	Nonpoint sources of water pollutants
1.		
2.		
3.		
4.		

3. Describe primary, secondary, and tertiary water treatment and the products of each.

	Description	Products
Primary water treatment		
Secondary water treatment		
Tertiary water treatment		

See the Ohio State University extension site: **ohioline.osu.edu/aex-fact/0768.html**

4. Compare and contrast the goals of the Safe Drinking Water Act and the Clean Water Act.
5. List six hazardous water pollutants and describe the primary sources and health effects of each.

	Hazardous water pollutant	Sources	Health effects
1.			
2.			
3.			
4.			
5.			
6.			

6. Describe how ancient civilizations rendered themselves extinct by long-term irrigation. See **Annenburg/CPB Learner.org** at **www.learner.org/exhibits/collapse/mesopotamia.html**

POP QUIZ

Matching

Match the following terms and definitions

____1. The level of this nutrient increases in water when sewage is present:	a. mercury
____2. Odors tend to be emitted from decomposing sewage if ______ levels are low.	b. selenium
____3. Very low BODs are found here:	c. dried sludge
____4. Very high BODs are found in the ______ off the coast of Louisiana.	d. eutrophication
____5. An algal bloom caused by sewage leaking into Lake Washington, WA, is an example of ______.	e. lead
____6. A new lake with low nutrient levels is called ______.	f. coliform bacteria
____7. A natural process of nutrient and sediment accumulation in a lake is termed	g. point source of water pollution
____8. Massive releases of sewage cause this to happen in the streams that immediately receive the pollution:	h. PCBs
____9. The wreck of the Exxon *Valdez* and subsequent oil spill is classified as ______.	i. water downstream of a mountain waterfall
____10. Agricultural soil runoff that contains fertilizers, pesticides and herbicides from southern Illinois is an example of a(n) ______	j. nitrogen, phosphorus
____11. The compounds of this liquid metal tend to bioaccumulate in tuna:	k. red tide
____12. Water pollution from western farm sediment runoff may contain toxic amounts of ______	l. cultural eutrophication
____13. Compounds of this metal were added to pre-1980s gasolines to raise octane and lubricate valves:	m. oxygen
____14. A general indicator of water quality is the *count* of ______	n. nonpoint source of water pollution
____15. Tertiary sewage treatment can make water usable if it removes ______	o. oligotrophic

____16. Secondary treatment of sewage produces ______ that can be used as fertilizer for flowers.	p. *dead zone*
____17. There are restrictions on eating large fish from Lake Michigan due to contamination by ______	q. fish kills

Multiple Choice

1. Which of the following organisms in polluted water does not cause disease?
   a. *E. coli*
   b. cholera
   c. *Giardia*
   d. *Cryptosporidium*
   e. rotavirus

2. An outbreak of which organism resulted when Milwaukee's water treatment was inadequate in 1993?
   a. *E. coli*
   b. cholera
   c. *Giardia*
   d. *Cryptosporidium*
   e. rotavirus

3. Arcata, CA, uses nature to treat sewage in ______
   a. fertilized vineyards
   b. estuaries
   c. reservoirs
   d. underground septic systems
   e. all of the above

4. The most important source of nonpoint pollution is which of the following?
   a. farming
   b. construction
   c. municipal sewage
   d. strip mining
   e. hog farming

5. Artificial eutrophication results when the level of which nutrient is high?
   a. nitrogen
   b. phosphate
   c. potassium
   d. all of the above

6. In the 1970s, the content of ______ in detergents was reduced for economic reasons and to improve water quality.
   a. nitrogen
   b. potassium

c. sodium
d. calcium
e. phosphate

7. Construction causes high levels of ______ in streams.
a. turbidity
b. sediment load
c. sulfuric acid
d. c only
e. a and b

8. When stream turbidity increases due to soil sediments, this occurs:
a. decrease in photosynthesis by aquatic plants
b. increase in populations of algae
c. BOD decreases
d. eutrophication increases in lakes downstream
e. all of the above

9. Pig farms have released sewage into streams. The most likely result would be ______
a. low BODs
b. algal blooms downstream
c. fish kills
d. c only
e. b and c

10. Which statement is correct concerning the red tide?
a. They tend to occur in the summer.
b. Dinoflagellate toxins can kill manatees.
c. Agricultural and sewage pollution contribute to red tides.
d. Oysters cannot be taken from areas affected by red tides.
e. All of the statements are correct.

11. Which group of water pollutants primarily cause liver damage?
a. heavy metals
b. industrial solvents
c. construction sediments
d. mine acids
e. high levels of nitrates and phosphates in drinking water

12. Which water pollutant primarily causes cerebral palsy, mental retardation, and kidney diseases?
a. lead
b. carbon tetrachloride
c. mercury
d. zinc
e. PCBs

13. Which water pollutant contributes to secondary hypertension and ADD syndrome in children?
a. lead
b. carbon tetrachloride

c. mercury
d. zinc
e. PCBs

14. Which of the following materials contributes to lung and intestinal cancer?
   a. PCBs
   b. mercury
   c. lead
   d. radon
   e. ethylene glycol (antifreeze)

15. Which chemical is best, in terms of toxicity or hazardous byproducts made, to use to purify city drinking water?
   a. ozone
   b. chlorine
   c. sodium hypochlorite (bleach)
   d. fluoride
   e. chloride

16. Which of the following chemicals would not be expected in agricultural runoff and aquifer leaching?
   a. pesticides
   b. herbicides
   c. nitrate
   d. phosphate
   e. Trick question alert! All of these would be expected.

17. Heavy metals can be removed from industrial wastewater by which process?
   a. filtration
   b. sedimentation
   c. ion exchange
   d. trickling filters
   e. bacterial decomposition

18. Aluminum sulfate is used in water treatment for which process?
   a. filtration
   b. sedimentation
   c. ion exchange
   d. trickling filters
   e. bacterial decomposition

19. The ships of which country, that suspended chlorination of drinking water, brought cholera bacteria to Mobile Bay, Alabama, in the 1990s?
   a. France
   b. U.K.
   c. Panama
   d. Peru
   e. China

20. Which of the following is a problem with combined sewer systems?
    a. Industrial wastes may add heavy metals and solvents to the mix.
    b. Heavy rains cause overflow into rivers.
    c. The cost of sewage treatment goes up.
    d. The costs of effective sewage treatment increase.
    e. All of the above are problems.

21. The principal contaminant of well water that causes blood hypoxia in children and contributes to carcinogen formation in the intestine is ______.
    a. nitrate
    b. phosphate
    c. insecticide residues
    d. herbicide residues
    e. chloroform

22. Which of the following is a carcinogenic oxygenating gasoline additive that leaks into groundwater from leaking filling station tanks?
    a. ethanol
    b. tetraethyl lead
    c. MTBE
    d. methanol
    e. STP

23. Which of the following materials is mostly removed by primary water treatment?
    a. nitrates
    b. dissolved phosphate
    c. suspended organic solids
    d. sediment
    e. digested sludge

24. Which materials are mostly removed by secondary water treatment?
    a. nitrates
    b. dissolved phosphate
    c. suspended organic solids
    d. sediment
    e. digested sludge

25. The percentage of U.S. cities not meeting EPA standards for maximum contaminant level is closest to ______.
    a. 0%
    b. 5%
    c. 15%
    d. 25%
    e. 75%

26. Which of the species below is invasive in the Great Lakes?
    a. sea lampreys
    b. zebra mussels
    c. smelt fish

d. yellow perch
e. a and b

27. Which geographical area is not appropriately matched with its primary water pollution problem?
    a. Po River, Italy/sediment
    b. Lake Maracaibo, Venezuela/fertilizer runoff
    c. Ganges, India/decomposing human bodies
    d. Mississippi River, U.S./fertilizer runoff
    e. Trick question alert! All of these are correct.

28. Which process results in salinization of soils?
    a. all irrigation
    b. overfertilization
    c. shallow irrigation
    d. organic fertilization
    e. all of the above

CYBER SURFIN'

1. From the World Health Organization, Water and Sanitation, check out: **www.who.int/water_sanitation_health/Documents/righttowater/righttowater.htm** and for advice on diseases that can be acquired by tourists, see the interactive map at **www.who.int/ith**
2. See the National Resources Defense Council site, Water Pollution: **www.nrdc.org/water/pollution/default.asp**
3. From the World Bank, New Ideas in Pollution Regulation, see an interactive map at **www.worldbank.org/nipr/Atrium/mapping.html** and Greening Industry: The New Model: **www.worldbank.org/research/greening/cha7new.htm.** To see the U.S. EPA overview of water quality and health, see **www.epa.gov/safewater/dwhealth.html**; for water pollution due to factory animal farming effects, go to **cfpub.epa.gov/npdes/faqs.cfm?program_id=7**
4. North Carolina State University has an informative site on the dinoflagellate *Pfiesteria* that is related to animal farming pollution, see **www.pfiesteria.org**
5. To read about *Listeria*, Salmonella, Shigella, hepatitis A, and other microorganisms that contaminate food or water, check out the USDA's *Bad Bug Book:* **www.cfsan.fda.gov/~mow/intro.html**
6. The University of Virginia has a stream study description: **wsrv.clas.virginia.edu/~sos-iwla/StreamStudy/StreamStudyHomePage/StreamStudy.HTML**
7. What you can do about protecting water quality? Check out the Water Keeper Alliance: **www.keeper.org/; http://www.epa.gov/safewater/dwh/getin.html;** The National Resources Council: **www.nrdc.org/water/pollution/gsteps.asp;** Lake Michigan Federation: **www.lakemichigan.org;** the Hudson River, NY; Los Angeles. CA: **www.riverkeeper.org/;** and ocean issues at **www.oceanconservancy.org**

'LIKE' BOOKING IT

1. Goldstein J. ***Demanding Clean Food and Water: The Fight for a Basic Human Right.*** Oxford, UK: Perseus Publications, 1990.
2. Wolverton BC, and Wolverton JD. ***Growing Clean Water: Nature's Solution to Water Pollution.*** Picayune, MS: Wolverton Environmental Services, Inc., 2001.
3. Van der Ryn S. ***The Toilet Papers: Recycling Waste and Conserving Water.*** White River Junction, VT: Chelsea Green Pub, Co., 1995.
4. Barlow M, and Clarke T. ***Blue Gold: The Fight to Stop the Corporate Theft of the World's Water.*** New York: New Press, 2001.
5. Reisner M. ***Cadillac Desert: The American West and Its Disappearing Water.*** New York: Penguin Putnam, 1993.

ANSWERS

Seeing the Forest

List eight categories of water pollutants, including principal sources and method of remediation for each.

	Water pollutant	Principal sources	Method of remediation
1.	Sewage	Leakage of municipal sewage systems, overflows of combined sewer system, and factory animal farms	Storm, industrial and organic sewage separated. Tertiary water treatment
2.	Disease-causing agents	Untreated sewage	Effective sewage treatment. Effective medical treatment of those infected or infested
3.	Sediment	Construction, soil erosion from farms	Silt fences at the construction site, contour plowing, shelterbelts
4.	Inorganic plant and algal nutrients	Farm runoff and leaching into groundwater	Use more organic fertilizer, direct fertilizer to root level instead of broadcast. Use less fertilizer
5.	Organic chemicals, solvents, pesticides, herbicides	Industrial sources and farm runoff	Use sustainable organic farming methods. Biologically control pests. Use crop diversity and natural predators of pests
6.	Inorganic chemical toxins, heavy metals	Industrial sources	Control the pollution at the source so that it does not contaminate organic waste.

7.	Radioactive isotopes—iodine, strontium, and radon	Iodine and strontium from radioactive waste from bomb making and nuclear explosions. Radon from natural rock formations in the eastern U.S. and Midwest	Prevent leakage from bomb making sites, nuclear safety in disposal of nuclear waste. Seal basements and don't use contaminated well water.
8.	Thermal pollution	Steam-generating power plants—both nuclear and conventional	Cooling towers

Questions to Ponder

1. Cattle and chicken feedlots and pig farms have major problems with water pollution. List four effects of manure releases or runoff into streams. See **www.bmts.com/~jkaminski/kincard/Spills.htm**, and **cfpub.epa.gov/npdes/faqs.cfm?program_id=7** and **www.aeconline.ws/pollution.htm** (Alabama Environmental Council).

	Effects of manure releases or runoff into streams
1.	Fish kills due to low oxygen levels created by bacterial decomposition of organic waste
2.	Algal blooms in slow moving areas of streams and lakes or ponds receiving that flow
3.	Increased infections of cattle and people by *E. coli* O157:H7 if the waste comes from cattle. Increased *Listeria* contamination of food products. Increased neurological disease due to *Pfeisteria* toxins.
4.	Fish kills in estuaries, coastal areas, and aquaculture operations. Toxic reactions in humans eating infected fish include blurred vision and dermatitis. Memory loss and slow mental development occurs in experimental animal models.

2. List and briefly describe four point and nonpoint sources of water pollution. See **www.epa.gov/owow/nps/qa.html** and the UN Food and Agricultural Organization: **www.fao.org/docrep/W2598E/w2598e04.htm**

Point source of water pollutants	Nonpoint sources of water pollutants
1. Mercury discharges from paper-processing plants (mercury is used to prevent fungal growth in the paper fiber slurry)	Fertilizer runoff from tilled soil. Mercury contamination from anti-fungal seed coatings.
2. Leaded gasoline (in many developing nations) leakage from an underground tank	Fertilizer leaching into groundwater causing increased nitrate concentrations in well water

3. Oil spills from ocean tanker accidents	Pesticides runoff from farming
4. Because there is so much underground and surface mining in eastern Kentucky and southwestern West Virginia, mine acid could be considered as nonpoint or as coming from many point sources.	Herbicides runoff from farming

3. Describe primary, secondary, and tertiary water treatment and the products of each.

	Description	**Products**
Primary water treatment	Removes larger suspended and floating materials that include sand, silt, and organic matter	Primary sludge
Secondary water treatment	Removes smaller suspended particles using trickling filters or aerated activated (with bacteria and oxygen) sludge process. Particles and microorganisms settle out to form secondary sludge.	Dried secondary sludge can be used as fertilizer for plants not intended for human consumption (U.S. law). Sludge can be further digested to make biogas. See Chapter 12. It can be incinerated also.
Tertiary water treatment	Removes nitrate and phosphate and other inorganic materials ions if ion exchange or electrolysis is employed.	Useable water if cleaning is thorough. Think of what astronauts in the space stations drink.

See the Ohio State University extension site: **ohioline.osu.edu/aex-fact/0768.html**

4. Compare and contrast the goals of the Safe Drinking Water Act and the Clean Water Act. *The Safe Drinking Water Act established maximum levels of contaminants in drinking water. Seventy-five percent of cities meet those standards. The Clean Water Act has successfully reduced point source water pollution. Although the fines for point source pollution are too low to be very effective, voluntary compliance in the U.S. has been good because of the bad publicity generated when toxic pollutants are discovered. Agencies enforcing both laws are understaffed.*
5. List six hazardous water pollutants and describe the primary sources and health effects of each.

Hazardous water pollutant	**Sources**	**Health effects**
1. Cadmium	By-product of lead and mercury production. Phosphate fertilizer residues.	Kidney failure due to destruction of tubules; skeletal failure due to interference with calcium metabolism—called “itai-itai” in Japan

2. Mercury	Municipal waste incinerators; electrical switches and thermometer production and disposal. Paper pulp waste, seed coatings.	Headache, depression, irregular heart beat, birth defects
3. Lead	Fertilizers, gasoline additive in many developing nations, paint additive.	Hypertension, birth defects, hearing loss, ADD
4. Organic solvents: carbon tetrachloride, vinyl chloride, formaldehyde, methyl ethyl ketone (paint stripper), benzene	Industries, leaking gasoline storage tanks, homes	Liver failure and cancer, kidney failure and cancer, bladder cancer
5. Radioactive iodine	Bomb processing, explosions of booms and nuclear fission reactors	Thyroid cancer
6. Radon	Naturally occurring in certain soils containing uranium-238	Cancer in the alimentary canal. The National Academy of Science estimates about 150 deaths a year due to stomach cancer caused by radon in water. Also a similar number due to lung cancer. Test for radon first at home

6. Describe how ancient civilizations rendered themselves extinct by long-term irrigation. See Annenburg/CPB Learner.org at **www.learner.org/exhibits/collapse/mesopotamia.html.**
Long-term shallow water irrigation in arid and semiarid regions deposits salts at root level as the water evaporates. The effect is more pronounced downstream because of the nonpoint runoff from irrigated areas upstream. Salts eventually build up so that standard crop plants grow poorly or not at all.

Pop Quiz

Matching

1. f
2. m
3. i
4. p
5. l

6. o
7. d
8. q
9. g
10. n
11. a
12. b
13. a
14. f
15. j
16. c
17. h

Multiple Choice

1. a
2. d
3. b
4. a
5. d
6. e
7. e
8. a
9. e
10. e
11. b
12. c
13. a
14. d
15. a
16. e
17. c
18. b
19. d
20. e
21. a
22. c
23. d
24. c
25. d
26. e
27. e
28. c

22

The Pesticide Dilemma

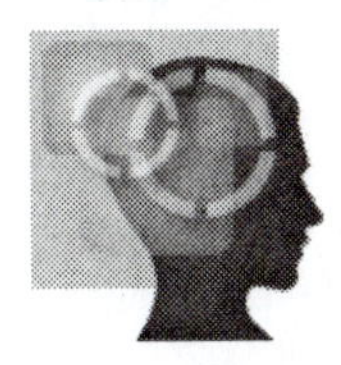

LEARNING OBJECTIVES

1. Define *pesticide*, distinguish among various types of pesticides (such as insecticides and herbicides), and describe major groups of insecticides and herbicides.
2. Relate the benefits of pesticides.
3. Summarize the problems associated with pesticide use, including development of genetic resistance, bioaccumulation, biological magnification and mobility in the environment.
4. Discuss pesticide risks to human health (including short-term and long-term effects), pesticides as endocrine disrupters, and risks to children.
5. Describe alternative ways to control pests including cultivation methods, biological controls, reproductive controls, pheromones and hormones, genetic controls, quarantine, integrated pest management, and irradiating foods.
6. Briefly summarize the three U.S. laws that regulate pesticides: the Food, Drug, and Cosmetics Act; the Federal Insecticides, Fungicide, and Rodenticide Act; and the Food Quality Protection Act.
7. Describe the purpose of the Stockholm Convention on Persistent Organic Pollutants.

SEEING THE FOREST

Fill-in the table below, referring to Table 22.1 and some Internet research at **pops.gpa.unep.org/04histo.htm** and Exotox Net from Oregon State U., Michigan State U., Cornell U., U. Cal. Davis, and Idaho State U.: **ace.orst.edu/info/extoxnet/pips/ghindex.html**

Name	Chemical classification	Specific use or use with the greatest benefit	Toxicity and regulatory status	Ecological and health effects	Persistence
DDT					
Chlordane					
Dieldrin, aldrin					
Malathion®					
Carbaryl (Sevin®)					
Pyrethrins					
Atrazine					
2,4 D					
Methyl bromide					

SEEING THE TREES

VOCAB

No man is justified in doing evil on the ground of expediency.

President Theodore Roosevelt

1. Pesticide—chemical that controls organisms that attack crops, ornamental plants, animals, or human or affect their health indirectly. Agriculture used 85% of the 2.9 million tons produced annually. They are divided into six groups:
   a. Insecticides—these consist of chlorinated hydrocarbons, organophosphates, and carbamates; botanical insecticides include pyrethrins extracted from marigold and chrysanthemum flowers, nicotine from tobacco, and rotenone (also used to kill fish). Synthetic pyrethrins are available for termite and other insect control.
   b. Herbicides—herbicides are used for weed control in farming.
   c. Fungicides—this is used to control fungal infections including various molds including powdery mildew, wheat rust, and corn smut.

d. Rodenticides—rodents are reservoir hosts for hantavirus (fatal respiratory infection) plague bacteria, Rocky Mountain spotted fever bacteria, and tapeworms that can infect humans.
e. Algicides—algae is controlled in pools by increasing acidity, chlorination, and copper sulfate pentahydrate. Copper sulfate is also toxic to snails in aquatic environments.
f. Molluskicides—for snails or slugs, they include arsenic (also an endocrine disrupter), metaldehyde (very toxic), iron phosphate, and copper sulfate.

2. Narrow spectrum pesticide—ideally pesticides and antibiotics should kill only the pest or bacterium that we target it against. Unfortunately, both are usually broad spectrum poisons that kill or inhibit the growth of many other organisms. Both kill or make sick many organisms not intended for control. Therefore, the malathion® you spray on your bean plants may save them but kill visiting bees. Similarly, killing the natural flora of the intestine with a broad spectrum antibiotic may lead to blood clotting problems, diarrhea, and yeast infections as the helpful bacteria are killed.

3. Chlorinated hydrocarbon insecticides—DDT (a first-generation insecticide DDT) was discovered in 1939 by Swiss chemist Paul Muller. It was a miracle pesticide for several reasons: it was broad spectrum in action, it had major benefits in controlling malaria mosquitoes in the tropics and human lice in war-torn Europe; it was persistent so that it did not have to be reapplied frequently; it was not water soluble, so when applied in oil or kerosene, it clung to the leaves; it could be made into a dry powder to control external parasites, such as lice; and it was inexpensive. It is an acetylcholinesterase inhibitor/antagonist that prevents muscle cells from responding to appropriate signals from the nervous system. DDT, chlordane, lindane, dieldrin (very toxic to vertebrates), endosulfan, and methoxychlor are chlorinated hydrocarbon insecticides. All except methoxychlor bioaccumulate. All are toxic to aquatic organisms, and DDT specifically interferes with eggshell formation due to its endocrine disruption of calcium metabolism. Chlordane was used for many years to control termites under houses and surrounding soils. The chlorinated hydrocarbons kill by contact.

4. Organophosphate insecticides—these are acetylcholinesterase inhibiting poisons discovered in German research on fluorophosphate nerve gases before World War II. Diazinon and chloropyrifos (dursban®) are organophosphates that were recently banned, the latter spurring the development of new termiticides that have long-lasting effects. Malathion® is used to control the Mediterranean fruit fly. They are highly toxic to bees, birds, and aquatic invertebrates. The organophosphates kill by contact.

5. Carbamate insecticides—Carbaryl (sevin®) and aldicarb are poorly soluble acetylcholinesterase inhibitors that are used as dusts or suspensions on crops. The insects die when they eat the insecticide. This is the single mostly widely used insecticide for the control of crop pests.

The twelve most contaminated fruits and vegetables

Rank	Crop
1	Strawberries
2	Bell peppers
2	Spinach
4	Cherries (U.S.)

5	Peaches
6	Cantaloupe (Mexico)
7	Celery
8	Apples
9	Apricots
10	Green beans
11	Grapes (Chile)
12	Cucumbers

Compiled from FDA and EPA data and the Environmental Working Group.

6. Bacterial insecticides and nematicides—bacteria can be employed to attack insect pests and root nematodes (roundworms) that damage crops. *Bacillus popillae* or milky spore can kill caterpillars such as cabbage moth larvae, Japanese beetle larvae, and other grubs underground. Grubs dine on plant roots and produce the Japanese beetle adults that eat bean, fruit tree, and grape leaves. Of importance to controlling West Nile encephalitis, malaria, and so on, *Bacillus thuringiensis*-toxin *(Bt)* is applied to water where mosquito larvae are present. It is also used to control pests on cotton and potatoes. Both *Bacilli* are harmless to humans and other vertebrates. Bt-toxin is activated only by the alkaline digestive secretions of insect larvae. Vertebrates have an acid-containing stomach that denatures the Bt-toxin.
7. Herbicides—chemicals used to control weeds that compete for water and nutrients in soils where crops are grown. Generally herbicides interfere with the hormonal-control plant growth. There are selective herbicides that kill specific types of plants: broadleaf (dicot) plants that compete with grass species (corn, wheat, and lawn grass); or grasses (monocot) that compete with broadleaf crops such as beans. Broad spectrum herbicides kill most plants (Roundup®). The jungles of the northern region of what was then South Vietnam were denuded to prevent the then North Vietnamese army from moving personnel and supplies to the south. Most of the tropical rain forest in that area has been replaced by shrubs and weedy grasses, similar to that in Brazil where deforestation has occurred. Dioxin contamination of the Agent Orange herbicide has poisoned the water in Vietnam and has moved through the food chain to humans. Many soldiers returning from Vietnam developed cancers due to dioxin exposure.
8. Evolution of genetic resistance—There is a striking parallel between the rapid evolution of bacterial resistance to antibiotics and the development of resistance by pests. Over the past 40 years, the incidence of postsurgical infections in hospitals has increased over 350% in the face of better knowledge of antiseptic technique and improved antibiotics. What do these disparate groups have in common? These organisms reproduce quickly. Natural selection will favor genes that confer pesticide or antibiotic resistance. Trillions of houseflies were killed by DDT in its early days, but some flies with special genes survived. Those reproduced to a large number of flies that were not susceptible to DDT. Five hundred twenty species of

insects have evolved resistance. The diamondback moth and palm thrips are resistant to all major classes of chemical insecticides. Resistance to *Bacillis thuringiensis* (Bt) toxin has emerged in the tobacco budworm, cotton bollworm, and the diamondback moth. Bt-toxin can have unintended effects by killing butterfly caterpillars and causing food allergies.

9. Treadmill effect—Both pesticides and antibiotics exhibit a *treadmill effect.* Over time, more powerful and more expensive chemicals have been developed to control bacterial infections and crop pests. These chemicals have to be changed regularly to keep up with pest resistance. Farmers are very conservative in adopting new approaches, so it is hard for many to get off the pesticide treadmill. Crop yields may actually decline as a result of using more toxic compounds or crop pests may actually increase in population over time due to the toxic effects on predators of pests, which coincidentally reproduce much slower than does the resistant pest species that they normally control. Instead of broadcasting pesticides over a longer time period, *calendar spraying* applies insecticide precisely at the time it is most effective. Also *scout-and-spray* works so that one knows exactly when the pests arrive. During the Chinese cultural revolution of the 1960s, citizens were instructed to search out flies and kill them. It is said that killing one fly in the winter is like killing 10,000 in the summer.

10. Resistance management—a new approach to delaying the spread of insecticide resistance genes to crop pests by timing the delivery of insecticide, interbreeding insects without resistant genes with resistance gene-carrying pests, thus diluting the resistance genes so that they are ineffective. Populations of pests that do not carry resistance genes are maintained in protected areas called *refugia.*

11. Imbalances to ecosystems—when insecticide is sprayed by crop duster aircraft over fields, pesticide drifts into patches of woods nearby. Birds can then be seen lying on the forest floor dying, not being able to breathe as the normal conduction of impulses from the nervous system to breathing muscles is interrupted. When insecticides kill the predators of crop pests, crop pest problems become worse. Why? The pests can reproduce much faster (*r* strategists) than the predators (the vertebrate predators are *K* strategists). Add to that the evolution of resistance.

12. Bioaccumulation and biomagnification—individual predators will *bioaccumulate* pesticides and their residues over time in fatty tissues. In *biomagnification,* when trophic levels of a food chain are considered, certain pesticide concentrations actually increase within the tissues of the organism in each succeeding trophic level. See the figure below. The fat content of organisms, by percentage of body mass, generally increases in each succeeding trophic or feeding level.

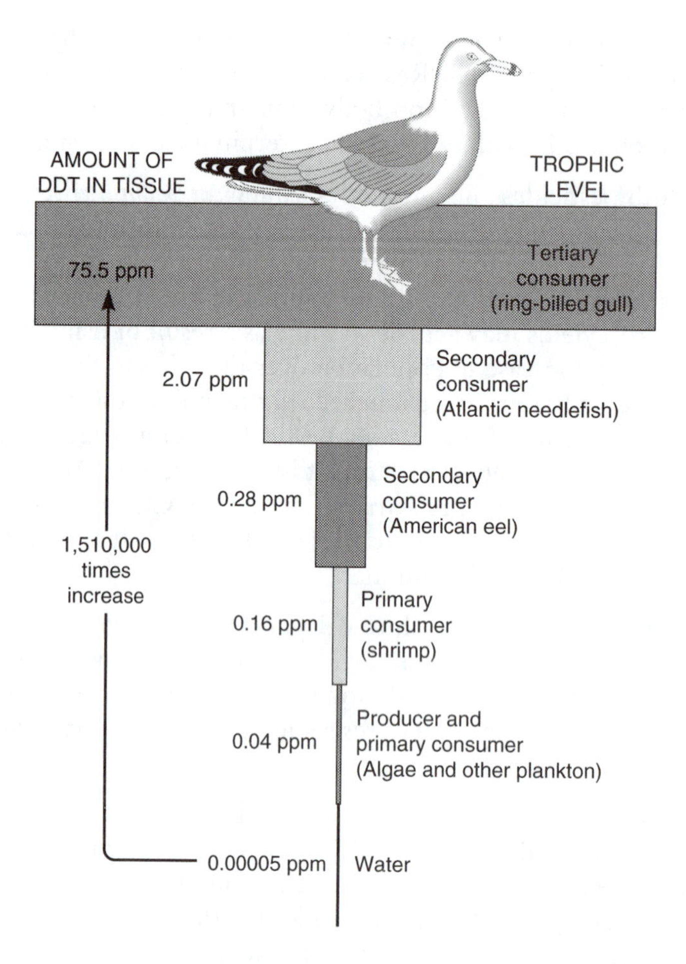

13. Endocrine disrupting pesticides—many pesticides and industrial chemicals are capable of interfering with the functioning of estrogens, androgens, and thyroid hormones in humans and animals. Exposures can cause sterility or decreased fertility, behavioral changes including the inability to pair bond with the opposite sex, changes in the timing of nesting and fertility, inability to produce a calcium carbonate eggshell (DDT and DDE effect), birth defects in the development of organs of reproduction, learning deficits, and metabolic disorders.
14. Alternatives to chemical pesticides—these include the following approaches:
    a. Cultivation methods—*interplanting* different species of crops discourages the increase of populations of crop-specific pests when that crop is grown in monoculture. Some plants attract wasps that then prey on the pests of crops (the molasses grass that is planted with corn). *Strip cutting* cuts only one strip of a crop, leaving habitat intact to provide cover for predators of pests. Greenbelt planting is even better, where scrubs and trees provide cover for predators of pests.
    b. Biological controls—wasps are grown and sold to farmers for a variety of pests of crops. The vedalia beetle was imported from Australia to control citrus scale insects. Even though it is an exotic species, it only eats scale insects. Pink bollworm of cotton is controlled by the wasp *Trichogrammatoidea bactrae.* Bt-toxin and *Bacillus pompillae* are examples of biological control described above. Gypsy moths, a voracious consumer of oak tree leaves, are an exotic species from Europe that has caused much damage in the eastern deciduous forest of North America. It has "busts" of population numbers due to virus and fungal infections.

c. Pheromones—pheromones are aromatic hydrocarbons that insects use to locate mates. They have been used to control cotton bollworms. So much pheromone is released that the scent of the female moths cannot be located by males. Japanese beetle traps were popular at one time, but unless everyone uses them, all they do is attract beetles to your plants. Their use has been officially discouraged by state agricultural extension services (Georgia). The gypsy moth, a pest of eastern deciduous forest, has been controlled with gyplure attractant-baited traps.
d. Hormones—synthetic ecdysone, a molting hormone, can disrupt the normal development of larvae if applied at a premature stage.
e. Reproductive controls—the screw-worm fly maggot that is very damaging to cattle and other livestock has been nearly exterminated from the Western Hemisphere by using the *sterile male technique.* The female flies only mate once. So if a female mates with a sterile male she produces no offspring. The *sterile male technique* is being used for the control of tsetse flies that transmit "sleeping sickness" to humans and cattle in Africa, Mediterranean fruit flies in Italy, and, surprisingly, sea lampreys in the Great Lakes.
f. Genetic control—genetic engineering or modification of crops can add genes that produce Bt or other toxins that kill or repel predators of certain crops. When crop species are discovered that are resistant to the attack of a particular pest, the genes providing the appropriate chemical can be isolated and moved to nonresistant seeds. Interbreeding can produce the same results.
g. Integrated pest management—when biological, chemical, and cultivation controls are combined, pesticide use can be reduced greatly. Indonesia is actually growing more rice at a greater profit margin and creating less toxic water and soil pollution due to integrated pest management. Greenbelt maintenance of the predators of pests can be combined with calendar or scout spraying, and soil conservation.

15. Food irradiation—gamma rays can sterilize bacteria and fungi that contaminate food from vegetables to meat, reducing waste due to spoilage and the risk of harmful bacterial infection. As we discussed in Chapter 11, irradiating food does not make it radioactive, but it may create free radicals. Currently there is no evidence that irradiation produces more free radicals than those present from the normal decomposition of food molecules.

16. U.S. laws regulating insecticide production and use include the following:

a. Food, Drug, and Cosmetics Act (FDCA) of 1938—a U.S. law, as modified by the Miller Pesticide Chemicals amendment in 1954 and the Delaney Clause in 1958, prohibits the presence of any chemical in processed food known to cause cancer in animal models or humans. It exempted fruits, vegetables, fish, shellfish, poultry, pork, lamb, and beef.
b. Federal Insecticide, Fungicide, and Rodenticide Act (FIFCA)—a U.S. act, passed in 1947, required testing and registration of pesticide ingredients. It was amended in 1972 to allow the EPA to regulate pesticide production and use. This led to the banning of many chlorinated hydrocarbon pesticides, such as chlordane, lindane, and DDT. Aldrin, dieldrin, kepone, and recently chlorpyrifos (dursban®) were also prohibited. In 1988, the law was amended to require the reregistration of older pesticides but did not address the issues of food laborer safety in handling pesticide-contaminated food or the residues of pesticides remaining on marketed foods. The FDA has advised that fruit given to small children be peeled before eating.
c. Food Quality Production Act—a 1996 U.S. law that amended both FDCA and FIFCA laws above by requiring that the increased susceptibility of infants and children be considered and that all health risks, not just cancer, be assessed. It also reduced the time that it requires for a pesticide to be banned from 10 years to only 14 months. The right to know the toxicity of

food ingredients or contaminants is important for informed choice by consumers. Pesticide residues may be of sufficient risk to avoid certain foods. The residues contaminating foods grown or processed in countries that still use certain pesticides of the "dirty dozen" or have as persistent soil and water chemicals should be assayed, and concentrations and risks made known to consumers of imported foods. The FDA recently banned the importation of British beef due to "mad cow" disease. Similar action can be taken in cases of pesticide contamination.

17. Stockholm Convention on Persistent Organic Pollutants—an international treaty adopted in 2001 that attempts to protect human and environmental health from 12 persistent organic pollutants or POPs. Nine of these are insecticides – see the first nine below. The "dirty dozen" includes aldrin, chlordane, DDT, diedrin, endrin, heptachlor (present in chlordane), hexachlorobenzene, mirex (once used to control fire ants in the southeastern U.S.), toxaphene (end of insecticides), PCBs, dioxins, and furans. The Stockholm Convention requires that signatories plan to eliminate the production and use of these chemicals, except DDT, that can be produced inexpensively by developing countries to control malaria and other mosquito or fly-vectored diseases.

Questions to Ponder

1. List and describe two important general benefits of pesticides.

Benefit	Description
1.	
2.	

2. Compare and contrast the specific benefits (specific disease or crop-eating pest controlled) and harmful effects of DDT use since its discovery. See **pops.gpa.unep.org/04histo.htm**

3. Describe the following phenomena and means of dispersal or movement by filling in the table below.

	Description	Means of dispersal or movement
Biological accumulation and magnification		
Mobility in the environment		

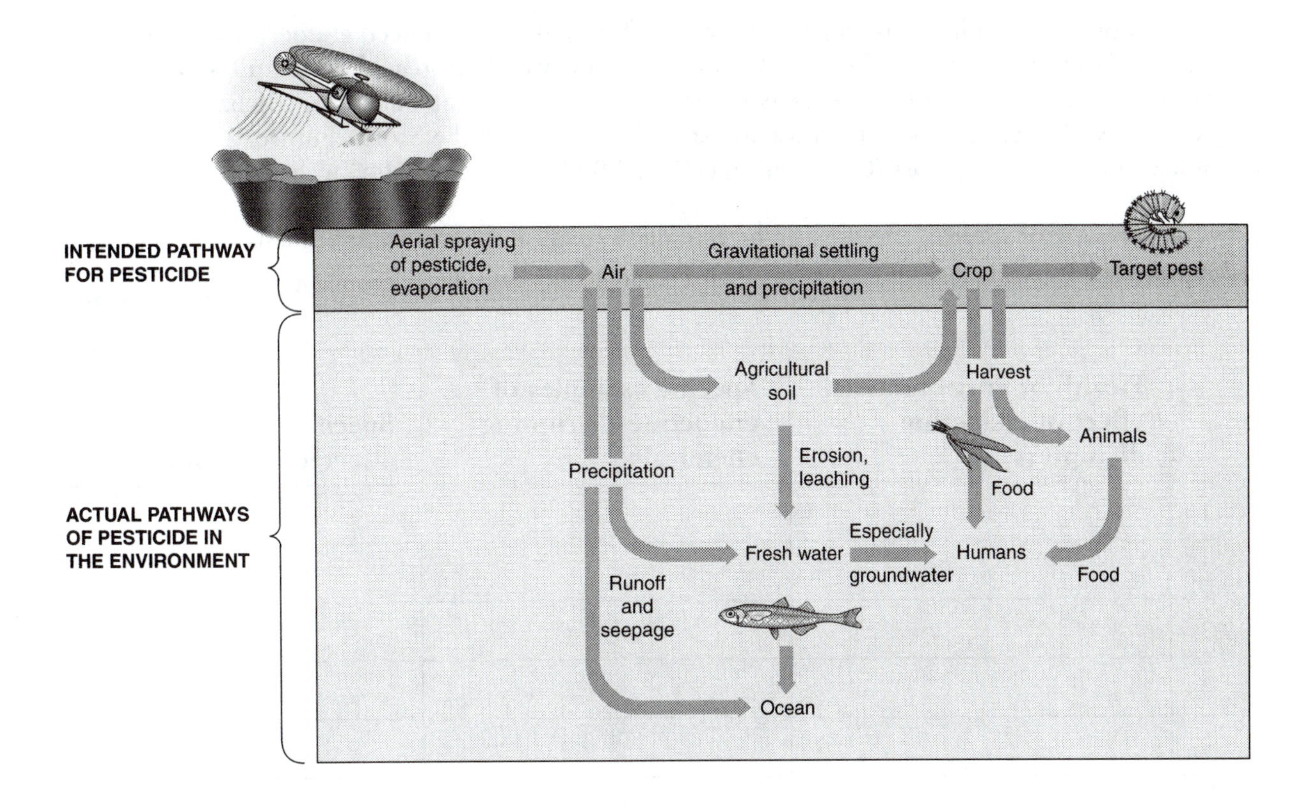

4. Explain how the use of a specific insecticide can actually make the farmer's insect pest problems worse over time.

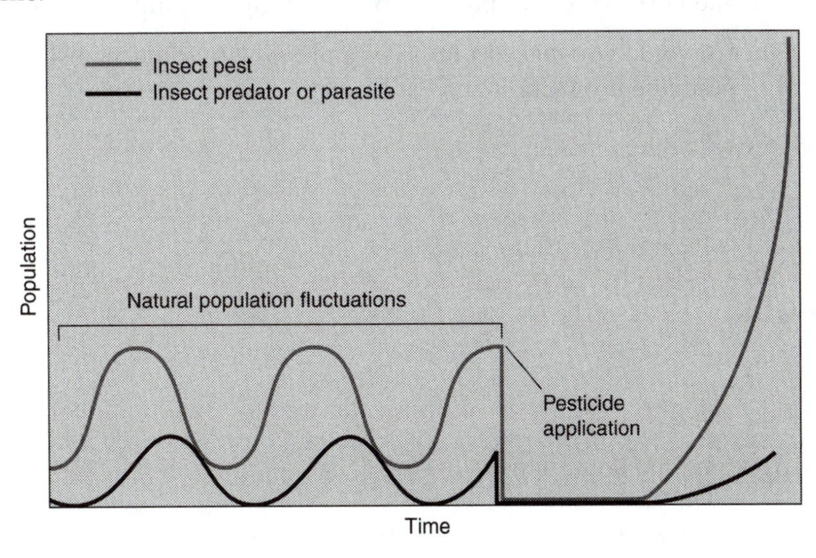

Health problems of humans and animals result from pesticide-induced endocrine disruption. See the National Resources Defense Council site: **www.nrdc.org/health/default.asp** and **www.nrdc.org/breastmilk/chems.asp**. See WWF environmental chemist, Michael Warhurst's website: **website.lineone.net/~mwarhurst/index.html**, and the Stolen Future.org site: **www.ourstolenfuture.org/Commentary/JPM/2002-1115scirevolution.htm.**

a. Describe five specific effects of endocrine disrupters on the health of humans and other organisms, list five chemicals that act in this way, and list five hormones that are disrupted.

	Health or environmental effect of endocrine disrupters	**Specific examples of endocrine-disrupting chemicals**	**Specific hormone affected and function**
1.			
2.			
3.			
4.			
5.			

b. Describe the suspected endocrine disrupter that is increasing in human breast milk.

5. Pretend you are a farmer who is going to have an ecologically sound pest management program for raising cotton and corn. What would you do differently as compared to the factory farm?
6. What improvements would you make in the laws and regulatory mechanisms currently in place for the control of pesticide production and use?

POP QUIZ

Matching

Match the following terms and definitions

____1. One of the "dirty dozen," this pesticide was at one time considered to be harmless to humans:	a. carbaryl (sevin)
____2. Most pesticides are used in this activity:	b. methyl isocyanate

____3. DDT, chlordane, and lindane are in this chemical classification:	c. atrazine
____4. Marigolds and chrysanthemums make this insecticide:	d. organophosphates
____5. The most commonly used single insecticide for the control of insects on crops is ______?	e. rice
____6. Malathion® is in the chemical classification of ______?	f. pyrethrins
____7. Chlorinated hydrocarbons and organophosphates kill insects by acting as inhibitors of the neurotransmitter ______.	g. aldrin, diedrin
____8. Indonesia has increased its production of ______ while decreasing use of pesticides.	h. strip cutting
____9. This is a broadleaf herbicide used to control weeds in wheat fields:	i. biomagnification
____10. This herbicide is an endocrine disrupter:	j. *K* strategists
____11. Top carnivores have greater concentrations of pesticides and pesticide residues than algae or minnows. This is termed ______.	k. agriculture
____12. The use of antibiotics or pesticides on ______ organisms invites the evolution of resistance.	l. Bt-toxin
____13. Vertebrate predators of pests are generally ______.	m. DDT
____14. An explosion of this material used to make carbamate insecticides, killed approximately 5,000 people in Bhopal, India:	n. 2,4-D
____15. Bacteria produce this chemical that kills when ingested by mosquito larvae and caterpillars:	o. chlorinated hydrocarbons
____16. Cover for predators of pests can be provided by greenbelts between fields or _____.	p. *r* strategists
____17. The FDCA and FIFRA laws enforced by the U.S. EPA banned these highly toxic chlorinated hydrocarbon insecticides in 1974 after 80% of all dairy products were found contaminated:	q. acetylcholine

Multiple Choice

1. Which chemical is not appropriately matched with its use?
   a. DDT/malaria mosquitoes
   b. 2,4-D/broadleaf weed killer
   c. Bt-toxin/larval mosquitoes
   d. Pyrethrins/grass herbicides
   e. Carbamates/dusts for crop pests

2. Regarding human health, insecticides have saved the most lives by preventing death from:
   a. hepatitis.
   b. yellow fever.
   c. malaria.
   d. equine encephalitis.
   e. West Nile encephalitis.

3. *Bacillus pompillae* and pheromone traps are used to control what common garden pest?
   a. Japanese beetles
   b. rats
   c. slugs
   d. cabbage moths
   e. cotton boll weevils

4. A farm manager has a crop. She sees corn borers in her stalks. She has a crop duster fly over her 100 acres of corn releasing carbamate dust. Next year she has a worse problem and spends more money for multiple dustings. Which statement below describes this situation?
   a. monoculture
   b. no predators available to eat pests
   c. pesticide treadmill
   d. she will have to switch to a more powerful and costly pesticide soon
   e. all of the above

5. The population of this bird recovered strongly after DDT was banned:
   a. brown pelican
   b. bald eagle
   c. passenger pigeon
   d. ring-billed gull
   e. a and b only

6. Bioaccumulation and biomagnification have the greatest effect on which trophic level?
   a. fish hawks
   b. sunfish (freshwater)
   c. herbivorous minnows
   d. decomposition bacteria
   e. single-celled algae

7. Which of the explanations below are correct concerning biomagnification?
   a. Pesticides affect single-celled organisms more dramatically.
   b. The food chain is interrupted at the producer level.

c. *K* strategists have more fat stores.
d. *r* strategists are more directly affected.
e. a, b, and d.

8. Aerial spraying of pesticides moves pesticides through which pathway?
   a. air to crop pest to soil
   b. air to surrounding woods to wasps and birds
   c. air to soil to freshwater to ocean
   d. air to crop to food to food laborers and consumers
   e. all of the above

9. The infamous Agent Orange contained ______.
   a. 2,4-D
   b. 2,4,5-T
   c. dioxins
   d. c only
   e. a, b, and c

10. Which statement is correct concerning genetic resistance?
    a. It develops relatively quickly in *r* strategists.
    b. Gene mutations produced variants that made chemicals that counteracted pesticides.
    c. Pests evolved.
    d. There is competition among chemical companies to create more powerful insecticides to keep up with genetic change.
    e. All of the statements are correct.

11. Which pest actually became one only after extensive spraying of DDT?
    a. cotton boll worm
    b. citrus scale insect
    c. cabbage moths
    d. Colorado potato beetle
    e. Japanese beetles

12. Genetically modified corn has a gene incorporated into its genome that makes ______ to resist attack by corn stem borers.
    a. *Bacillus thuringiensis* toxin
    b. nicotine
    c. pyrethrins
    d. rotenone
    e. caffeine

13. Endocrine disrupters cause which of the following hormones?
    a. thyroid hormones controlling metabolism
    b. estrogens increased in males
    c. pancreas makes too much insulin
    d. c only
    e. a and b

14. Pheromones have been used to control which major pest of oak trees in the eastern deciduous forest?
    a. Japanese beetles
    b. Dutch elm beetles
    c. gypsy moth
    d. diamondback moth
    e. all of the above

15. Which pest has been successfully controlled using the sterile male technique?
    a. sea lamprey
    b. Mexican screw-worm maggot
    c. tsetse flies
    d. b only
    e. a, b, and c

16. Which of the following chemicals would not be expected in agricultural runoff and aquifer leaching?
    a. pesticides
    b. herbicides
    c. fungicides
    d. nitrates enhancers of toxicity
    e. Trick question alert! All of these would be expected.

17. This pest has developed resistance to Bt-toxin:
    a. European corn borer
    b. diamondback moth
    c. cotton ringworm
    d. tomato pinworm
    e. all of the above

18. Which of the following statements is an accurate description of resistance management?
    a. Genetic modification incorporates genes for crop resistance.
    b. Interbreeding moves genes from resistant plants to crops.
    c. Areas are maintained where nonpesticide pests are bred.
    d. Hybridization of pest species produces all offspring with pesticide resistance.
    e. Trick question alert! All of these are accurate.

19. The widespread use of Bt-GM crops could cause a catastrophic fall of the populations of which creature?
    a. butterflies
    b. robins
    c. bald eagles
    d. brown pelicans
    e. all of the above

20. Which of the following actions is not required by U.S. laws regulating pesticide production and use?
    a. testing and registration of all pesticides
    b. the banning of any substance that causes cancer

c. the determination of possible effects on children
d. the quantities of pesticide residues on or in imported food are published
e. old pesticides were retested to determine their toxicity

21. Which of the following foods have the greatest cancer risk from pesticide residues?
    a. tomatoes
    b. apples
    c. potatoes
    d. beef
    e. peaches with those little hairs that pesticides cling to

22. Which of the following statements is accurate concerning irradiation of food?
    a. Irradiating food increases its level of radioactive isotopes and therefore its level of radioactivity.
    b. Gamma rays knock electrons off small molecules, creating free carcinogenic radicals and peroxides at dangerous levels.
    c. Gamma rays knock electrons off small molecules, creating free radicals and peroxides at very low levels.
    d. Ionizing radiation turns food into mush.
    e. Mutant bacteria and fungi are likely to become pathogens.

23. Irradiating food will produce this health effect:
    a. decrease in *E. coli* O157:H7 infections
    b. decrease in hepatitis A infections from strawberries grown in Mexican feces-fertilized soils
    c. increased risk of cancer in workers in the irradiating plants
    d. increased nuclear waste generated
    e. all of the above

24. Which of the following insecticides is not on the "dirty dozen" list of persistent organic pollutants?
    a. DDT
    b. carbamates
    c. chlordane
    d. dieldrin
    e. mirex

25. Integrated pest management controls pests by:
    a. increasing predator populations.
    b. increasing crop diversity.
    c. using sterile male matings.
    d. c only.
    e. a and b.

26. The Delaney clause of the FDCA had to be modified in 1996 for this reason:
    a. Older pesticides were not restricted as food residues.
    b. Newer chemical assay techniques revealed that there were contaminants of nearly all foods.
    c. Food producers successfully lobbied to eliminate all pesticide residue testing.
    d. Adult exposures to pesticides were significantly more important because of bioaccumulation.
    e. All of the above were notable reasons.

CYBER SURFIN'

1. See the University of Reading (U.K.) primer on pesticides: **www.ecifm.rdg.ac.uk/pesticides.htm.**
2. From the Global Plan for Action on the Protection of Marine Environments site, see the page on the history of persistent organic pollutants: **pops.gpa.unep.org/04histo.htm**
3. The Environmental Working Group—Richard Wiles, Kert Davies, Susan Elderkin—have a very informative "Shoppers Guide to Pesticides in Produce": **www.ewg.org/pub/home/Reports/Shoppers/Chapter1.html.** Also check out POPs in human breast milk from the Natural Resources Defense Council: **www.nrdc.org/breastmilk/chems.asp**
4. See the U.S. Fish and Wildlife page on controlling sea lampreys: **midwest.fws.gov/marquette/etc/slm2.html#sterile** and their site on the effects of pesticides on wildlife: **contaminants.fws.gov/Issues/Pesticides.cfm**
5. Broward County, Florida, Pollution Prevention and Remediation Division has a primer called "Common Sense Pest Control." Alternative pest control measures are described for the home and yard. See **www.co.broward.fl.us/ppi02100.htm.** See also the offering from San Luis Obispo County, California: **www.sloag.org/Bugs/alternative%20pest%20index.htm**
6. The University of Minnesota has an introduction to pheromone insect control, "Understanding Semiochemicals with Emphasis on Insect Sex Pheromones in Integrated Pest Management Programs'" by Hollis Flint and Charles Doane at **ipmworld.umn.edu/chapters/flint.htm**
7. Check out the Pesticide Action Network—North America presentation on Genetically Modified Crops and Foods at **www.panna.org/resources/geTutorial.html**

'LIKE' BOOKING IT

1. Carson R. ***Silent Spring.*** Boston: Houghton Mifflin, 1962.
2. Winston M L. ***Nature Wars: People Versus Pests.*** Cambridge, MA: Harvard University Press, 1997.
3. Lapierre D, and Moro J. ***Five Past Midnight in Bhopal.*** New York: Warner Books, 2001.
4. Colbun T, Dumanowski D, and Myers J. ***Our Stolen Future: Are We Threatening Our Fertility, Intelligence, and Survival? – A Scientific Detective Story.*** New York: Plume Books, 1997.
5. Fagin D, and Lavelle M. ***Toxic Deception: How the Chemical Industry Manipulates Science, Bends the Law, and Endangers Your Health.*** Center for Public Responsibility.
6. Ellis B W, Bradley F M, Atthowe H, and Yepsen R. ***The Organic Gardener's Handbook of Natural Insect and Disease Control: A Complete Problem-Solving Guide to Keeping Your Garden & Yard Healthy Without Chemicals.*** Emmaus, PA: Rodale Press, 1996.

ANSWERS

Seeing the Forest

Fill-in the table below, referring to Table 22.1 and some Internet research at **pops.gpa.unep.org/04histo.htm** and Exotox Net from Oregon State U., Michigan State U., Cornell U., U. Cal. Davis and Idaho State U.: **ace.orst.edu/info/extoxnet/pips/ghindex.html**

Name	Chemical classification	Specific use or use with the greatest benefit	Toxicity and regulatory status	Ecological and health effects	Persistence
DDT	Chlorinated hydrocarbon insecticide	Control of malaria	Moderately toxic, banned in 1972	Toxic to birds, fishes, and aquatic invertebrates. Prevents eggshell formation in falcons, eagles, ospreys, brown pelicans. Has endocrine disruption properties	Moderate persistence, but bio-accumulates in fatty tissues
Chlordane	Chlorinated hydrocarbon insecticides	Termite treatments for buildings, crops	Moderately toxic, banned in 1988	Toxic to humans, birds, fishes, and aquatic invertebrates. It has endocrine-disruption properties leading to impaired reproduction.	Long-term persistence lasting over 20 years; bio-accumulates in fatty tissues.
Dieldrin, aldrin	Chlorinated hydrocarbon insecticides	Termites of homes and crops	Banned in 1975 in the U.S.	Toxic to humans, birds, fishes, and aquatic invertebrates. It has endocrine-disruption properties leading to impaired reproduction.	Persistence of 5+ years, bioaccumu-lates in fatty tissues.
Malathion®	Organo-phosphate insecticide	Sucking insects on crop leaves, flies, mosquitoes	Licensed for use	Highly toxic to aquatic wildlife, such as amphibians and invertebrates.	Low persistence, one month approx.
Carbaryl (Sevin®)	Carbamate insecticide	Caterpillars on crops, leaf-chewing insects in general	Most commonly used insecticide	Moderately toxic to fish, highly toxic to aquatic invertebrates and bees. Toxic to soil microorganisms including *rhizobia* (bad) and root-eating nematodes (good). Neurotoxic, personality changes noted in animals dusted with carbaryl.	Low persistence
Pyrethrins	Botanical insecticides made by marigolds and chrysan-themums	Quick "knockdown" of flying insects by contact.	Most common ingredients of fly sprays	Highly toxic to fish and aquatic wildlife	Low persistence

Atrazine	Herbicide	Broad spectrum herbicide kills broadleaf weeds and grasses	Licensed	May cause frog developmental deformities	Very high persistence in soils
2,4-D	Herbicide	Broadleaf herbicide	Licensed	Highly toxic to fish and aquatic invertebrates. Endocrine disrupter. Products may be contaminated by dioxins.	Short persistence in soils and aquatic ecosystems but longer when leached into groundwater.
Methyl bromide	Fungicide, nematicide, insecticide	Fumigant	Restricted use	Highly toxic to humans. Destroys nerve cells. Moderately toxic to aquatic organisms.	Evaporates to become an ozone-layer-destroying compound.

Questions to Ponder

1. List and describe two important general benefits of pesticides.

1. Increased factory farm productivity	Large-scale farming operations have increased productivity of food crops, but at high costs to the environment and human health.
2. Decreased disease	*Plasmodium* (malaria) is a disease organism that has killed and debilitated more humans in tropic and subtropical regions than any other organism. If insecticides were eliminated, mosquito-vectored diseases, such as yellow fever and various viral encephalitis infections, would increase; biting tsetse and sleeping sickness would increase; and lice and lice-vectored diseases, such as typhus, would increase.

2. Compare and contrast the specific benefits (specific disease or crop-eating pest controlled) and harmful effects of DDT use since its discovery. See **pops.gpa.unep.org/04histo.htm**. *DDT was considered to be the ideal pesticide—it was inexpensive to produce, and it was considered to be "harmless" to humans and animals. During World War I there were five million deaths due to louse-vectored typhus (rickettsia bacteria) infection. In post–World War II Europe DDT powder was sprayed into the clothes of refugees and prison camp survivors. During World War II, soldiers were exposed to insect-vectored diseases (such as malaria, yellow fever, kala azar, etc.) when fighting in tropical and subtropical areas in North Africa, South Europe, Southeast Asia, and the South Pacific islands. The U.S. army had insect control scientists who accompanied the soldiers, and they applied DDT, successfully controlling these diseases. Farmers began to use DDT after the war. Insecticide was applied to fruit trees, vegetable crops,*

and cotton. Sprays were developed for home use, first in pump spray devices and later in pressurized cans. Homeowners controlled flies, mosquitoes, and roaches. The pesticide treadmill effect then occurred: first the farmers began to use larger and larger quantities of DDT to control the cotton pest, and finally resistance was noted in house flies. The chemical companies responded by developing new, more toxic insecticides to counter the evolution of resistance. Rachel Carson's book in 1962 opened the public's eyes to the problems of insecticide toxicity and persistence in the environment. Later the endocrine-disrupting properties were described biochemically to explain the falling populations of American robins, falcons, eagles, and brown pelicans. The bioaccumulation and biomagnification of DDT were described by scientists. Large amounts of DDT were used in forests and on city streets in the 1960s to control the widespread outbreak of Dutch elm fungal disease of trees in the eastern deciduous forest of North America, vectored by bark beetles. DDT was banned from production and use in 1972. At that time human breast milk had more DDT and its residues than allowed by the FDA in processed food.

3. Describe the following phenomena and means of dispersal or movement by filling in the table below.

	Description	**Means of dispersal or movement**
Biological accumulation and magnification	Because each succeeding trophic level of a food chain has generally greater fat stores, the total content of fat soluble chlorinated hydrocarbons and other organic chemicals increases with each succeeding feeding level.	Water in an aquatic ecosystem to algae, which are eaten by shrimp, which are eaten by eels, which are eaten by Atlantic needle fish, which are eaten by ring-billed gulls.
Mobility in the environment	Pesticides reach nontargeted soils and waters by overspray or drift, precipitation, and leaching from soil into groundwater, handling the crop in picking, processing or consumption; or by excretion from humans and animals. From soil, pesticides and their residues can move again into crops and back into humans and animals. Similarly, pesticides and residues can move through an aquatic food chain back into humans and animals, with bioaccumulation occurring with certain pesticides.	Overspray and drift, precipitation, leaching into groundwater to aquatic ecosystems, food handling, and ingestion.

4. Explain how the use of a specific insecticide can actually make the farmer's insect pest problems worse over time. *When an insecticide is used to control insect pests over a large area of a monoculture, the insecticide not only kills the targeted pests but kills the predators of the pest in the fields and nearby woods or tall natural vegetation. The wasp and bird predators cannot recover as fast by reproduction, because they generally are* K *strategists that reproduce slowly. Because most insecticides affect their target insects by interference with neurotransmission, the same chemicals will affect vertebrates. Depending on the dosage, death may result or endocrine disruption that interferes with fertility. The pests are* r *strategists that reproduce quickly and in large numbers to overcome high death*

rates. Therefore their populations recover faster and the pest problem persists. Refer to the figure above in "Questions to Ponder." The first response that farmers have tried historically is to increase application. The next response, like a physician trying to cure a persistent infection with antibiotics, is to try a different and more powerful insecticide. An insecticide treadmill results with accompanying high overhead expenses for the farmer.

5. Health problems of human and animals result from pesticide-induced endocrine disruption. See the National Resources Defense Council site: **www.nrdc.org/health/default.asp** and **http://www.nrdc.org/breastmilk/chems.asp**. See WWF environmental chemist, Michael Warhurst's website: **website.lineone.net/~mwarhurst/index.html** and the Stolen Future.org: **www.ourstolenfuture.org/Commentary/JPM/2002-1115scirevolution.htm.**

   a. Describe five specific effects of endocrine disrupters on the health of humans and other organisms, list five chemicals that act in this way, and list five hormones that are disrupted.

	Health or environmental effect of endocrine disrupters	**Specific examples of an endocrine-disrupting chemicals**	**Specific hormone affected and function**
1.	Feminizing effects in males due to blocking of testosterone receptors in target cells of the male reproductory tract, resulting in decreased production of sperms and undescended testes. Delayed puberty.	Phthalates in plastics, DDE, dioxins, PCBs	Testosterone normally stimulates the descent of testes in the fetus and sperm production after puberty.
2.	Advance of puberty in females due to estrogen mimicry, enhanced breast cancer tumor growth	DDT and its metabolites DDD and DDE	Estrogen affects mating behavior, causes mineral eggshell formation in reptiles and birds.
3.	Delay of puberty of females due to estrogen-blocking action	Atrazine (herbicide)	Estrogen normally causes puberty and regulates cyclic changes of the uterus during the estrous/menstrual cycle.
4.	Decreased metabolism due to the blocking of thyroid hormone reception by target cells	PCBs industrial wastes	Thyroid hormones control the activity and energy use of cells.
5.	Affects of hypothalamus-pituitary-gonad axis of fertility	Triazine herbicide	Hypothalamic-releasing hormones and pituitary gonadotropins are reduced, resulting in sterility.
6.	Antagonistic binding of chemical to glucocorticoid and mineralocorticoid receptors.	Arsenic compounds used in wood treatment	Glucocorticoid or cortisol is important to stress resistance. Decreased resistance to stress results, so that the affected human may suffer from stress-related heart disease and be more likely to die from shock due to loss of blood pressure.

b. Describe the suspected endocrine disrupter that is increasing in human breast milk. *The flame retardant chemical polybrominated diphenyl ethers (PBDEs) are a class of widely used flame retardants used in plastics (television cabinets and computers) and are also found in construction materials, furniture, and textiles. The chemical structure of PBDEs mimics that of PCBs and dioxins. PBDEs are persistent in the environment and have been found in marine mammals. Its effects have yet to be determined, but PBDE is likely to bioaccumulate and disrupt hormones due to its chemical similarities noted above.*

6. Pretend you are a farmer who is going to have an ecologically sound pest management program for raising cotton and corn. What would you do differently as compared to the factory farm?

   I would take the following actions:

   a. *I would first apply nonpersistent pesticides using the calendar and scout methods. I would not broadcast insecticides over a large area. I would not use crop duster aircraft.*
   b. *I would maintain greenbelts or strip cropping for predator shelters. The greenbelts will border every large field.*
   c. *I will raise insect predators, such as wasps.*
   d. *Each field will be planted with a different crop—no adjoining fields will have the same crop or related one that is attacked by the same pest.*

7. What improvements would you make in the laws and regulatory mechanisms currently in place for the control of pesticide production and use? *I would institute a strict* ***Precautionary Principle Testing Program*** *for any exotic compound introduced into the environment in large quantities. I would strengthen the* ***Right-To-Know*** *laws so that each ingredient in a product must be disclosed.*

Pop Quiz

Matching

1. m
2. k
3. o
4. f
5. a
6. d
7. q
8. e
9. n
10. c
11. i
12. p
13. j
14. b
15. l
16. h
17. g

Multiple Choice

1. d
2. c
3. a
4. e
5. e
6. a
7. c
8. e
9. e
10. e
11. b
12. a
13. e
14. c
15. e
16. e
17. b
18. c
19. a
20. d
21. a
22. c
23. e
24. b
25. e
26. b

23

Solid and Hazardous Wastes

LEARNING OBJECTIVES

After you have studied this chapter you should be able to:

1. Distinguish between *municipal solid waste* and *nonmunicipal solid waste.*
2. Describe the features of a modern sanitary landfill and relate some of the problems associated with sanitary landfills.
3. Describe the features of a mass burn incinerator and relate some of the problems associated with sanitary landfills.
4. Summarize how source pollution, reuse, and recycling help reduce the volume of solid waste.
5. Define *hazardous waste* and briefly characterize representative hazardous wastes (dioxins, PCBs, and radioactive wastes).
6. Contrast the Resource Conservation and Recovery Act and the Comprehensive Environmental Response, Compensation, and Liability Act (the Superfund Act).
7. Discuss the scope, accomplishments, and shortcomings of the Superfund Program.
8. Define *environmental justice* and discuss the issue of international waste management as it relates to environmental justice.

SEEING THE FOREST

a. It has been said that old sanitary landfills are like *ticking time bombs* or on the other hand, *gold mines.* If cities would dig these up, what hazardous chemicals and valuable recyclables could be found? Fill in the table below.

Hazardous Chemical	Sources of Hazardous Chemical	Recyclables

b. Describe the principal environmental problems of old landfills and open dumps. See **http://www.epa.gov/region02/soer/r2soe/r2soe98wasteframe.htm.**

SEEING THE TREES

VOCAB

At first, the argument was that citizens would not go to the trouble to sort recyclable items from their trash. We now know that well-designed and publicized curbside collection programs in typical American suburban communities routinely achieve participation rates of 80% and higher. Skeptics also said that markets for recovered materials would not absorb all the new materials being collected. But since 1985, consumption of recovered metals, glass, plastic and paper by American manufacturers has grown steadily, even as commodity prices for virgin and recycled materials naturally fluctuate.

John Ruston and Richard Denison

1. Municipal solid waste—a mixture of solid materials of every description that are disposed of in municipal trash pickup from houses or businesses. Materials may be recycled, buried in landfills, placed in open dumps, dumped into the ocean in some areas of the world (New York City dumped garbage into the ocean until 1935), or incinerated. Some solid waste is trucked from the northeastern U.S. for disposal in Appalachian Kentucky.
2. Nonmunicipal waste—mining wastes (75%) including overburden from mining of waste rock after mineral extraction, slag from metal smelting, industrial wastes including chemicals and trimmed materials (9.5%), and agricultural wastes (13%) including broken equipment, manure, and unused plant materials.

3. Sanitary landfill—sanitary landfills differ from open dumps currently because they compact the garbage and cover it each day with soil. Each landfill is lined by a supposedly impenetrable concrete, clay/plastic or other barrier material to prevent leakage into groundwater.

4. Incineration of waste—incineration of combustible solid waste can reduce the amounts by 90%. The heat produced can make steam that generates electricity. Ideally, municipal incinerators should be equipped with *lime scrubbers* to remove acid rain gases and *electrostatic precipitators* to remove large particles of *fly ash*. See Chapter 19. Some fly ash particles may be contaminated by heavy metals (such as mercury, cadmium, and lead) or residue organic compounds such as dioxins if incineration temperatures are too low.

   Incinerator types include:
   a. Mass burn incinerator—large furnaces that burn all materials, including combustible and noncombustible materials. Noncombustible materials leave a solid *bottom ash* or *incinerator slag* that must be buried.
   b. Modular incinerator—smaller units that burn combustible wastes that are contaminated by noncombustible materials.
   c. Refuse-derived incinerators—these burn only combustible materials as all solid materials have been sorted mechanically and by hand before incineration.

5. Source reduction—solid wastes can be reduced by designing less waste in manufacturing processes, and capturing and reusing wastes at the factory where they are produced.

6. The Pollution Prevention Act of 1990—a U.S. law required the EPA to develop models for industrial waste recycling.

7. Dematerialization—many household appliances and even cars have gotten lighter because of the replacement of metals by plastics.

8. Product stewardship—the idea that a corporation would take responsibility for the disposal of its used products. For instance, oil producers would recycle the used oils and liquid greases, manufacturers would include more recyclable products in their vehicles and purchase those materials from the market of recyclables and reprocessed materials.

9. Fee-for-bag approach—this is a measure taken by many municipalities to reduce solid waste by encouraging recycling. In Charlottesville, VA (home of the University of VA) they achieved reductions in the weight (14%) and volume (37%) of solid waste. The weight did not go down as much because people became experts at compacting their garbage.

10. Principle of inherent safety—industrial processes are designed to use less toxic materials.

11. Solid Waster Recovery Act of 1965 and the Resource, Conservation, and Recovery Act (RCRA) amendments of 1970 and 1976—laws enacted to improve solid waste disposal methods and the management of hazardous and nonhazardous wastes. RCRA also promotes resource recovery and reduction of hazardous waste generation from point sources, such as open dumps and landfills.

12. Comprehensive Environmental Response, Compensation, and Liability Act (CERLA)—commonly known as the Superfund Act it designated funds to clean up abandoned and illegal toxic waste sites. The corporate cleanup tax was removed in 1995, and the Bush administration recently removed 33 sites from the Superfund allocation lists.

13. Superfund National Priorities List—Twenty percent of Superfund sites are old open dumps or landfills, the rest are military bases, nuclear materials processing sites-making sites, mining sites, and industrial sites.

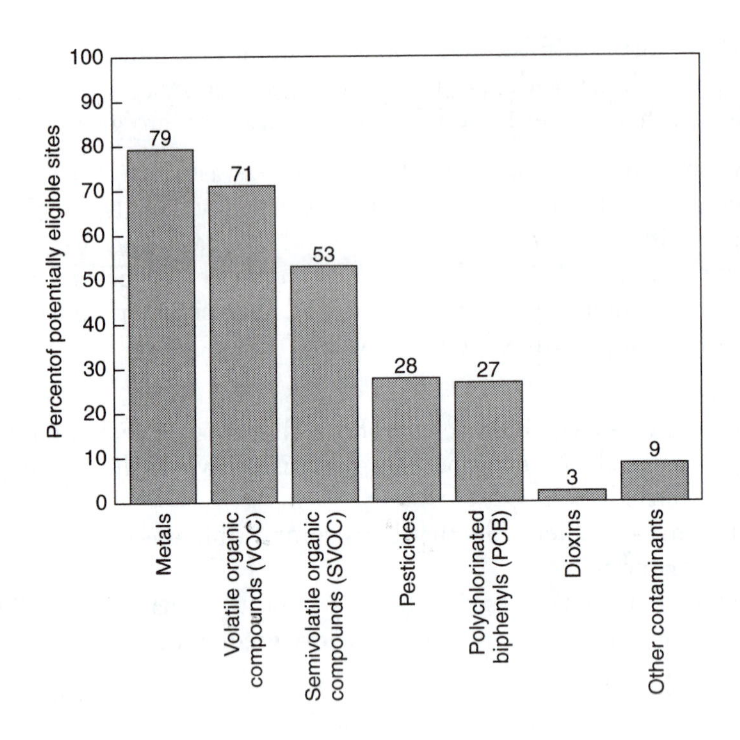

14. Bioremediation and phytoremediation—some plants, bacteria, and algae not only tolerate certain toxic metals and other elements but concentrate them in their cells. This bioaccumulation process allows toxic metal removal from the land. The twist flower, *Streptanthrus polygaloides*, can remove nickel from the soil: the fern, *Pteris vittata* can remove arsenic. The plants can be processed to remove the metals for industrial and other use. Nickel, for instance, was used in World War II to plate automobile bumpers, thus conserving the more valuable chrome for military use.
15. Environmental chemistry—also known as *green chemistry*, it is a subdiscipline of chemistry that identifies chemicals in wastes and determines their concentration. Green chemistry can reduce or eliminate hazardous chemicals at their source.
16. Environmental justice—the principle that every citizen has a human right to be protected from environmental hazards.
17. Basel Convention—this is an international treaty developed by the U.N. Environmental Program that attempts to restrict the transport of hazardous waste. It was signed in 1989, but not ratified by the U.S. Congress. An amendment made in 1995 bans the movement of any hazardous wastes between countries.
18. Integrated waste management—the *three R's* of waste management—reduce, reuse, and recycle—are combined to manage solid waste.
19. Voluntary simplicity—simpler is better or the KISS method can lead people to the understanding that "less can be more." Such an understanding is the foundation of any conservation effort.

Questions to Ponder

1. Describe two problems that open dumps have that sanitary landfills do not have to the same extent.

Problem	Description
1.	
2.	

2. Describe two materials that represent special problems for sanitary landfills.

Material	Description
1.	
2.	

3. Describe three benefits of recycling old cars.

	Description
1.	
2.	
3.	

4. List six materials that can be recycled in large amounts from municipal solid wastes, an explanation of why each is not more thoroughly exploited as a source of materials, and examples of the products that can be made from them.

	Recycled Material	Explanation Why More Is Not Recycled	Products
1.			
2.			
3.			
4.			
5.			
6.			

5. List six hazardous materials in municipal solid waste and several sources for each.

	Hazardous Material	Examples of Sources
1.		
2.		
3.		
4.		
5.		
6.		

6. Go to **map3.epa.gov/enviromapper/index.html.** Find and describe three Superfund sites nearest to your location. Get the description from **www.epa.gov/superfund/sites/npl/npl.htm.**

	Superfund Site and Location	Description
1.		
2.		
3.		

7. Explain how Hanford, WA, nuclear reservation can potentially pollute wide areas and will be one that is very difficult to cleanup. See **seattlepi.nwsource.com/local/66920_hanford18.shtml.**

8. List and describe three improvements that you would make to existing laws and treaties on the disposal of solid waste.

	Improvement	Description
1.		
2.		
3.		

POP QUIZ

Matching

Match the following terms and definitions

____1. The most toxic throwaway commercial items in municipal solid waste are:	a. aluminum
____2. Most solid waste is placed into a ______.	b. phytoremediation
____3. The most serious pollution problem related to solid waste disposal is ______.	c. dematerialization
____4. Because the land is cheap, many new ______ are located near or on landfills.	d. source reduction
____5. The chief reason explaining why cities use waste incineration is:	e. sanitary landfill
____6. The waste heat from incineration of solid waste can be used to produce ______?	f. Superfund sites
____7. Bottom ash and fly ash must be disposed of in ______.	g. plastic
____8. Designing manufacturing processes so that they produce less waste is an example of:	h. PCBs
____9. Refrigerators and cars are much lighter today and produce much less solid waste because of:	i. electricity
____10. The recycling of _______ saves large amounts of natural resources, mining damage and energy.	j. integrated waste management
____11. The recycling of ______ conserves oil resources.	k. groundwater leaching
____12. A hazardous waste from electrical transformers is ______.	l. computers
____13. The combustion of chlorinated organic compounds make this hazardous byproduct:	m. hazardous waste landfills
____14. _______wastes are stored near Hanford, Washington.	n. schools
____15. The funds that were accumulated by taxing polluting industries were set aside for cleaning up at ______.	o. 90% reduction of volume
____16. Cattails concentrating selenium in their cells describes:	p. dioxin
____17. "Reduce, reuse, and recycle"	q. radioactive

Multiple Choice

1. Electrical connections are soldered together with ______. There are at least 3 pounds of it in a desk model computer.
   a. dioxins
   b. PCBs
   c. lead
   d. mercury
   e. cadmium

2. Which of the following materials is most importantly reclaimed by the recycling of automobile and truck catalytic converters?
   a. steel
   b. platinum
   c. gold
   d. aluminum
   e. Trick question alert! All of these are equally important.

3. In automobile recycling, the "fluff" generated includes:
   a. glass
   b. antifreeze
   c. brake fluid
   d. oil
   e. all of the above

4. Beverage deposit laws encourage which of the following responses by the public and bottlers?
   a. Less litter is produced.
   b. Acrylic or glass containers are used by beverage producers.
   c. Beverage containers are recycled.
   d. Oil is conserved.
   e. All of the above are correct descriptions.

5. The country that produces the most solid waste is ______.
   a. U.K.
   b. France
   c. Canada
   d. Russia
   e. U.S.

6. The country that recycles the greatest percentage of its solid waste is ______.
   a. U.K.
   b. France
   c. Canada
   d. Russia
   e. U.S.

7. The U.S. could do a much better job of recycling and exporting ______ as compared to other countries.

a. paper
b. glass
c. copper
d. coal
e. steel

8. The explanation of why more plastic is not recycled is which of the following statements?
   a. It cannot be compressed.
   b. It is more valuable to burn.
   c. Oil is cheaper than reclaimed plastic.
   d. The chemical processes for recycling plastic are not known.
   e. All of the above are correct.

9. This material in landfills tends to migrate upward because it is less dense than surrounding materials:
   a. refrigerators
   b. tires
   c. paper
   d. ethylene glycol in antifreeze
   e. all of the above

10. Not including municipal waste, which activity produces the most solid waste?
    a. automobile manufacture
    b. appliance manufacture
    c. agriculture
    d. mining
    e. the military

11. Which of the following materials is produced by decomposing organic matter in landfills, could be claimed, but is usually "flamed off"?
    a. ethyl alcohol
    b. methane
    c. cellulose fiber
    d. rubber
    e. oil

12. Which of the following describes a use for recycled tires?
    a. as an ingredient in road surfaces/asphalt
    b. sound tire carcasses for recapped tires
    c. burned to generate steam for electricity generation
    d. reef-building
    e. all of the above

13. Thermostats, electrical switches, and thermometers cannot be incinerated because they contain:
    a. platinum
    b. dioxins
    c. mercury
    d. tetraethyl lead
    e. all of the above

14. Incineration of solid wastes in Japan produces 10 times the amount of ______ in the air as compared to other highly developed countries.
    a. lead
    b. dioxins
    c. PCBs
    d. mercury
    e. cadmium

15. The main problem in cleaning up cancer-causing PCBs in the environment is which of the following?
    a. They are leached into groundwater.
    b. There are large quantities buried in landfills.
    c. They are very persistent.
    d. They can only be cleaned up by bacteria on the surface of soil.
    e. All of the above are correct.

16. The Rocky Mountain Arsenal is contaminated by which hazardous chemical?
    a. nerve gas and its residues
    b. diedrin
    c. mustard gas and its residues
    d. jellied leaded gasoline/napalm
    e. all of the above

17. Which of the following materials cannot be presently lessened by phytoremediation or biological accumulation?
    a. selenium
    b. PCBs
    c. nickel
    d. strontium-90
    e. Trick question alert! All of these can be biologically remediated.

18. Solid waste incineration requires the use of which pollution control technique?
    a. fly-ash precipitators
    b. lime smokestack scrubbers
    c. burning of solid incinerator slag and fly ash in leak-proof landfills
    d. c only
    e. a, b, and c

19. Which of the following proposals would reduce municipal solid waste the most?
    a. make fewer new cars and repair the old cars
    b. make cars entirely of plastic
    c. have national beverage container deposit laws
    d. dematerialize all appliances
    e. make compost piles of yard clippings

20. Using waste paper and tires to produce energy for making electricity saves which natural resource?
    a. coal
    b. natural gas

c. trees
d. c only
e. a and b

21. The Resource Conservation and Recovery Act designated which agency of the U.S. government responsible for determining whether a solid waste chemical is hazardous?
    a. U.S. Department of Agriculture
    b. U.S. Environmental Protection Agency
    c. U.S. Food and Drug Administration
    d. U.S. Department of Defense
    e. U.S. Fish and Wildlife Service

22. Which act or treaty legislated against the transport of hazardous waste across national boundaries?
    a. Comprehensive Environmental Response, Compensation, and Liability Act
    b. the Basel Convention
    c. the Stockholm Treaty
    d. Superfund Amendments and Reauthorization Act
    e. a and b

23. The most enviromentally sound way to dispose of municipal and agricultural organic food and vegetation wastes is which of the following?
    a. composting
    b. burning to produce electricity
    c. phytoremediation to remove selenium that is sold as vitamins
    d. feed it to pigs
    e. bury it in landfills

24. Which of the following materials would not be expected in incinerator solid waste?
    a. mercury
    b. lead
    c. dioxins
    d. cadmium
    e. arsenic

CYBER SURFIN'

1. Read a History of Waste Management in New York City at **www.johnmccrory.com/bags/history/historyl.html**
2. The Sierra Club has an informative site on Superfund sites: **www.sierraclub.org/toxics/superfund**
3. The U.S. EPA has a lot of information on solid waste:
    a. An interactive toxic waste and Superfund map at **http://map3.epa.gov/enviromapper/index.html**
    b. A general overview of solid waste management at **www.epa.gov/osw**

c. Information on specific types of hazardous waste at **www.epa.gov/epaoswer/osw/hazwaste.htm**
d. "Reduce, Reuse, and Recycle Waste": **www.epa.gov/epaoswer/osw/rrr.htm**

4. Check out websites on the infamous Love Canal in Niagara Falls, NY: The Center for Health, Environment and Justice, **www.chej.org/lovecanal.html**; and ATOMcC Science News at **web.globalserve.net/~spinc/atomcc/lovecana.htm**
5. To see a discussion and lists of toxic hazardous wastes from the Communicable Disease Centers (CDC), go to **www.atsdr.cdc.gov/toxhazsf.html**
6. From Environment Canada, see The Environmental Implications of the Automobile at **www.ec.gc.ca/soer-ree/English/products/factsheets/93-1.cfm**

'LIKE' BOOKING IT

1. Colten C E, and Skinner P N. ***The Road to Love Canal: Managing Industrial Waste Before EPA.*** Austin: U. of Texas Press, 1996.
2. Mazur A. ***A Hazardous Inquiry: The Rashomon Effect at Love Canal.*** Cambridge, MA: Harvard U. Press, 1998.
3. Barlow M, and May E. ***Frederick Street: Living and Dying on Canada's Love Canal.*** New York: HarperCollins, 2000.
4. Ackerman F. ***Why do we Recycle?: Markets, Values and Public Policy.*** New York: Island Press, 1996.
5. Strasser S, McMillan M, and Austen A. ***Waste and Want: A Social History of Trash.*** New York: Owl Books/Henry Holt, 1999.
6. Rathje W L, and Murphy C. ***Rubbish!: The Archaeology of Garbage.*** New York: HarperPerennial Books, 1993.
7. Lilienfeld R M, and Rathjie W L. ***Use Less Stuff: Environmental Solutions for Who We Really Are.*** New York: Fawcett/Ballantine Books, 1998.

ANSWERS

Seeing the Trees

a. It has been said that old sanitary landfills are like *ticking time bombs* or on the other hand, *gold mines.* If cities would dig these up, what hazardous chemicals and valuable recyclables could be found? Fill in the table below.

Hazardous Chemical	Sources of Hazardous Chemical	Recyclables
Lead	Car batteries, TVs, computers, paint	Iron
Mercury	Electrical switches, car batteries, paint	Aluminum
Arsenic	Pesticides, paint, and treated wood	Chromium
Cadmium	Rechargeable batteries, paints	Zinc (used in alloys)
PCBs	Older appliances, radios, TVs, computers, electrical transformers and capacitors	Glass or plastics, depending on market value

b. Describe the principal environmental problems of old landfills and open dumps. See **www.epa.gov/region02/soer/r2soe98wasteframe.htm.** *Old landfills are notoriously leaky, polluting groundwater, streams, and other bodies of water nearby. Water pollutants include microorganisms and toxic chemicals. The ground sinks as the landfill settles and materials in it decompose. Odors emanate from the site. Methane concentrations may be high in the air above the site.*

Questions to Ponder

1. Describe two problems that open dumps have that sanitary landfills do not have to the same extent.

Problem	Description
Rats, mice, birds	Open dumps have disease-carrying rats and mice that breed in garbage dumps and move to nearby communities.
Smells	Rotting food smells travel downwind to communities. Open dumps many times have fires.

2. Describe two materials that represent special problems for sanitary landfills.

Material	Description
Tires	Tires tend to be less dense than other materials, rising to the surface.
Plastics	Most plastics do not decompose within 20 years.

3. Describe three benefits of recycling old cars.

	Description
1.	Reduction of environmental damage done by surface and deep mining of coal (for smelting energy) and metallic mineral ores
2.	Conservation of metals resources so that our supplies will extend to future generations
3.	Conservation of energy supplies, most of which are nonrenewable

4. List six materials that can be recycled in large amounts from municipal solid wastes, an explanation of why each is not more thoroughly exploited as a source of materials, and examples of the products that can be made from them.

	Recycled Material	Explanation Why More Is Not Recycled	Products
1.	Paper	It is difficult, but not impossible, to separate contaminated paper from clean.	More paper for news and the office
2.	Cardboard	Cardboard is valuable enough to be exported. But Asian rice paper cardboard has brought its price down recently.	More cardboard for boxes

3.	Metals	If more communities would recycle, we could conserve more metal.	Metal for appliances, building tools, etc.
4.	Glass	The price of glass is low right now because there is so much sand available. Glass may be broken in pieces that are difficult, but not impossible, to separate from other trash.	Window glass, glass containers
5.	Plastic	Oil is still too cheap. Also plastic has high volume for a small weight, increasing the cost of transport, and it consists of several types that must be separated.	More plastic containers, also "plastic lumber" for building
6.	Organic wastes	People think it's yucky. Organic wastes can be composted to fertilize soil or digested to make biogas.	Fertilized soil and biogas for heating and the generation of electricity

5. Go to **map3.epa.gov/enviromapper/index.html.** Find and describe 3 Superfund sites nearest to your location. Get the description from **www.epa.gov/superfund/sites/npl/npl.htm.**

	Superfund Site and Location	**Description**
1.	Luminous Processes Inc. U.S. Hwy 78 and 29, Athens, GA 30622	"Between 1952 and 1978, the company used radioactive material in the production of luminous watch and clock dials. The building and surrounding grounds, including adjacent property, are contaminated with radium-226 and tritium in excess of generally accepted permissible levels of radiation or radioactive materials in unrestricted or noncontrolled area."
2.	Diamond Shamrock Corp Landfill, W. Girard Ave., Cedartown, GA 30125	"Residues from herbicide manufacturing and latex waste from the carpet-manufacturing industry were buried in unlined trenches. According to records obtained from Velsicol Chemical Corp.'s plant in Chattanooga, Tennessee, the wastes contained arsenic and organic chemicals, including benzonitrile and herbicides. About 3,000 tons of hazardous waste were buried during the operational period."
3.	Savannah River Nuclear Reservation Site, Aiken, S.C.	"SRS operations generate a variety of radioactive, nonradioactive, and mixed (radioactive and nonradioactive) hazardous wastes. Past and present disposal practices include seepage basins for liquids, pits and piles for solids, and landfills for low-level radioactive wastes. According to a 1987 USDOE report, shallow ground water on various parts of the site has been contaminated with volatile organic compounds (degreasing solvents), heavy metals (lead, chromium, mercury, and cadmium), radionuclides (tritium, uranium, fission products, and plutonium), and other miscellaneous chemicals (e.g., nitrates)."

7. Explain how Hanford, WA, nuclear reservation can potentially pollute wide areas and will be one that is very difficult to cleanup. See **seattlepi.nwsource.com/local/66920_hanford18.shtml.** *Radioactive materials have been poured into pits in the ground. Pools and vats of slowing decaying radioisotopes boil and condense in aging containers. Groundwater pollution advances toward streams in the Columbia River drainage. Stontium-90 and iodine-129 enter food chains. We will need a massive cleanup or commitment of two hundred years of governmental maintenance of the site until it cools off by radioactive decay.*

8. List and describe three improvements that you would make to existing laws and treaties on the disposal of solid waste.

	Improvement	Description
1.	National deposit law	This has been very successful in Michigan and other states in reducing litter and conserving resources.
2.	The elimination of non-recyclable containers dispensing more than 6 ounces of beverage.	Combined with product stewardship by the manufacturer and sellers of beverages, it would conserve resources.
3.	Require all cities and counties to recycle glass, plastic, paper and metals.	Solid waste disposal should be subsidized by the citizens. We are running out of landfill space and the NIMBY response will make us switch to incineration or recycling. As virgin natural resources decline, market forces that will make us recycle.

Pop Quiz

Matching

1. l
2. e
3. k
4. n
5. o
6. i
7. m
8. d
9. c
10. a
11. g
12. h
13. p
14. q
15. f
16. b
17. j

Multiple Choice

1. c
2. b
3. a
4. e
5. e
6. e
7. a
8. c
9. b
10. d
11. b
12. e
13. c
14. b
15. e
16. e
17. e
18. e
19. a
20. e
21. b
22. e
23. a
24. c

24

Tomorrow's World

LEARNING OBJECTIVES

After you have studied this chapter you should be able to:

1. Define *sustainability* and explain the very different reasons why people in highly developed countries and developing countries are not living sustainably.
2. Discuss how the natural environment is linked to sustainable development.
3. Explain why we should respect and care for the community of life.
4. Describe some of the challenges confronting our efforts to improve the quality of human life worldwide.
5. List two major steps we must take to stay within Earth's carrying capacity.
6. Describe the role of education in changing personal attitudes and practices that affect the environment.
7. Discuss at least two important environmental goals that can be accomplished most effectively at national and international levels.
8. Write a one-page essay describing what kind of world you want to leave for your children.

SEEING THE FOREST

This edition has considered issues in air quality, water quality, soil quality, biodiversity, economics, and consumption of natural resources. How would the concept of sustainability affect the current state of each? Who would be held accountable for harmful changes in each?

Environmental Aspect	Effects of Application of a Principle of Sustainability	Accountability: Who or What Will Be Held Accountable for Harmful Changes?
Air quality		
Water quality		
Soil quality		
Biodiversity		
Economic health of the world		
Consumption behaviors		

SEEING THE TREES

VOCAB

We shall require a substantially new manner of thinking if mankind is to survive.

Albert Einstein

1. *Caring for the Earth*—this is a report published in 1991 by the World Conservation Union, the United Nations Environment Program, and the Worldwide Fund for Nature.
2. Sustainable development—a level of economic development that does not further compromise the environment or the availability of natural resources in the future. *Caring for the Earth* concluded: (1) We should be living on the interest of the *natural resources bank account* rather than on the principal and (2) the power to affect the environment of increasing numbers of humans has increased, the rate of change may be so great, that scientific and technological remedies may fail to keep pace. Kai N. Lee of Williams College in *Compass and Gyroscope: Integrating Science and Politics for the Environment* concluded that "sustainability is like freedom and justice, something that may not be universally attainable, but nevertheless is a noble goal." Dr. Lee also wrote, "Human action affects the natural world in ways we do not sense, expect, or control." (book synopsis, see Cyber Surfin' below).

3. Nine principles to build a sustainable society follow:

   I. *Build and maintain a sustainable society:* Behavioral attitudes must be changed in the following ways: Nature, unaltered or managed, must be valued at emotional and economic levels. The advice of scientists must be taken into equal account with economics. We must consider consumption and environmental exploitation from moral and ethical perspectives. We must value residents of under developed countries as we value ourselves.

   II. *Respect and care for the community of life:* The ecosystem services provided by intact or managed ecosystems must be valued emotionally and economically to attain sustainable development. We must consider the renewability of natural resources. We must be concerned about plant and animal extinctions caused by human activities. We must realize our absolute dependence on nature.

   III. *Improve the quality of human life:* The ultimate goal of economic development should be improving the quality of human life rather than simple profit. One person in every four in the world is living in poverty. One in seven poor people is receiving less than 80% of the required calories recommended for adequate growth, maintenance of organs, and mental function. World poverty, inadequate education of professionals, and the status of women must be addressed by economic development.

   IV. *Conserving the Earth's vitality and biological diversity:* Eighty percent of the world's species live in tropical regions and underdeveloped countries. The science of *restoration ecology* must be practiced along with the conservation and preservation of the remaining natural resources. We must preserve the soil and water that it takes to grow crops and animals. Human ingenuity can be applied to make agriculture sustainable, to reduce fertilizers and the bee- and bird-killing pesticides used, and responsibly introduce varieties of naturally and genetically engineered crops to increase food quality and offer built-in resistance to pests. Soil erosion and aquifer depletion trends must be reversed.

   V. *Keep within the Earth's carrying capacity:* The concept of carrying capacity of an environment is equivalent to sustainability. An ecosystem can absorb and transform wastes into nutrients for organisms in food chains that support humans. To attain a sustainable biosphere requires the control of excessive consumption, waste and human population. *Full-cost accounting* should be applied to account for the expenses of excessive consumption and the waste of natural resources. Capitalists accuse environmentalists of wanting to consider all organisms as having equal rights to existence and of assuming that newborns are consumers instead of producers. Nature supports production by plants, herbivory, and predation in ecological balance but does not favor overexploitation—humans are consumers of natural resources and producers of wealth. The emphasis of environmentalism is survival, not the individual accumulation of wealth.

   VI. *Changing personal attitudes and practices:* "Less is better," apathy, ignorance, or incentives for wasteful consumption (the tax deductions of $25,000 in business expenses given to those who purchased SUVs in 2002). An educated public will be more responsible than an uneducated one. Therefore increase environmental education; encourage local conservation and preservation organizations; support natural history museums, zoos, aquaria, and botanical gardens. Write to your representative and complain about the poor quality of the U.S. Department of Commerce's "National Aquarium." Encourage the inclusion of environmental material in churches and other groups.

   VII. *Enabling our communities to care for their own environments:* Garbage is truly not "throwaway," it has to go somewhere. Large cities have been trucking garbage to sites far away from metropolitan centers. They are filling up, and new landfills are now encroaching on developing communities. Reduce, reuse, and recycle is fast becoming an economic imperative. Community action follows from community knowledge. Modern telecommunications can get communities the information they need on weather and pest trends. Energy production must be decentralized, giving each community the ability to produce their needs from solar, wind, and water power.

VIII. *Building a national framework for integrating development and conservation:* Many agencies of the U.S. government must cooperate to produce a sustainable economic growth. Scientists, citizens, lawyers, politicians, accountants, and businesspeople must collectively come to a consensus that leads to political action. "The economy is a wholly-owned subsidiary of the environment," said U.S. Senator Tom Wirth (CO).

IX. *Creating a global alliance:* As has been noted in several chapters, international consensus on environmental action can lead to treaties that legalize that consensus with regulations. The new *global economy* links nations together more closely than at any time in the history of humankind. *Economic imperialism*, or the flow of money out of developing countries toward the developed countries, has been recognized and partly reversed within the global community. The U.S. government in recent times has been notably lax in its consideration of the welfare of the world, as evidenced by the lack of ratification of many international treaties. It seems contradictory that on one hand an open world economy is promulgated, and on the other hand isolationism holds on to economic prerogatives (the right to produce huge amounts of carbon dioxide, deep sea mining, biological diversity in developing countries, banning production and use of persistent organic compounds, and the international transport of hazardous waste).

Table 24.1 **Selected International Environmental Treaties**

Treaty	*What it Does*	*U.S. Ratification*
Kyoto Protocol	Addresses the problem of global climate change	No
U.N. Fish Stocks Agreement	Regulates marine fishing	Yes
Madrid Protocol	Protection of Antarctica's unique environment	Yes
U.N. Convention of the Law of the Sea	Regulates deep-sea mining	No
Convention on International Trade in Endangered Species of Wild Flora and Fauna	Protects endangered plant and animal species	Yes
U.N. Convention on Biological Diversity	Protects biological diversity	No
Montreal Protocol	Protects the ozone layer	Yes
Stockholm Convention on Persistent Organic Pollutants	Bans or restricts the production and use of 12 extremely toxic chemicals	No
Basel Convention	Restricts the international transport of hazardous waste	No

4. What kind of world do we want?—Are our attitudes about nature and material values out of touch and out of balance with the biosphere? Can the divisions within our society be healed so that we can truly cooperate to achieve a better, sustainable world? Revisit Einstein's quote that began this chapter of the study guide. Can we intelligently deal with each other and nature in a way that will preserve all? Will an environmental collapse teach us the error of our ways? Few of us are any good at predicting the future. Predictions of environmental collapse made thirty years ago can be attacked for a lack of accuracy. Why? Human ingenuity can postpone inevitable changes. However, the population of the Earth cannot double indefinitely before homeostatic mechanisms will try to correct for the imbalance. Moreover, we cannot destroy plant life and expect it to compensate for an unlimited amount of carbon dioxide. We choose the future for ourselves and our children and their children.

Questions to Ponder

1. List four ways that people in developing countries and developed countries are not living sustainably. Which resources are being depleted?

Practice that Is Not Sustainable in Developing/Underdeveloped Countries	Practice that Is Not Sustainable in Developed Countries
1.	
2.	
3.	
4.	

2. Describe how ecosystem services can be valued economically.
3. List five conditions that work against the improvement of the quality of life in developing countries.

	Description
1.	
2.	
3.	
4.	
5.	

4. List and describe two goals that must be achieved for humans to stay within Earth's ecological carrying capacity.

   1.

   2.

5. List four local community, national, and international goals that would help sustain the biosphere.

	Local Community Goals	National Goals	International Goals
1.			
2.			
3.			
4.			

6. In which five ways can attitudes in developed countries such as the U.S. be changed to achieve a sustainable biosphere?

	Helpful Attitude Change
1.	
2.	
3.	
4.	
5.	

7. What kind of world do you want to leave to your children? Please explain.

POP QUIZ

Matching

Match the following terms and definitions

____1. After most people agree that environmental sustainability is necessary, this must proceed:	a. ecosystem restoration
____2. The fact that the U.S. has not ratified many international treaties on the environment reflects:	b. energy
____3. During most of the 20th century, more money flowed out of developing countries into developing countries than the reverse. This is an example of ______.	c. full-cost accounting
____4. This segment of the population of developing countries suffers most from poverty and lack of resources:	d. unlimited
____5. Most species of plants and animals are found here:	e. political isolationism
____6. A "throwaway mentality" assumes that the Earth's resources and space are ______.	f. developing/under-developed countries
____7. In the U.S., tax deductions were given in 2002 for the purchase of ______ by business.	g. human ingenuity
____8. The economy of the U.S. is largely based on the artificially low cost of ______.	h. ecosystem services

____9. In some areas of Brazilian rain forest, small islands of rain forest are preserved in areas of deforestation. This is an attempt to do:	i. education
____10. This process values the ecosystem services of developed land:	j. economic imperialism
____11. The process of learning that causes a change in behavior as a result of the learning is termed ______.	k. political action
____12. The reason that environmental degradation and natural resource depletion have not proceeded at the rates predicted circa 1970 is ______.	l. women and children
____13. Sewage treatment, the production of oxygen, the lowering of carbon dioxide are ______.	m. SUVs and vans

Multiple Choice

1. The U.S. with 4.6% of the world's population controls ______ % of its economy?
   a. 4.6
   b. 10
   c. 25
   d. 40
   e. 80

2. Which environmental change in the 20th century is mismatched with its percentage difference?
   a. Population increased 240%.
   b. Agricultural land in cultivation decreased 20%.
   c. We are feeding 100% more people that we did in 1950.
   d. Since 1950 we have cut 1/3 of the world's forests.
   e. Trick question alert! All of these are correct.

3. The per capita income of people in developed countries is approximately ______ times higher than in developing countries.
   a. 10%
   b. 2×
   c. 4×
   d. 6×
   e. 10×

4. Of those living in poverty, approximately how many are undernourished?
   a. 5%
   b. 10%
   c. 15%
   d. 50%
   e. 100%

5. Which of the following is an effect of poverty?
   a. death of adults due to starvation
   b. infant deaths
   c. women forced to contribute financially as well as care for the family's clothing and nutrition
   d. increase in deaths due to communicable diseases
   e. all of the above

6. Which of the following international treaties did the U.S. ratify?
   a. Kyoto Protocol on carbon dioxide emissions
   b. U.N. Convention of the Law of the Sea regulating deep sea mining
   c. U.N. Convention on Biological Diversity
   d. Stockholm Convention on Persistent Organic Pollutants
   e. Trick question alert! The U.S. did not sign any of these treaties.

7. Population control is generally associated with which of the following conditions?
   a. high poverty rates
   b. fully industrialized economies
   c. lack of education
   d. environmental catastrophe
   e. all of the above

8. Which of the following problems works most against the achievement of sustainable economic development?
   a. waste
   b. overconsumption
   c. poverty
   d. overpopulation
   e. greed

9. Which of the following resources has not been seriously depleted in the 20th century?
   a. forests
   b. oxygen
   c. soils
   d. stratospheric ozone
   e. commercial whale and fish stocks

10. To stay within the Earth's carrying capacity, we must stabilize human population and ______.
    a. eliminate communicable diseases
    b. feed all starving people
    c. reduce consumption and waste
    d. increase land area in cultivation
    e. all of the above

CYBER SURFIN'

1. Read works by Dr. Kai Lee online,
   a. *Our Common Journey: A Transition Toward Sustainability* by Kai Lee, see the National Academies Press website, **http://lab.nap.edu/catalog/9690.html**
   b. *Appraising Adaptive Management.* Conservation Ecology 3(2): 3. [online]: **www.consecol.org/vol3/iss2/art3**
   c. *Upstream: Salmon and Society in the Pacific Northwest,* Committee on Protection and Management of Pacific Northwest Anadromous Salmonids, National Research Council, National Academies Press: **www.nap.edu/catalog/4976.html**
   d. A synopsis of *Compass and Gyroscope: Integrating Science and Politics for the Environment,* by Kai Lee, **see www.williams.edu/CES/people/klee/compass.html**
2. For one religious/ethical perspective, the United Methodist Church has statements in its social principles about environmental degradation and wasteful consumption of natural resources. See **www.umc.org/abouttheumc/policy/natural**
3. Check out Northern Arizona University's site on Sustainable Environments of the Colorado Plateau at **environment.nau.edu** and the Harvard U. Forum on Science and Technology for Sustainability, **sustsci.harvard.edu**
4. A comprehensive set of environmental sustainability links can be seen at Ecosustainable Hub.com: **www.ecosustainable.com.au/links.htm#10.**
5. From the U.N. Food and Agriculture Organization, various topics on environmental sustainability, biodiversity and women's issues, see **www.fao.org/waicent/faoinfo/sustdev/index_en.htm**
6. Norm Chomsky has a comment about U.S. policies aimed at nonsustainable development of underdeveloped countries, see **zmag.org/ZSustainers/ZDaily/2000-05/30chomsky.htm**

'LIKE' BOOKING IT

1. Wondolleck J M, and Yaffee S L. ***Making Collaboration Work: Lessons from Innovation in Natural Resource Management.*** New York: Island Press, 2000.
2. Daniels S E, and Walker G B. ***Working Through Environmental Conflict: The Collaborative Learning Approach.*** New York: Praeger Publications, 2001.
3. Lee K N. ***Compass and Gyroscope: Integrating Science and Politics for the Environment.*** New York: Island Press, 1995.
4. White R. ***Land Use, Environment, and Social Change: The Shaping of Island County, Washington,*** Seattle, WA: U. of Washington Press, 1992.
5. Meadows D H, and Meadows D. ***Global Citizen,*** New York: Island Press, 1991.
6. Meadows D H, Meadows D L, and Randers J. ***Beyond the Limits: Confronting Global Collapse, Envisioning a Sustainable Future.*** White River Junction, VT: Chelsea Green Publishing Co., 1993.

ANSWERS

Seeing the Forest

This edition of *Environment* has considered issues in air quality, water quality, soil quality, biodiversity, economics, and consumption of natural resources. How would the concept of sustainability affect the current state of each? Who would be held accountable for harmful changes in each?

Environmental Aspect	Effects of Application of a Principle of Sustainability	Accountability: Who or What Will Be Held Accountable for Harmful Changes?
Air quality	A rule could be established that no further increases in fossil fuel burning would occur. Then, treaties controlling carbon dioxide emissions would attempt to prevent further increases. The restoration of ecosystems would mitigate further increases. Finally, free enterprise will come up with long-awaited technological innovations to produce energy locally so as to reduce costs in transmission.	Governments, industries, communities, and individuals
Water quality	A rule could be first established that no more water than can be restored naturally will be consumed by farms, communities, and industries. Aquifer depletion will be arrested and restoration begun. Farmers will use root-level irrigation and no-till methods. Shallow broadcast irrigation will generally cease. Broadcast use of insecticides and herbicides will cease. Standards will be established for chemical residues in groundwater, streams, lakes, and the ocean, the latter by international treaty.	Governments, agribusinesses, farmers, communities, industries, and individuals
Soil quality	First, chemical use in farming and in yards must be reduced. No-till and other methods will conserve soil. We will cease to remove large amounts of soil from production by paving over it for cheap parking lots. Standards will be established for chemical residues in soil.	Governments, agribusiness, farmers, businesses, institutions, communities, industries, and individuals
Biodiversity	First, we need an attitudinal change to respect other forms of life. We need full-cost accounting to value loss of air, water, soil, minerals, and species. We need careful testing of GE/GM crops to determine that no genes will "escape" into natural plant populations or that GE/GM chemicals will not have unintended effects on nonpest species.	Governments, agribusiness, farmers, businesses, institutions, communities, industries, and individuals

Economic health of the world	Economic imperialism will cease by international treaty. The concept of free enterprise will be available to all people. Undue political influence by large corporations will be mitigated by democratic action. We will respect people in developing/underdeveloped nations as ourselves. Individuals and institutions will increasingly invest in socially responsible ways.	Governments, corporations, institutions, communities, industries, and individuals
Consumption behaviors	The wasteful overconsumption of resources will not be encouraged by manufacturers, businesses, or politicians. Religious authorities will see nature as a Creation that must be protected. Others will recognize the moral and ethical imperatives of the preservation of nature.	Political leaders, leaders of corporations, institutions, and industries; religious leaders, and individuals

Questions to Ponder

1. List four ways that people in developing countries and developed countries are not living sustainably. Which resources are being depleted?

Practice that Is Not Sustainable in Developing/Underdeveloped Countries	Practice that Is Not Sustainable in Developed Countries
Clearing of forests, particularly tropical rain forest, for grazing and crop land, firewood, and building materials. We have lost roughly 1/3 of our tropical rain forests in the 20th century.	Quickly using up fossil fuels for cheap energy and plastics manufacturing
Destruction of ecologically and agriculturally productive land by surface mining operations.	Sanitary landfills used as dumps for disposable solid waste
The operation of automobiles and industries without pollution controls and accumulation of pollutants in soils, such as lead in gasoline. Human health is affected.	Increasingly large amounts of water used by farms, industries, businesses, and communities, large amounts of which are wasted in broadcast irrigation and to dilute pollutants.
Uncontrolled population leading to depletion of natural resources within the borders of the developing country.	Excessive consumption and waste of valuable minerals and metals

2. Describe how ecosystem services can be valued economically.

In the late 1960s there was an issue in Georgia that clarified that point. Real estate developers and industries wanted to fill in estuaries and use the readily available water for production, dilution of pollutants, or recreation. State legislators touted the benefits of increased tax revenues and jobs. Ecologists, notably Eugene Odum of the University of Georgia, explained the value of marshland in sewage treatment and recycling materials into food chains that supported local fisheries and tourism. Some state legis-

lators cried that to limit their activities was a violation of "free enterprise." Ultimately and perhaps fortunately, the issue was decided not by the legislature, but by a decision of the State Attorney General that the land "between the tides" was deeded to the state of Georgia by the King of England at the time of colonization. The state then acted to preserve marshland, thereby preserving the fishing enterprise for the economic and health benefit of future generations of consumers, tourists, and fisheries workers. Another principle was honored in this agreement: All land and terrestrial ecosystems are valuable in an unaltered state. Full-cost accounting must take this into consideration.

3. List four conditions that work against the improvement of the quality of life in developing countries.

	Description
1.	Economic imperialism. As long as more wealth leaves the country than enters, people in developing countries will remain in poverty and their numbers will increase.
2.	The low status of women and female children: slavery—trading in women and children, genocide due to tribal or religious differences
3.	Lack of laws regulating environmental destruction
4.	Apathy in the developed countries in regards to the living conditions in developing countries
5.	The human population exceeding its environmental carrying capacity due to prolonged drought or other natural catastrophe, or overpopulation due to family pressures to have more children to help with raising food and gathering firewood

4. List and describe two goals that must be achieved for humans to stay within Earth's ecological carrying capacity.

	Goals for Humans to Stay within Earth's Carrying Capacity
1.	Stabilize population with human ingenuity rather than waiting for the Earth's negative feedback homeostatic mechanisms to respond. Massive starvation will cause much suffering and death. Wars will occur as nations and tribes fight over scarce water and soil.
2.	Control overconsumption and waste of natural resources. Use full-cost accounting to measure actual economic benefit or harm.

5. List four local community, national, and international goals that would help sustain the biosphere.

	Local Community Goals	**National Goals**	**International Goals**
1.	Reduce use of outdoor and indoor water	Cease the practice of environmental isolationism and ratify international treaties	Protect developing countries against the economic imperialism of nations and multinational corporations

2.	Support environmental education in public schools	Fund and support environmental education in public schools	The developed nations should promote, fund, and provide education for the people of the third world as a means of repayment for the environmental damage of overconsumption and waste
3.	Promote environmental awareness in social groups, institutions of higher learning, and religious denominations	Cease the practices of economic imperialism by not enacting laws or regulations that protect corporate patent rights to the extent that the manufacture of cheaper products within the country that needs them is prohibited	Multinational corporations should follow the same pollution and land disruption/reclamation standards as that required in most developed countries
4.	Localize the production of energy as it was 80 years ago. Many sources of hydroelectric power were abandoned in favor of large, centralized power generators that waste much energy in transmission. Use more solar and wind power	Reject the attempts of national political parties to divide the people into probusiness and antibusiness camps, or proenvironment and antienvironment camps	Prevent the export of genetically modified crops and animals to underdeveloped countries where their effects on local populations of organisms are not known
5.	Preserve some areas in their natural state. Manage others for recreation: parks, lakes, etc.	Establish the ecological principle that natural resource use of water and wood products must equal the mass of materials restored by nature (forest, biomass harvested = biomass restored by growth)	Promote the rights of each individual to live in a healthy environment. Promote the rights of people to know what chemicals that they are being exposed to and what environmental damage will occur when their natural resources are extracted

6. In which five ways can attitudes in developed countries like the U.S. be changed to achieve a sustainable biosphere?

	Helpful Attitude Change
1.	Have empathy or sympathy for those people living in poverty. Fight poverty with education
2.	View overconsumption and environmental degradation as moral and ethical problems. Reflecting on a recent campaign to discourage the purchase of large SUVs, "What would Jesus drive?" Or would he walk?

3.	View ourselves as being in more control over the production and consumption of energy
4.	View our opinions as solutions that proceed from a universal thirst for knowledge. Think of people who disagree as potential allies rather than as incorrigible enemies
5.	Be self-aware. We live a busy life, so we should not be destructive or wasteful just because we are in a rush.

7. What kind of world do you want to leave to your children? Please explain.

   My opinion: I want my children to ride their bikes in our community without fear from rushing, distracted motorists. I want my children to see the North Rim of the Grand Canyon, Rocky Mountain National Park, Banff in Canada, Yosemite, etc., in some state of preservation. I want my children to feel the riches of experience of nature: to hear the songs of hermit thrushes, to wonder about the ages of rocks in mountains, to be able to move about without bumping into other people constantly, and to be able to find solitude in nature. I want them to know a father or mother who does not spend three hours a day commuting to work, instead spending that time with a family. I want my children to be consumers without feeling that what they consume has been derived from the suffering of others—I want them to have light to read great books without having images of mountain top removal in Appalachia haunting them. I want my children to know that they are not merely means of national production, but a spiritual manifestation of a biosphere or a creation—a self-aware creature that cares for the environment that created her or him.

It has been my pleasure to share this cup of tea to you. Like the rufous sided towhee (bird) says, "Drink your teeee!"

John V. Aliff

Pop Quiz

Matching

1. k
2. e
3. j
4. l
5. f
6. d
7. m
8. b
9. a
10. c
11. i
12. g
13. h

Multiple Choice

1. c
2. e
3. d
4. c
5. e
6. e
7. b
8. d
9. b
10. c

Appendix

Elementary Environmental Chemistry

ENVIRONMENTAL MODELING AND HOMEOSTASIS — See pg. 103 of Chapter 6 of the textbook

Homeostasis (maintaining a steady state) is a fundamental concept for the understanding of physiology and disease that can be applied to planet Earth. The "Gaia hypothesis" assumes that the Earth is a living organism that maintains a steady state of atmospheric gases, temperature, and geological materials. Balance of these factors is maintained by cause-and-effect cycles. There are two types of homeostasis: **negative feedback** and **positive feedback.**

Negative feedback homeostasis produces a response or output opposite to the input stimulus or stress (see figure below). For instance, you go outside on a cold January day in Canada without enough clothes on. Your skin and the blood in it are cooled, the cooled blood flows to the brain, where the hypothalamic thermostat detects the change. Then the thermostat sends a command to constrict the skin's blood vessels, which, in turn, will reduce heat loss. The central nervous system stimulates muscles to contract or shiver, releasing heat. Notice that in negative feedback homeostasis, the sign, positive or negative, is opposite for the input and output.

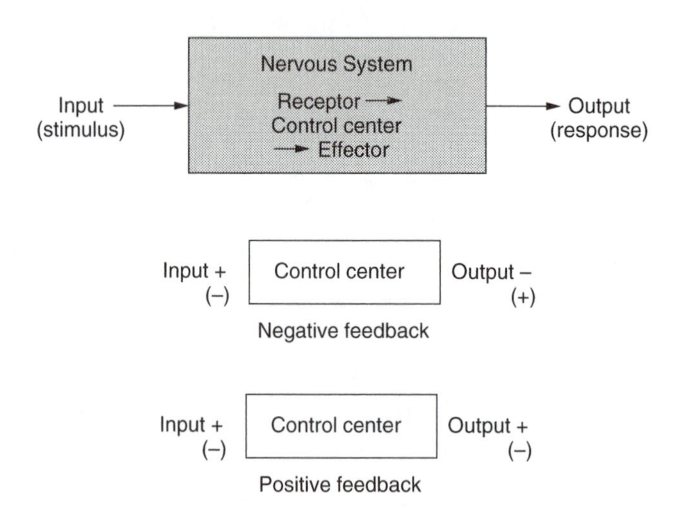

Negative feedback in environmental systems (ecosystems) is seen in the biosphere when carbon dioxide increases in the atmosphere. More carbon dioxide encourages more **photosynthesis** (see below) that uses up carbon dioxide. Also the formation of carbonate and carboniferous (coal) rocks speeds up. Both photosynthesis and carbonate rock formation will lower carbon dioxide. Over time the input, increased carbon dioxide, is followed by an output decreased carbon dioxide.

Positive feedback homeostasis is characterized by the same sign in the input and output and is many times related to disease processes (the vicious cycle of high blood pressure and arteriosclerosis/hardening of the arteries by calcium deposits). Arteriosclerosis makes blood vessels less flexible and narrow. Arteries harden in response to high blood pressure. High blood pressure, as chronically endured by uncontrolled sugar diabetes causes arteriosclerosis, arteriosclerosis causes more high blood pressure, and so on in a vicious cycle. An increase in blood pressure causes a further increase in blood pressure—this is positive feedback or an increase results in a further increase feedback cycle.

A positive feedback loop is usually destructive to the environment. For instance, as the birth and survival rates of human populations increase, resources, such as farm lands, are used up at a greater rate. The feedback can be negative at this point, and starvation will reduce population by negative homeostasis. This has been happening in Ethiopia, the Sudan, and Eritrea in Africa for many years. The farm area that supported an average family has been reduced by more than 75%. But what if food is brought in from the outside?

This has had the effect of maintaining the high birth rates that led to the reduction of farm land. Therefore, increased birth rate → increased resource loss → increased birth rate is a positive feedback.

The two great unknowns in greenhouse climate modeling are whether increased heat resulting in increased evaporation will cause low clouds to form that decrease surface temperatures (negative feedback) or high clouds that serve to reflect heat back to the surface (Venus)—positive feedback.

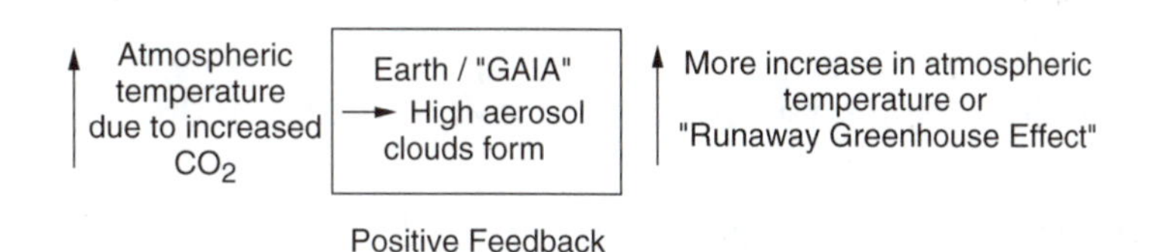

Positive Feedback

Compounded positive feedback due to population increase: Increasing human population leads to more burning of fossil fuels and wood (see combustion on pg. 427 of this chapter), which leads to more carbon dioxide in the atmosphere, which leads to more heat being trapped in the atmosphere and global warming and more acid rain. Meanwhile, the economic benefits of the use of combustion allow for a higher number of human beings to be supported that increases demand for fossil fuels, therefore even more carbon dioxide added to the atmosphere. The latter will be explained further as you read.

Chemistry

Every student of a science or those desiring to achieve good citizenship should have a basic knowledge of chemistry. Most of the issues that we will deal with in this study guide depend on understanding cycles of chemical and energy changes that accompany these cycles.

Matter

Matter occupies space or volume and has mass—the latter we call *weight* where gravity pulls the mass toward the center of the earth.

Matter is constructed of tiny atoms, indivisible by chemical means. Although the ancient Greeks first coined the term, they thought there were only four types of elements, Earth, air, fire, and water. Men had more fire, and women had more water. In the 19th century, Roentgen discovered X-rays, Rutherford discovered protons, and Thomson described electrons. In the 1930s,

Chadwick proved the existence of neutrons predicted by Rutherford, and it was discovered that atoms could be split, according to Einstein's theory, to release a large amount of energy. When large atoms were split (nuclear fission) by Hahn and Strassmann, smaller atoms and subatomic particles were produced.

Matter is composed of what is currently recognized as 103 different kinds of naturally occurring elements with different numbers of protons, neutrons and electrons. See **www.webelements.com**

VOCAB

Atoms are the building blocks of matter, the smallest units of matter indivisible by chemical means.

Atomic mass is the weight (on Earth) of all the subatomic particles of a given atom.

Atomic mass number is the number of protons and neutrons added together. The number of neutrons may be calculated by subtracting the atomic number from the atomic mass number.

Atomic number is the number of protons in the nucleus of a given atom: it is always equal to the number of electrons orbiting at near the speed of light around the atomic nucleus in an unreacted atom.

Elements are substances composed of like atoms, e.g., aluminum foil.

Molecules are composed of two or more atoms. H_2 is a diatomic or homoatomic molecule, but also units of the element Hydrogen (H). NaCl, sodium chloride or table salt, is a heteroatomic molecule.

Molecular weight is the sum of the atomic weights of all the atoms in a molecule—H_2O, water, 2 H (1) + O (16) = 18. What is the molecular weight of octane, one of the gasoline mixtures of molecules, C_8H_{18}?

A **Mole** is a standard unit of measurement for the qualities of reactants and products of a chemical reaction. If one carefully measures out the molecular weight of a compound, one will have almost exactly 6.023×10^{23} particles—a very large number! Concentration in chemistry uses **molarity** or the number of Moles of chemical per liter of solution. See "Acids and Bases," on pg. 424.

Compounds are composed of like molecules, each of which has two or more different kinds of atoms.

A **true solution** is a mixture of a smaller amount of chemical, the solute, with a larger amount of solvent (a teaspoon of pure salt mixed with a quart of water). If the solute particles are very small, kinetic energy (see below) will keep them mixed up as long as the solvent amount does not decrease.

A **suspension** is composed of larger solute particles that will settle out of a mixture if left standing. However, smoke, dust, and sulfur aerosols are suspended by moving air, and soil particles or silt are suspended in moving water (a muddy stream).

Atoms

Subatomic Particles Atoms are composed of four larger particles and many smaller particles that will not be discussed: **protons** (positive in charge), **neutrons** (neutral), **electrons** (negative in charge) that revolve around the atomic nucleus and their equal in mass but opposite in charge **positrons** found in the nucleus. One proton has a mass of about 1 dalton or atomic mass unit. The neutron is also approximately 1 dalton. Electrons and positrons are much smaller in size—about 1/1837 dalton.

Hydrogen

The simplest atom and what astronomers regard as the "building block of the universe" is common **hydrogen,** or **protium.** Hydrogen has three forms or **isotopes** that have the same number of protons and electrons (1) but a different atomic mass.

The atomic mass is due to the addition of neutrons to the nucleus. See below. **Protium,** atomic mass or weight of 1, has 1 proton and one electron. There are two other, rarer, forms of hydrogen: **deuterium** (mass = 2) and **tritium** (mass = 3). All isotopes of hydrogen have one positive proton and one negative electron. All ***unreacted*** **or uncombined atoms have # protons = # electrons to create a 0 electrical charge.**

How many neutrons do protium, deuterium, and tritium have? What is the difference between iodine-131 and I-127? Carbon-12 and C-14? See the periodic table and webelements below. Atoms with more neutrons than the most stable form (isotope of the atom) tend to be radioactive—they emit radiation and are called **radioisotopes.**

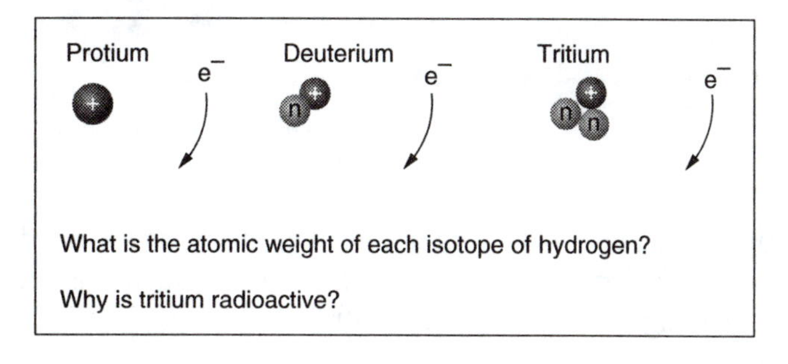

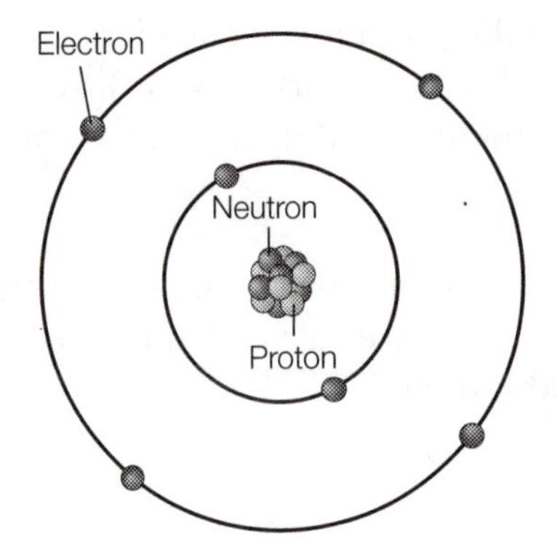

The atomic nucleus is composed of neutrons and protons. Atomic energy holds nuclear particles together.

Atoms and Energy

Physicists theorized that hydrogen atoms were being fused in the sun to release atomic energy, the energy that holds atoms together. Hydrogen fusion (simplified proton chain) powers the sun:

$$= {}^{1}_{1}H + {}^{1}_{1}H + {}^{1}_{1}H + {}^{1}_{1}H \text{ yields} \rightarrow {}^{2}_{4}He\text{, Helium + radiation}$$

The atomic number is at the left top of the atomic symbol and the atomic weight to the left bottom.

Einstein discovered that mass (an amount of matter, on earth designated as weight) can be converted to energy. Some mass is converted to radiant energy in a nuclear fusion reaction. Because measurable mass apparently disappeared, this was called the **mass defect.**

$$\mathbf{E = mc^2}$$

In a hydrogen bomb, deuterium and tritium are fused.

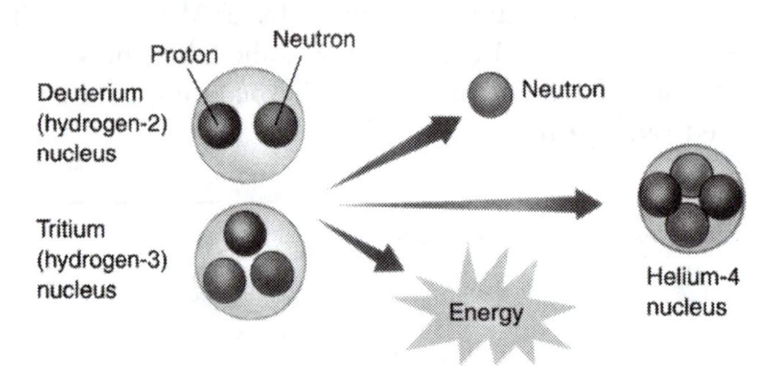

According to chemists, physicists, and astronomers, heavier atoms are cooked up in stars (carbon, iron, nitrogen, sulfur, phosphorus, zinc, copper, etc.). Because these atoms are in our tissues, the well-known astronomer Carl Sagan called us "star stuff."

Periodic Chart: Look up element #19 in the periodic chart.

Here is #12 as an example.

2	12
8	Mg
2	24.3

The numbers at left represent electron configuration in the first three shells.

The first shell of an atom is filled with 2e⁻, the second with 8, the third with 18. Argon with 8 in the third shell is inert.

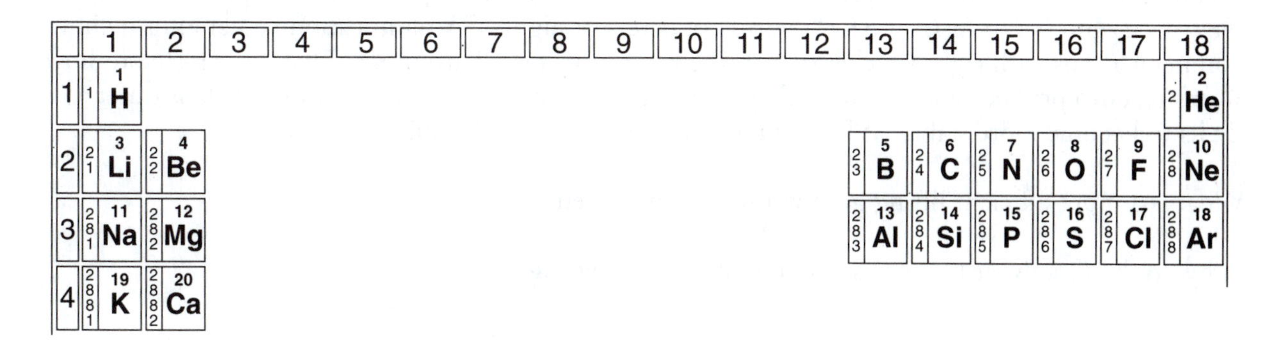

Periodic chart of elements #1–#20, showing electron shells and numbers of electrons in each shell.

Also see **www.chemicalelements.com/show/electronconfig.html** for the electron configurations of other atoms.

We will deal with small-medium size atoms. With the periodic chart link or a general biology or chemistry book as an aid, you should be able to draw hydrogen (H); helium (He); sodium (Na); chlorine (Cl); carbon (C) and oxygen (O). Notice that a vertical column in the periodic table has elements with similar electron configurations in the outermost shell. That similarity leads to similarities in the chemical properties of the atoms in a vertical column, particularly in chemical bonding and ion formation. This will be explained as you read.

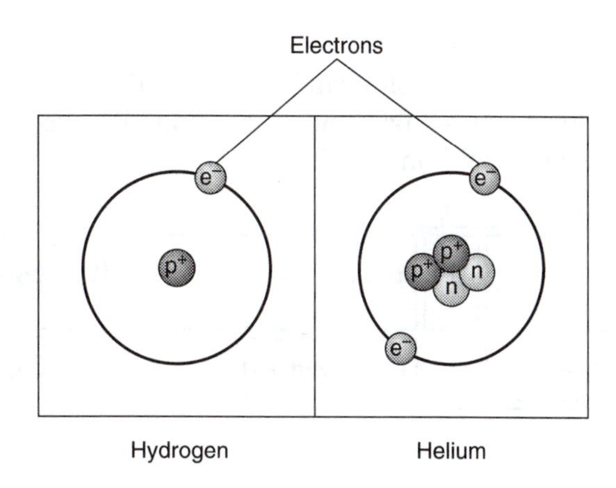

Hydrogen Helium

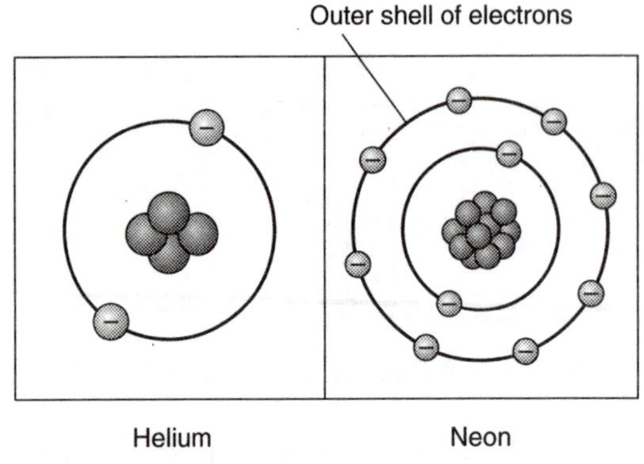

The Noble or Inert Gases

Mendeleev concluded that the properties of the smaller elements repeated every eighth one—Li to Ne (row 2) and Na to Ar (row 3)—"the rule of octets or eights." A ramification is that after helium, nonmetal atoms with eight (8) electrons in their outermost shell are **inert**. That gives a clue to how atoms react to produce compounds—they rearrange their electron configurations of their outer shells to become more stable. What is more stable Cl_2 gas or Cl^- ion?

What is a more efficient lifting gas for a blimp, hydrogen or helium?

Why do the Goodyear blimps use helium instead of hydrogen?

CHEMICAL BONDS AND MOLECULES

The reason why atoms react or combine to form molecules is that, in reacting, they reach a more stable outer shell electron arrangement, 8, the magic number for atomic stability and lower energy level. Which electrons have more energy? Those closer to the nucleus, or those farther away? Which electrons do chemical bonding? The answer is the outermost electrons. Because the positive nucleus pulls a negative electron toward it, the electron must have enough rotational or centrifugal energy to continue to orbit the nucleus. Therefore, electrons further away from the nucleus have more energy than those close in.

Chemical Reactions

Chemical reactions are expressed in equation form. This is similar in meaning to a recipe. Take this and add that and make something new—baking powder + flour + leavening + heat make bread. When atoms react, they form chemical bonds.

Reactant A	+	**Reactant B**	yields →	**Product(s) C**
C, carbon in coal	+	O_2, oxygen	→	CO_2

Types of Chemical Reactions

Endergonic (endothermic) reactions store potential energy in chemical bonds. The process of photosynthesis makes food molecules that store energy this way by converting the electromagnetic energy of the sun into chemical bond energy.

Exergonic (exothermic) reactions release energy, usually the greatest part in heat. When ATP breaks down to move muscle fibers, much heat is released. This explains why shivering warms you up.

All chemical reactions need **activation energy** added to the reactants in order to start a reaction. Types of activation energy include these that are of interest for Environmental Sciences:

a. *Kinetic Energy*—chemical reactions don't occur at a temperature of absolute zero (–273 degrees C). Some kinetic (motion) energy is needed to bring reactants (to left of chemical reaction equation) together. Reaction rates generally double with each increase of 10 degrees C. Therefore **heat** is the typical energy of activation for chemical reactions.

b. *Sunlight*—electromagnetic radiation (ultraviolet) is needed in the making of photochemical smog. See pg. 437.

Anabolic reactions build larger units from smaller atoms or molecules (when simple sugars are strung together to make a starch stored in plant roots). Catabolic reactions break larger molecules into smaller ones, such as when digestion breaks down glucose into carbon dioxide and water or when gunpowder explodes.

The **kinetic theory of matter** also explains the process of **diffusion.** Molecules tend to move from an area of higher concentration to an area of lower concentration. Why? The molecules or atoms (gas or liquid are good examples) move in straight lines and in random directions. Even solid metal atoms have some movement. Why does metal expand when it is heated? When water diffuses into the cells of plant roots, it is a special case of diffusion called **osmosis.**

Why is it easier to dissolve corn starch or table salt in warm water rather than cold water? The salt ions are moved into the solvent by water molecules that surround them. All particles move in straight lines and random directions until they collide with each other or the walls of a container—then they carrom like pool balls.

Chemical Bonds

Atoms are bound closely together to form a molecule in two ways.

1. *Ionic bond*—If a sodium atom, Na gives away an electron to a chlorine atom, the atom Cl keeps it, notice that both atoms have 8 electrons in their outer shell.

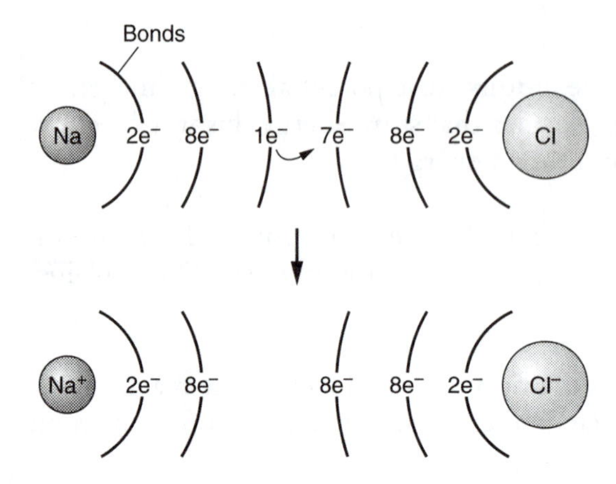

Now, the reacted Na^+ ion has a +1 electrical charge because it lost one negative electron and Cl^- has an extra electron and has a –1 negative charge. Charged atoms, Na^+, or groups of atoms, NH_4^+ are called **ions.** Polyatomic ions have more than one atom, e.g., NH_4^+, ammonium ion, or SO_4^{2-}, sulfate ion.

Positively charged ions are called **cations:** e. g., Na^+, sodium ion; Cu^+ cuprous ion; Ca^{2+}, calcium ion; Fe^{3+}, ferric iron atom; NH_4^+, ammonium ion.

Negatively charged ions are called **anions:** Cl^-, chloride; F^-, fluoride; SO_4^{2-}, sulfate; CO_3^{2-}, carbonate; and HCO_3^- bicarbonate.

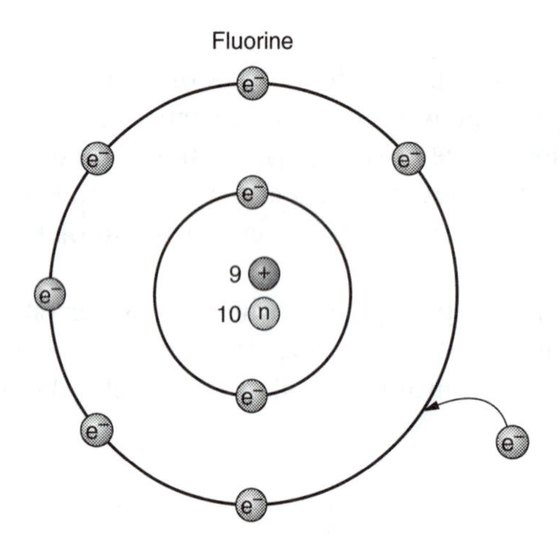

What is the difference between fluorine gas and fluorine ion? A big one! Fluorine ion has that extra electron it wanted for stability. Now it has 8 in the outermost shell of electrons.

Atoms or polyatomic ions that are less oxidized are designated with the **-ous** or **-ite** suffixes, e.g., H_2SO_3, sulfurous acid; Na_2SO_3, sodium sulfite; those that are more oxidized are named with the **-ate** suffixes; e.g., Na_2SO_4, sodium sulfate.

Ionic bonds are formed when atoms give away or receive electrons. When atoms lose electrons, an **oxidation** results, if atoms gain electrons, they undergo **reduction; e.g.,**

$$Na^0 \rightarrow Na^+ (+)\ 1e^- \text{(Oxidation of Na)};$$

$$Cl^0 + 1e^- \rightarrow Cl^- \text{(Reduction of Cl)}$$

Consider this example, closer in meaning to the term *oxidation.*

Ca^0 + O (one atom) → CaO, calcium oxide. The Ca in CaO is now Ca^{2+} ion, it lost two electrons. The oxygen went from 0^0 to 0^{-2}, it was reduced as it gained two electrons.

Or, when coal is combusted, the sulphur is oxidized to make sulfur dioxide, an acid rain gas.

$$S + O_2 \rightarrow SO_2$$

Nonmetals generally have **4–8 electrons** in the outer shell. Those with 4–7 hold on to their electrons and generally "grab for more" when bonding with metal atoms. Those with eight, 2 for helium, are inert. Their electron needs are satisfied because their outermost shell is full. Metals are weakly electronegative atoms. That means they give up electrons readily: **Metals** generally have **1–3 electrons** in their outer shells.

The electron needs of atoms are stated as **valence.** If a metal atom such as sodium gains stability by giving away one electron, its valence is +1: it is stable because its underlying shell has 8 electrons. The oxygen atom with 6 electrons in the outermost shell gains stability by receiving 2 electrons to fill its shell at 8. Therefore, the valence of oxygen is –2.

Look at the periodic chart. Are Li, Na, K, Mg, Al and Au (gold) metals? ______ How about S, O, Cl and Ar. What are they, metals or nonmetals? ______ Hydrogen may exist as a metal on the planet Jupiter when pressures are very high, 2000 times earth atmospheric pressure, or in other locations where temperatures are near absolute zero (–273 °C) and pressures are also high.

If you dissolve the ionic compound NaCl in water, Na^+ and Cl^- ions will float free and be surrounded by spheres of water molecules.

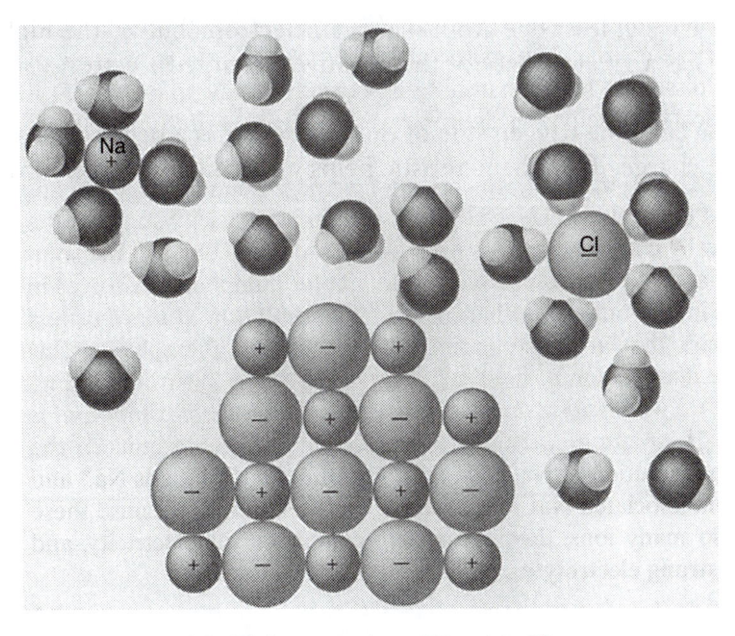

NaCl (in water) → Na^+ (+) Cl^-

Water solutions of ions are called **electrolytes,** essential for the operation of muscle and nerve cells, and equally important for conducting minerals, such as nitrates, phosphates, potassium, and calcium into plant roots.

2. *Covalent bonds* share electrons. Two electrons form a pair and share an **orbital** path; there are two shared electrons in one covalent **bond** represented with a dash.

The diatomic gas molecules are covalently bonded:

H-H or with electron dots (Lewis symbol), **H** · + **H** · → **H : H**

How many electrons does carbon have in its outermost shell? Notice all carbon atoms share 4 bonds ($8e^-$)—carbon dioxide, O=C=O, **O** :: **C** :: **O** ; and nitrogen gas N≡N, **N** ⋮⋮ **N.**

There are two kinds of **covalent** bonds:

Nonpolar covalent, where electrons are shared equally by all atoms and the electrical changes are therefore evenly distributed (Methane, CH_4).

```
      H                 H
      |                 ..
  H – C – H    or    H : C : H
      |                 ..
      H                 H
```

Oils are nonpolar compounds.

Polar covalent bonds have at least one atom that is an electron "bully," the highly electronegative atom (usually oxygen) has a greater share of the negatively charged electrons.

H–O–H is a polar molecule, the oxygen side of the molecule has a negative charge and the hydrogen side has a positive charge. This characteristic helps water dissolve NaCl or table salt.

Water Molecule:

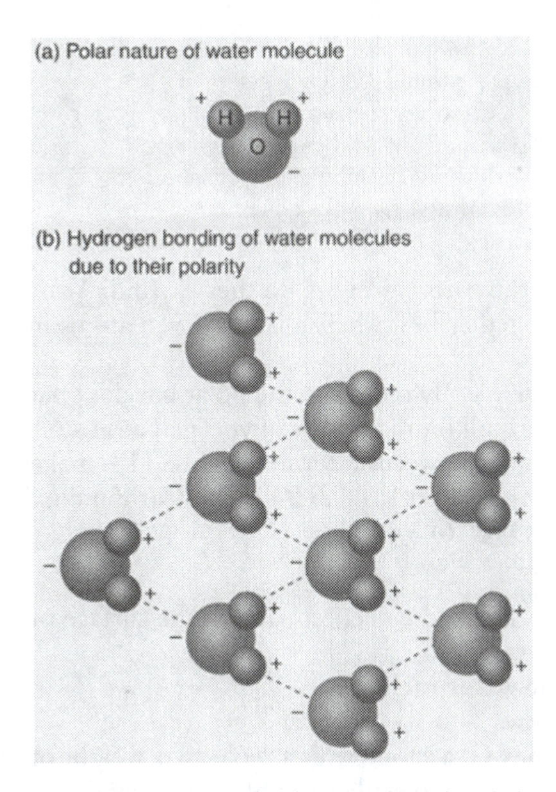

Water molecules stick to each other with weak **hydrogen bonds** of electrical attraction. The positive and negative ends of water molecules draw them together, forming oceans, lakes, etc. Hydrogen bonds may draw sections of large molecules together (protein helixes and DNA double helixes). These hydrogen bonds are broken when water evaporates.

Water as a Substance

What Are the Properties of Water?

Living organisms are mostly composed of water. We are approximately 76% water; jellyfish are over 90% water. Life may have begun in water. Why is water so important to life existing on Earth?

1. *Water is Wet! It will* ***dissolve or disperse*** *many substances.* Sea water is an ionic solution. Fresh water has much less salt dissolved in it. **True solutions,** like sea water, have ions and small molecules of **solute** dissolved in a water **solvent.** The solutes will never settle to the bottom because their molecules continually move and mix themselves up with the water molecules that also continually move. All atoms and molecules move as long as there is heat—this is called the **kinetic theory of matter** and the movement "Brownian motion."

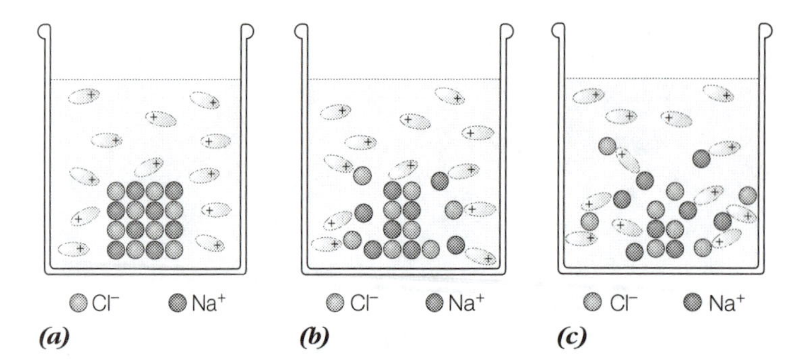

Consider your car engine. As the parts warm up, do they expand? Yes! The engine is designed to run when hot—the parts fit together best when it is hot. Even atoms in solid metals move slightly.

2. *Water sticks to itself: It is **cohesive.*** By reason of the polar covalent nature of water, electronegative oxygen atoms will attract or pull on the positive hydrogen atoms of adjacent water molecules, creating weak hydrogen bonds between water molecules. This makes water hard to boil because of its low molecular weight of 18, 1O(16) + 2H(1) = 18. Carbon dioxide has a heavier molecular weight of 44, 1 C (12) + 2 O (@16) = 44.

Why is water found in oceans and in the deepest parts of the surface of the earth's crust?

Some important facts about water include:

A. **Water boils at 100 degrees C,** acetone with a molecular weight of 58, boils at 58 degrees C! You expect compounds with low molecular weight to be heavier and more dense than those with high molecular weights.

B. To raise the temperature of 1 gram of water, 1 degree C, requires **1 calorie of energy** (heat). Food calories are "big" kilocalories, 1000 times more than the standard "small" calorie.

C. To vaporize 1g of water at 100 degrees C requires an additional 540 calories **(heat of vaporization)** to break hydrogen bonds. Water requires a lot of energy input in order to boil (vaporize).

D. Water has a **high specific heat** (1 calorie/1 degree C/1 gram) which means that it will store heat and thus it acts as a **moderator of climate** in the summer. The Earth is also warmer in the winter because oceans release stored heat.

Explain why Bismarck, ND, has more temperature extremes as compared geographically to Seattle, WA. Look at a map or globe.

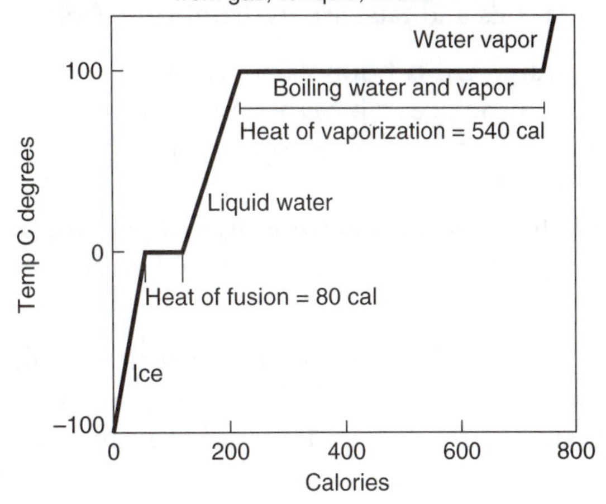

5. Because water molecules bond to each other by hydrogen bonding, the molecules on the surface are draw inward toward their fellow molecules down below. This creates a film-like surface or **surface tension** on which certain water bugs walk.

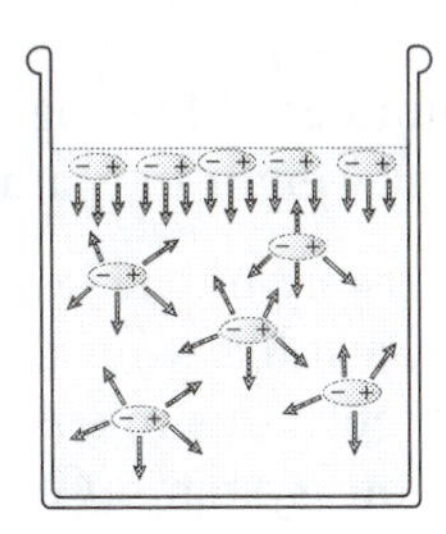

6. Water is **adhesive;** it will stick to certain other substances. Water will **adhere** to the inside walls of small hollow tubes, such as xylem tubes in plants and rise against the force of gravity. This is called **capillarity.** It helps water climb the trunks of trees hundreds of feet high in xylem tissue.

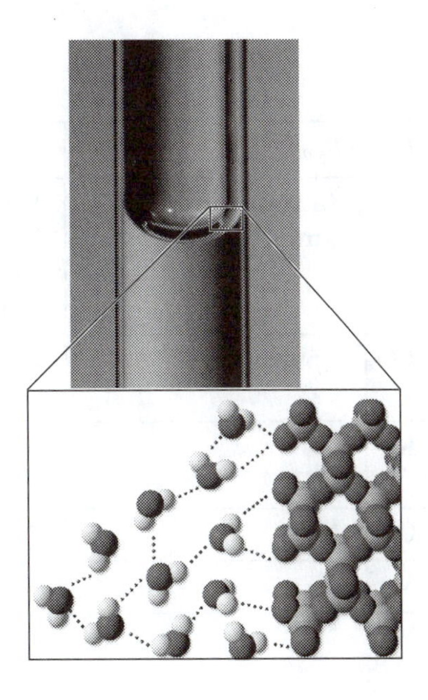

Water will adhere to **hydrophilic** (water loving) atoms or atom groups, such as ions, strongly electronegative atoms, such as oxygen, nitrogen, etc. Some substances that are filled with nonpolar covalent bonds, such as oils and fats, are **hydrophobic** (water fearing).

Why do water and oil separate? Oil is nonpolar covalently bonded, water is polar. The two repel each other. Oil spills on the ocean have been vacuumed off the surface. Also the density of oil (g/ml) is less that water, so the oil rises to the top. The rule is "like dissolves like." Does water in your gas tank mix with the gasoline (nonpolar)? No, it goes to the bottom of the tank and is more likely to cause engine roughness if you are near empty. Do gasoline and motor oil mix? Yes, both are nonpolar. You have a problem if your motor oil smells like gasoline.

7. At declining temperatures less than 4 degrees C, the water molecules begin to move away from each other. Therefore, solid water formed at 0 degrees C, ice, has less **density** (weight/volume) than liquid water and it floats. In the deepest layers of the ocean, the water temperature is 4 degrees C.
8. When ice forms over a northern coniferous forest lake or in the Arctic Ocean, the ice acts as an insulator for the living organisms below. What would happen if ice were more dense than liquid water when it freezes? Freezing water from the bottom would push all the fishes up to be killed by the cold air.
9. Water makes a very small but equal number of hydrogen, H^+ and hydroxide, OH^- ions [10^{-7} or one 10 millionth of a mole (a molecular weight in grams, e.g., 18 g of water].
10. Pure water is a poor conductor of electricity unless you dissolve ions in it. The positive and negative charges act as electrical poles, just like your car battery that has a positive pole and a negative pole to conduct electrical current. Ions in solutions are called **electrolytes** and are necessary for the operation of the nervous systems. Generally, covalent compounds are poor conductors.

Which of the preceding characteristics of water can be explained by its polar nature and hydrogen bonding? Check out 1–7, above.

Major Water Soluble Plant Nutrients

Nutrient	Ionic Form	Function in Plants
Nitrogen as nitrate (first number on a bag of general purpose fertilizer) and ammonia, NH_3	**NO_3^- (nitrate) and NH_4^+ ammonium**	Very important for stem growth. Needed to make plant proteins, DNA (genetic chemical), RNA (gene copies), and ATP (energy chemical)
Phosphorus as phosphate (2nd number)	**PO_4^{3-} (phosphate)**	Encourages root growth. Needed for DNA and ATP production
Potassium (3rd number)	**K^+**	Important for retaining water inside cells. Ionic balance
Calcium in limestone, $CaCO_3$	Ca^{2+}	Part of cell walls
Magnesium in dolomite, $MgCO_3$	Mg^{2+}	Needed to make chlorophyll

Iron	Fe^{2+}	In photosynthesis and cellular respiration chemicals. Many shrubs need extra amounts.
Sulfur	SO_4^{2-} (sulfate)	In proteins and vitamins
Molybdenum	MoO_4^- (molyboenate)	Needed for *rhizobia* bacteria to do nitrogen fixation. See textbook pg. 106.
Copper, Manganese, Zinc	Cu^+, Mn^{2+}, and Zn^{2+}	In photosynthesis and cellular respiration enzymes
Chloride	Cl^-	Needed to make oxygen in photosynthesis. Balances cations in cells

You may have heard of ammonium nitrate. It is an excellent fertilizer for spring grass. The high (35-0-0) nitrogen content encourages growth. Why? The **"lime"** spread on grass is ground **limestone,** calcium carbonate and magnesium carbonate. It not only raises the pH of the acid soils but also adds calcium and magnesium nutrients.

The most important nonsoluble material in soils is **humus,** an amorphous, dark material that results from the breakdown of plant debris and other organic detritus. It is very important for holding water and mineral nutrients at root level. Folks add peat moss, cow manure, and other products to increase the humus content of Georgia soils. Humus is low in southeastern clay soils because of a combination of high year-round temperatures that increase microbiological action breaking it down. Also, high rainfall leaches the minerals into lower soil levels away from the roots.

CHEMICAL REACTIONS

Note: to keep things simple, we have not balanced the reactions that follow by indicating the relative amounts of reactants and products involved.

The general form of a chemical reaction:

Reactant A + Reactant B → Product C + Product D

For example, Acid **(A)** and Base **(B)** → Salt **(C)** and Water **(D)**

This is called a neutralization reaction as will be explained below, or more precisely,

$$HCl + NaOH \rightarrow NaCl + H_2O$$

This reads: **hydrogen chloride** (in water, this is called hydrochloric acid) plus **sodium hydroxide** (base) makes **sodium chloride** (one of many salts, depending on the reactants) and **water.**

Chemical reactions have to be activated by energy in the form of heat or other forms of electromagnetic radiation. According to the **kinetic theory of matter,** molecules and atoms move. Gas molecules are an ideal example. They move in straight lines and in random directions, so that the carbon dioxide we exhale becomes evenly distributed in the room we occupy. Heat increases kinetic energy so that when reactant chemicals collide, they can undergo a reaction. Theoretically, no chemical reactions could occur at absolute zero, –273 degrees C.

ACIDS AND BASES

Acids are substances which produce H^+ ions or protons in a water solution that bind to water molecules, they are proton donors. For example, HCl is a polar covalently bonded gas molecule, but when dissolved in water makes H^+ ions and Cl^- ions. (Actually the hydrogen ions bind to water molecules creating hydronium ions, H_3O^+.)

H_2CO_3 is carbonic acid, H_2SO_4 is sulfuric acid, H_2SO_3 is sulfurous acid, HNO_2 is nitrous acid, HNO_3 is nitric acid.

Why are some acids strong and others weak at a given concentration?
For example, typically **100 HCl** molecules (dissolved in a large amount of H_2O) yield → **96 H^+** ions + 96 Cl^- (Chloride) ions (96% of the HCl molecules are ionized). 96 of 100 HCl molecules break into ions, and 4 HCl molecules exist at any point in time. On the other hand, 100 H_2CO_3 (in H_2O) yields → **$3H^+$** ions + $3HCO_3^-$ (bicarbonate) ions: 3 of 100 H_2CO_3 molecules break up (3% ionize).

Which acid is the strong one? The weak one?

When beverages are carbonated and in acid rain formation, $CO_2 + H_2O \rightarrow H_2CO_3 \rightarrow H^+ + HCO_3^-$. The difference in taste between a fresh soda pop and a flat one is carbonic acid.

Bases are the opposites of acids, they have OH^- ions, and will bond to H^+ ions/protons. For example, if a solution had a lot of H^+ ions, you could add OH^- ions, react the two and form HOH or water. NaOH or lye is a common base used for making soap and cleaning sink drains—NaOH (in H_2O) → Na^+ (+) OH^-. Any compound which produces OH^- is a base. But is NH_3 (ammonia) a base because $NH_3 + H^+ \rightarrow NH_4^+$ (The answer is yes!)

Bases react with acids to produce water or Acid + Base → Salt and H_2O

This is called neutralization:

Acid + Base	→	**Salt and H_2O**
$H^+ + OH^-$	→	H_2O
HCl + NaOH	→	NaCl + H_2O

Buffers are chemicals that resist change in acidity. For example, the mineral limestone contains the basic salts $CaCO_3$, and $MgCO_3$, calcium carbonate and magnesium carbonate, respectively. Calcium carbonate reacts with acid and neutralizes it as follows

$$H_2SO_4 + CaCO_3 \rightarrow CaSO_4 \text{ (calcium sulfate)} + H_2O + CO_2.$$

This can be demonstrated by pouring vinegar or battery acid over limestone or carboniferous rocks. Bubbling indicates the release of carbon dioxide as the acid is neutralized. Lakes or soils that have little **buffering capacity** (without limestone or dolmite) are more susceptible to the effects of acid rain. See Chapter 20.

Acids are very damaging to many plants, especially evergreens at high altitudes where the acidic cloud moisture soaks them daily. Acids and water will dissolve chemicals better than water alone. Research has proven that as acids bathe evergreen needles (leaves), the macronutrients Ca^{2+} and Mg^{2+} are leached out. The tree is then weakened. Final death of the weakened tree may be enhanced by the attack of insects (parasitic diseases are more likely to attack weakened organisms, see chapter 5). For years the explanation of schools of Agriculture/Forestry was that the trees were dying of insect outbreaks that they wanted to control by spraying insecticide. Schools of Arts and Sciences/Ecologists said that the acid was at fault. Similar reactions of the two camps were noted at the time Rachel Carson's book *Silent Spring* came out in the 1960s.

Many researchers of the 1960s who were supported by grants from major chemical companies, hated the book's thesis that the exotic compounds that we were releasing into nature would have drastic health and environmental effects. So who was right? We are living the future predicted by Carson, but the cause can be attributed more to habitat destruction than to pesticide contamination. However, pesticide contamination in still a problem for soils, aquatic ecosystems and certain birds.

Measuring Acidity

pH is an inverse logarithmic scale of H^+ ion concentration in moles per liter of solution. Mathematically, pH = 1/log H^+ or – log H^+ ion concentration in Moles/liter. Because the formula is 1 over the concentration of H^+ ion, as H^+ ion conc. increases, pH decreases and vice versa. The pH scale runs from 0 to 14. A pH of 0 to just below 7, means the solution is acidic and has more H^+ ions than OH^- ions. A pH above 7.00 means a basic solution where OH^- is more concentrated than H^+. *Beware: the scale at first look, appears backwards*—if the pH number is high, H^+ concentration is low; if the pH number is low, H^+ concentration is high! A simple rule that relates to a study of the metric system is a pH = 0 means H^+ concentration is 10^0 or 1 mole/liter. A pH of 1 means $H^+ = 10^{-1}$ mole/l. A pH of 7 means $H^+ = 10^{-7}$ mole/l. Is 10^{-7} a big quantity or a very small quantity? It is 1/10,000,000 of a mole. Each higher pH number has 10 times less H^+ ions than the lower pH number, or conversely a pH of 0 is ten times more concentrated in H^+ ions than a pH of 1.

Remember: Acid (+) base yields salt (+) water is neutralization. The resulting pH will be 7 if it was a complete neutralization of equally strong acid and base, with equal amounts (10^{-7}M) of H^+ and OH^- resulting. A pH of 7 is therefore termed neutral.

At a **pH = 7, H^+ concentration = OH^- concentration,** or 10–7 Moles H+ = 10–7 Moles OH–

You take an antacid for you hyperacidic stomach. The reaction is:

HCl (stomach acid) + $CaCO_3$ (calcium carbonate) → $CaCl_2$ (a salt) + H_2O + CO_2 (gas)

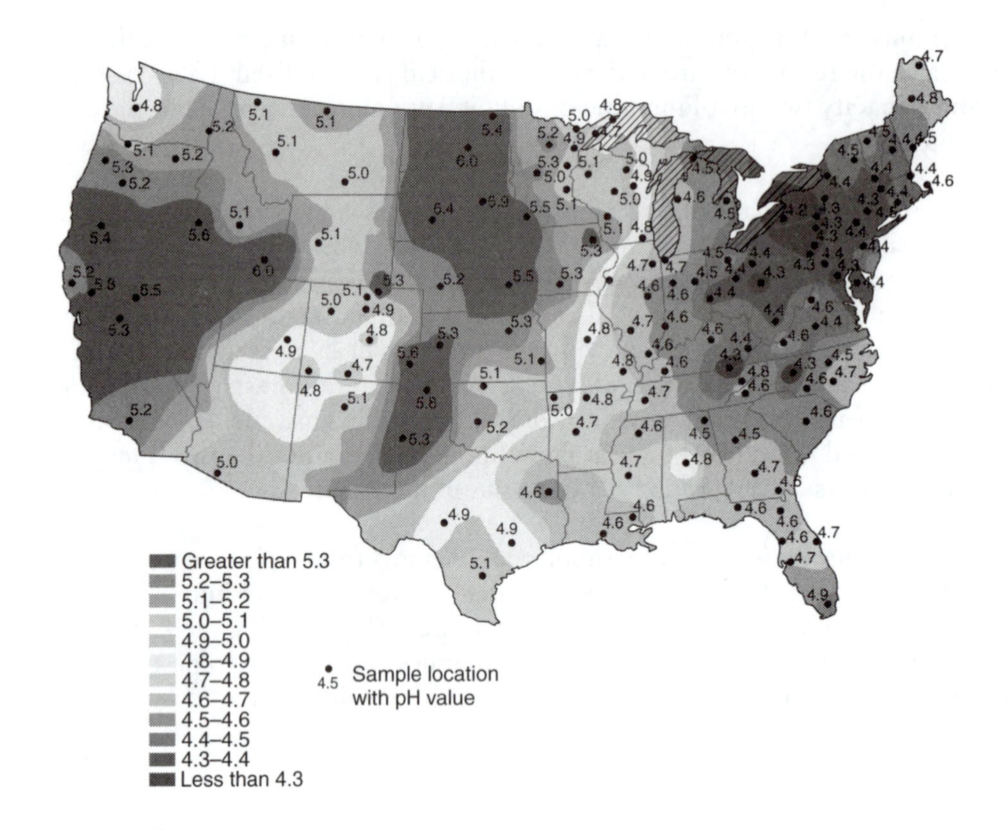

pH of acid rain by location, why are the lowest pHs in the northeast U.S.?

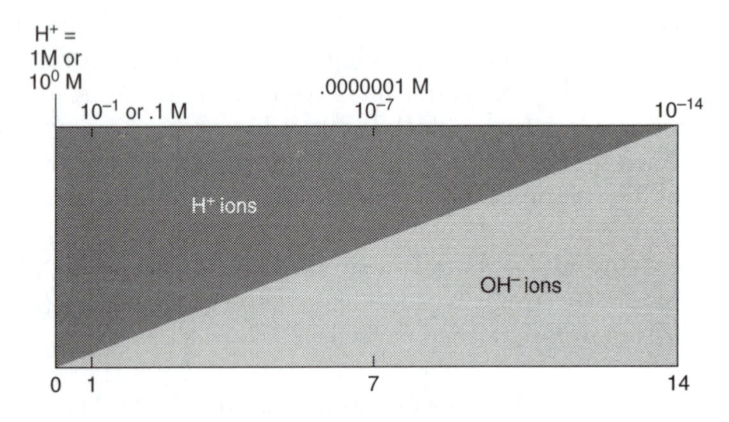

(H^+) = hydrogen ion concentration; (OH^-) = hydroxide ion concentration.

If the pH changes from 2.7 to 4.7, how much change occurred? One hundred times less acid or 10^{-2} less.

Hint: Because pH numbers are logarhythmic, log of the number 1 = 0, log of 0.1 = –1; log 0.01 = –2, log .001 = –3, etc.

AIR POLLUTION CHEMISTRY

You have heard of acid rain: It includes carbonic acid, sulfuric acid, nitric acid, and others. These acids are formed by the reaction of polluting gases with water and oxygen in the air.

Acid rain gas	Reaction to make acid	Principal sources
CO_2, carbon dioxide	$CO_2 + H_2O \rightarrow H_2CO_3$ (carbonic acid)	Combustion of hydrocarbon fossil fuels and wood. $C_xH_x + O_2 \rightarrow CO_2 + H_2O$
SO_2, sulfur dioxide	$SO_2 + H_2O \rightarrow H_2SO_3$ (sulfurous acid, also in onions)	$S + O_2 \rightarrow SO_2$. Coal-burning power plants, autos (this stuff wrecks our exhaust systems over time and decreases the effectiveness of your catalytic converter—Sulfur coats the platinum catalyst)
SO_3, sulfur trioxide	$SO_3 + H_2O \rightarrow H_2SO_4$ (sulfuric acid) $SO_2 + H_2O + O_2 \rightarrow H_2SO_4$	
N_2O, NO, and NO_2, nitrogen oxides—nitrous oxide, nitric oxide, and nitrogen dioxide respectively	N_2O or $NO + H_2O \rightarrow HNO_2$ (nitrous acid) $NO_2 + H_2O + O_2 \rightarrow HNO_3$ (nitric acid)	$N_2 + O_2$ (high temp. and pressures) $\rightarrow$ N_2O, $NO + NO_2$. Especially, high-compression diesel and gasoline engines and coal-burning power plants make nitrogen oxides, NO_x. The higher the combustion pressure and the higher the temperature, the more oxidation of nitrogen occurs and the more oxygen atoms attached to nitrogen atoms—NO_2. Nitrogen oxides contribute to ozone and smog formation in cities

Carbon Oxides

Combustion of organic or carbon compounds is an important process to know.

Carbon dioxide is produced by the burning of carbon (as in coal) or hydrocarbons in fuels or tobacco.

$$C + O_2 \rightarrow CO_2$$

This reads: Carbon and oxygen (plus heat to activate) yields carbon dioxide

$$C_XH_X + O_2 \rightarrow CO_2 + H_2O$$

This reads: hydrocarbons (e.g., propane, C_3H_8, octane, C_8H_{10}) plus oxygen yields carbon dioxide and water. Carbon monoxide is produced if less oxygen is reacted.

$$C_XH_X + \text{less } O_2 \rightarrow CO \text{ (carbon monoxide)} + H_2O$$

Smoke and **soot** are unburned/uncombusted carbon or carbon compounds and aerosols.

The **catalytic converter** on your car takes hydrocarbons and carbon monoxide and oxidizes them to CO_2.

$$C_XH_X + CO + O_2 \rightarrow CO_2 + H_2O$$

It reduces the nitrogen in the nitric oxide pollutant produced as a byproduct of gasoline combustion in nitrogen gas.

$$NO_x \rightarrow N_2 + O_2$$

Common hydrocarbons found in liquid petroleum gas are illustrated below.

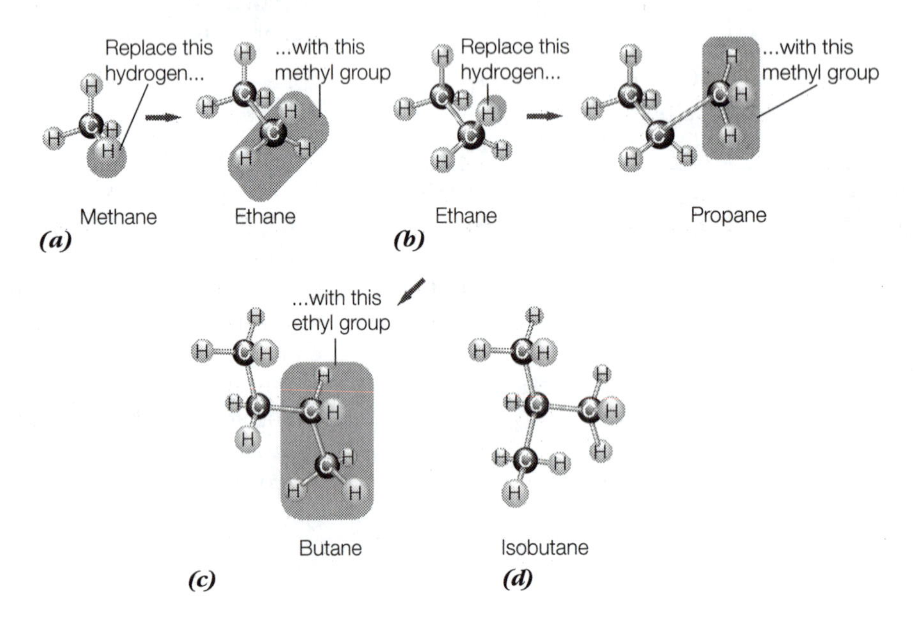

Sulfur Oxides

Sulfur contaminates in coal, gasoline, and other fuels oxidize to form sulfur dioxide, SO_2 (yellow in color), and sulfur trioxide, SO_3. Respectively, these gases dissolve in water to form sulfurous acid, H_2SO_3, and sulfuric acid, H_2SO_4. The acids dissolve the limestone of monuments and buildings and dissolve the calcium and magnesium out of conifer needles and soils—causing plant roots to soak up toxic aluminum. See pg. 427.

Nitrogen Oxides

Because combustion engines compress high octane gasoline and air previous to an ignition spark, atmospheric nitrogen is oxidized to NO_X (nitric oxides) during the higher temperature explosion.

Naturally, high compression engines make a lot of nitrogen oxide gases. Remember the air we are breathing contains 78% nitrogen gas. When the first atomic bombs were exploded in 1945, some feared that the atmosphere would catch fire. It did, but just in the fireball, fortunately.

$$N_2 + O_2 \rightarrow NO_X \text{ gases (nitrogen oxides: } N_2O\text{, nitrous oxide; NO, nitric oxide; } NO_2\text{, nitrogen dioxide)}$$

Forty years ago we had high compression/high horsepower auto engines—muscle cars like Pontiac GTOs. Standard engines have 8–9:1 compression ratios, high compression engines 10–12:1.

Why aren't these engines made today? They required highly leaded gasoline for valve lubrication and octane to prevent preignition due to the heat of compression, and they produced high amounts of nitrogen oxides.

Gasoline without high levels of lead, bromine, or benzene additives will explode prematurely in an engine due to the **"heat of compression."** When a gas is compressed, it heats up. In a diesel engine, the heat of compression makes the fuel and air mixture explode (combust)—it needs no spark plugs! If gasoline explodes prematurely it causes "ping" or "knock" that is damaging to it. The exploding gas pushes down on a piston that is coming up. This is not a good combination for performance or engine life.

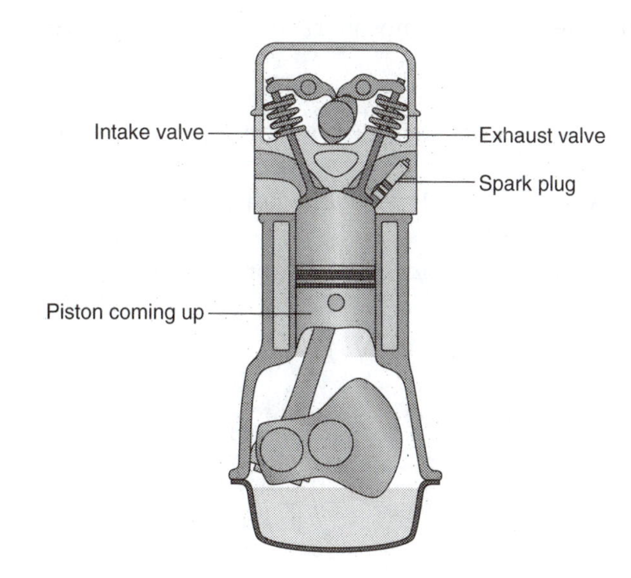

The Ozone Layer

The **ozone layer** is found in the upper atmosphere about 15 miles high or nearly twice as high as standard airline jets cruise (40,000 ft. or 8 miles high). Ultraviolet electromagnetic waves cause O_2 to break down into 2 separate oxygen atoms. One of these will combine with an O_2 to make an O_3 or ozone molecule, which then may break down again. These reactions are activated by ultraviolet energy, and convert UV to infrared (heat) radiation.

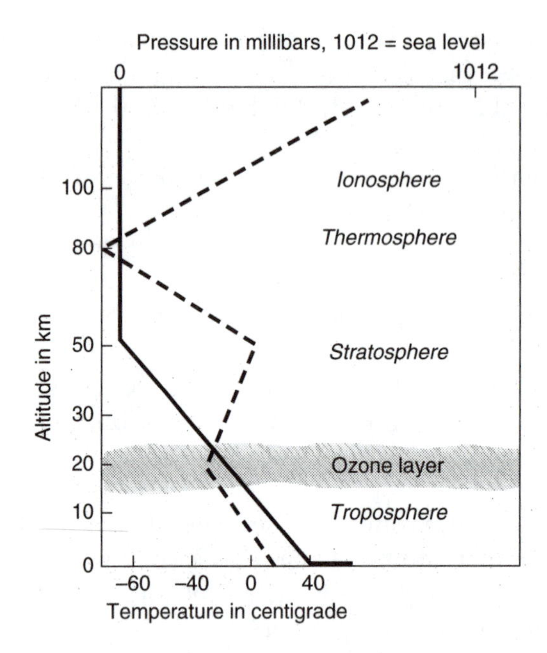

Temperature in centigrade = dashed line. Air pressure in MB = solid line.

Why does temperature increase in the thermosphere? Why does the aurora borealis occur in the ionosphere? See below.

$$O_2 + \text{UV photons} \rightarrow O_1 + O_1 + \text{Infrared photons (heat)}$$

$$O_1 + O_2 + \text{UV} \rightarrow O_3 \text{ (ozone)} + \text{Infrared photons (heat)}$$

$$O_3 + \text{UV} \rightarrow O_2 \text{ and } O_1 + \text{Infrared photons (heat)}$$

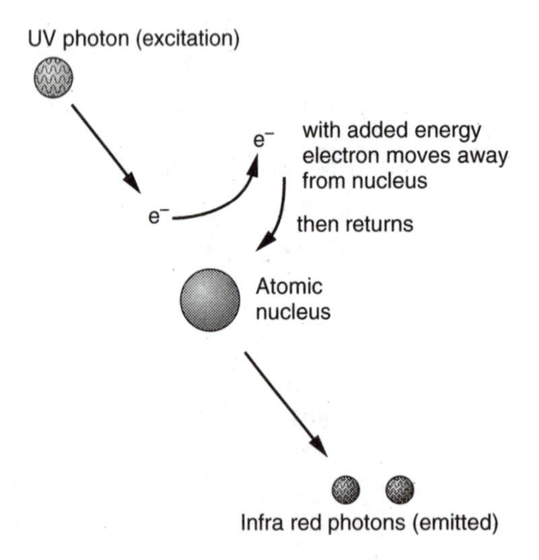

Why do temperature inversions (temperature suddenly increases with increasing altitude) happen in the stratospheric ozone layer? Because heat is released as ozone is created and breaks down cyclically.

Various compounds can destroy stratospheric ozone. Refrigerants or chlorinated fluorocarbons (CFCs) from spray cans or refrigeration/air conditioning units, carbon monoxide and unburned hydrocarbons from high altitude jet engine exhausts, methane from large herbivores, and chlorine gases can destroy the ozone causing **"ozone holes"** in the atmosphere that leak in increased amounts of ultraviolet radiation to the earth's surface. If ultraviolet radiation increases, an increase in the occurrence of skin cancers will be seen in unfortunate humans, more commonly in light skinned people.

Study Question: In 1973, the U.S. government decided not to encourage the building of a high flying supersonic transport aircraft (SST). What did the ozone layer have to do with this decision? The SST could have flown at altitudes in the stratosphere where ozone is present.

The *ozone hole* is near the South Pole over Antarctica. There are areas of thinning over the industrial northeast in the U.S. and Canada, and similarly from England over northern Europe in France and Germany. The ozone hole occurs annually over Antarctica between September and November, when the Antarctic spring (sunlight) returns.

The circumpolar vortex of cold air forms ice crystals to which chlorine and bromine adhere. When the circumpolar vortex breaks up, the chlorine and bromine are dispersed northward to New Zealand and South Australia. The resultant thinning of ozone over New Zealand has caused increased rates of melanoma and other skin cancers there.

Melanoma is the most aggressive and fatal form of skin cancers and is caused by a series of mutations to melanocytes that make the pigment melanin at the base of the upper cellular layer of the skin (epidermis). Once the cancer cells have penetrated into the dermis of the skin where blood vessels and lymphatic vessels are located, it quickly moves to lymph nodes, and to other tissues and organs. Squamous cell carcinoma and basal cell carcinoma are also caused by UV-B exposure. The latter is usually not fatal.

In the troposphere, ozone is a secondary air pollutant produced by photochemical reactions that produce smog, see below. At street level, ozone causes or worsens lung diseases, particularly in those who have compromised lung health by smoking.

Greenhouse Effect

Just like glass, the troposphere is transparent to visible sunlight, but when that sunlight strikes the Earth, the photons of violet-indigo-blue-yellow-orange-red are transformed into infrared photons or heat. Heavier tropospheric gases such as nitrogen oxides, carbon dioxide, methane, CFC, and water, reflect these infrared photons back toward the surface, creating the greenhouse effect.

The combustion of wood, grass, etc., and fossil fuels has dramatically increased the CO_2 level of the atmosphere as determined in south polar ice samples dating back 158,000 years. There have been periodic rises and falls of CO_2 levels that are related to cold glacial periods (low CO_2) and warm interglacial periods (high CO_2).

If you want to see what high CO_2 will do for a planet, look at the planet Venus. The sulfur oxidized from the rocks on the planet's surface is now suspended as a aerosol (droplets) of sulfuric acid in thick clouds from 68 to 100 km above its surface. Venus has a **runaway greenhouse effect** that has produced temperatures high enough to evaporate all the plant's liquid water to form a thick cloud layer that hides the surface from optical view. Venus has to be mapped with radar. The surface temperature is approximately 460 degrees C, and the atmospheric pressure is 100 times more than Earth.

By reason of the carbon dioxide currently in the atmosphere and more expected in the future (see Chapter 20), most climatologists predict that the Earth will experience an **enhanced greenhouse effect** after 2050.

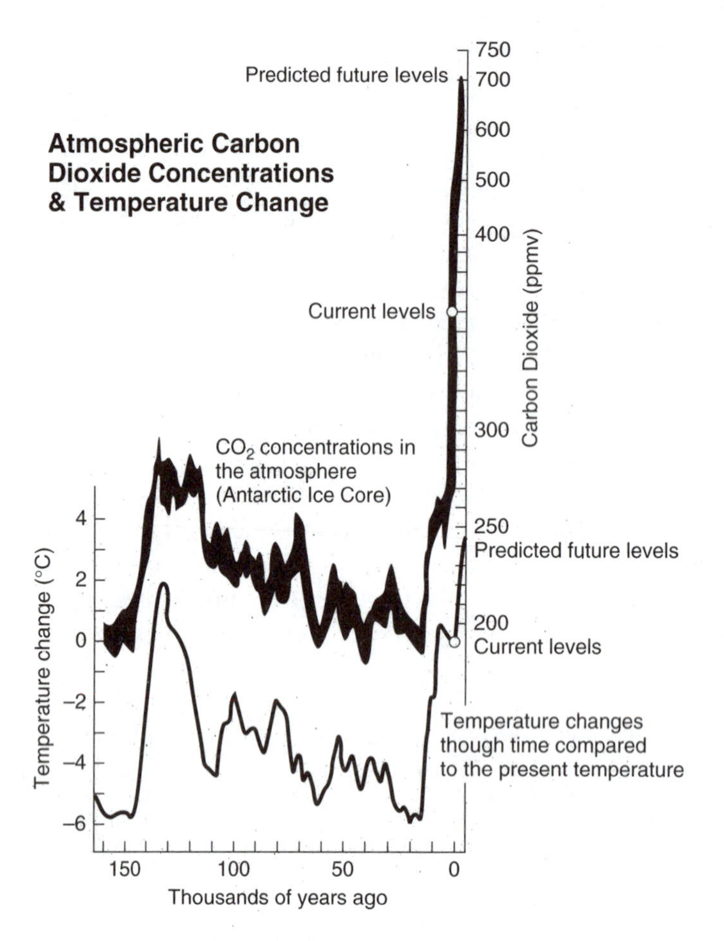

Methane is a greenhouse gas produced by rotting vegetation in marshes, swamps, and bogs, and **methanogen** bacteria in the guts of herbivores. You can see the bubbles of methane popping to the surface at the Okefenokee Swamp. Can the herbivorous dinosaurs' extinction be attributed to the "passing of gas?" Extensive volcanic eruptions can increase dust, cooling the Earth for the short term, but their release of carbon dioxide and sulfur dioxide could increase greenhouse effect for the longer term as it requires thousands of years to mitigate increased carbon dioxide.

There are many theories for dinosaur extinction. One outcome of the greenhouse effect is warming of the atmosphere that may have caused increased evaporation of water, which increased cloud cover, which then had a negative feedback effect causing atmospheric cooling, which caused the collapse of plant populations and the food chains that supported dinosaurs.

See **www.knowledge.co.uk/frontiers/sf080/sf080g12.htm** and **/www.epa.gov/globalwarming/index.html**

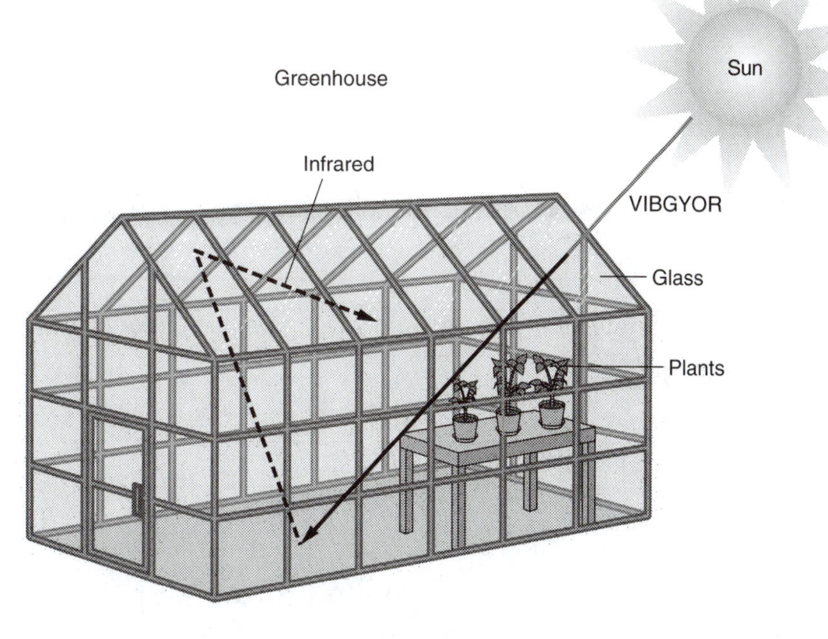

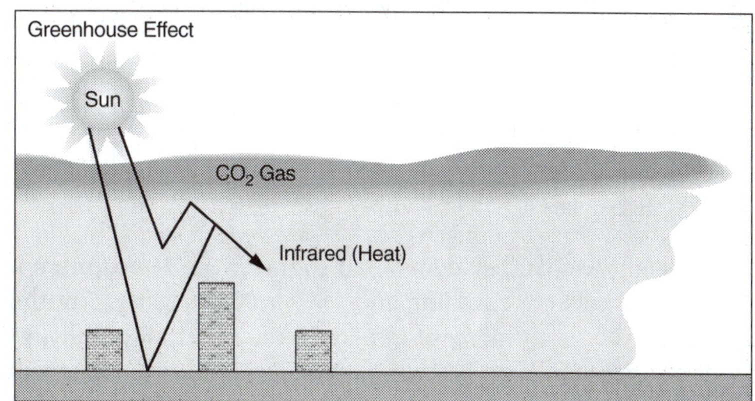

The Greenhouse Effect—Carbon dioxide and methane reflect infrared radiation (heat) back to the surface, causing increased temperatures at the earth's surface. Notice how UV-violet-indigo-blue-green-yellow-orange-red light wavelengths come through the atmospheric gases, but infrared radiation is reflected back to the surface.

Global warming has not increased as dramatically as carbon dioxide in the atmosphere.

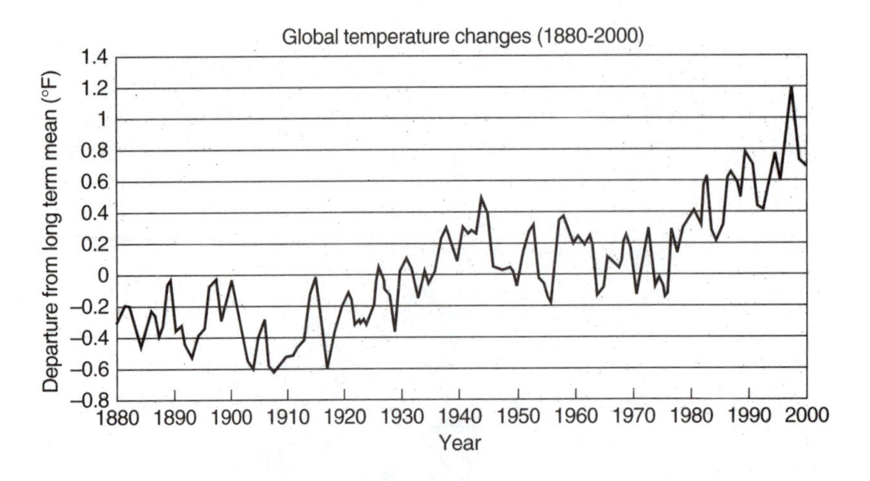

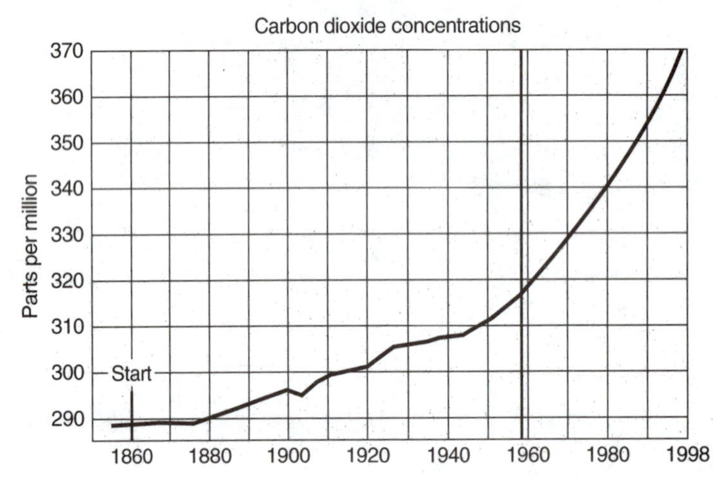

Other than the surface temperature data, evidence that suggests the atmosphere is warming is the fact that most of the world's glaciers are receding and sea levels are rising. Another possible outcome of increasing carbon dioxide is an increase in plant biomass. If, hopefully, we don't destroy the plants, their increase will mitigate the high carbon dioxide levels. If not, algae will take over, growing in nutrient-rich offshore waters. That could turn into a stinking mess and destroy commercial fisheries.

Experiments have indicated that increasing biomass is accompanied by a relative decrease in nitrogen content of the plant food. Nitrogen would not increase, it would stay the same. Therefore, herbivores would have to eat more plant material to get the same amount of nitrogen-containing amino acids to make their cellular proteins. Could this be another dinosaur extinction theory?

The glacier that resided here in Colorado is gone!

How will plant communities adjust carbon dioxide levels?

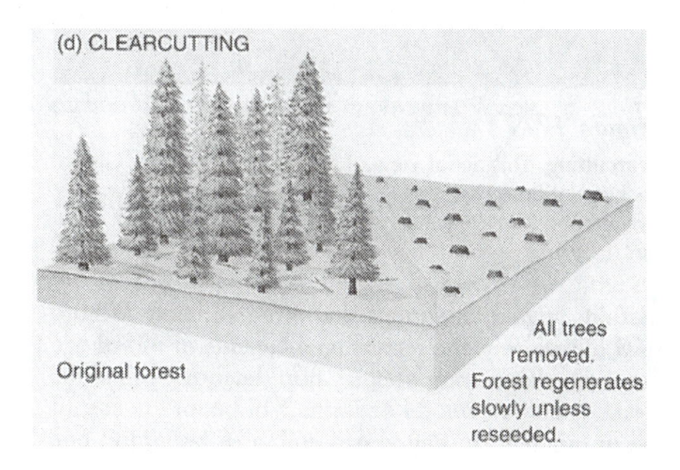

Adjusting Carbon Dioxide

Review the carbon cycle in Chapter 6. Nature can dispose of carbon dioxide in two ways:

1. Photosynthesis incorporates carbon into the bodies of plants and that carbon is passed to animals. When the plants are buried, the carbon may be stored geologically as coal or oil.

The question for the contrarians to answer is this: "Where does the carbon dioxide go if we are destroying plants?

2. When carbon dioxide dissolves in the ocean, it forms carbon acid that then reacts with calcium ions to precipitate limestone. Also certain protozoa and most mollusks make shells that contain carbonates. The most important actions humans can take in reducing carbon dioxide is to stop burning fossil fuels and allow plant populations to increase.

Temperature Inversions—Smoke, Sulfur Aerosols, and Dust as Air Pollutants

The normal circulation pattern is for air to be heated at the earth's surface and then rise as convection currents. Hot air, being less dense, rises; cold air, being more dense, sinks. However, if there is a layer of warmer air over a layer of colder air (see below) will the layers mix? No! Air pollution will fill the lower cold air layer and people with lung problems will begin to feel the effects. This is a common condition in Denver, Los Angeles, New York, and Mexico City. The trapped air will remain until a front or otherwise strong winds blow away the pollution, then the air will clear for a while.

A typical temperature inversion consists of a warm air layer trapped beneath a cold air layer. Above the inversion is what pilots call the "haze layer" that usually tops out at 5,500 ft. or more in the summer, the temperature lapse rate initially increases instead of showing the normal decrease (temp. lapse rate) with height.

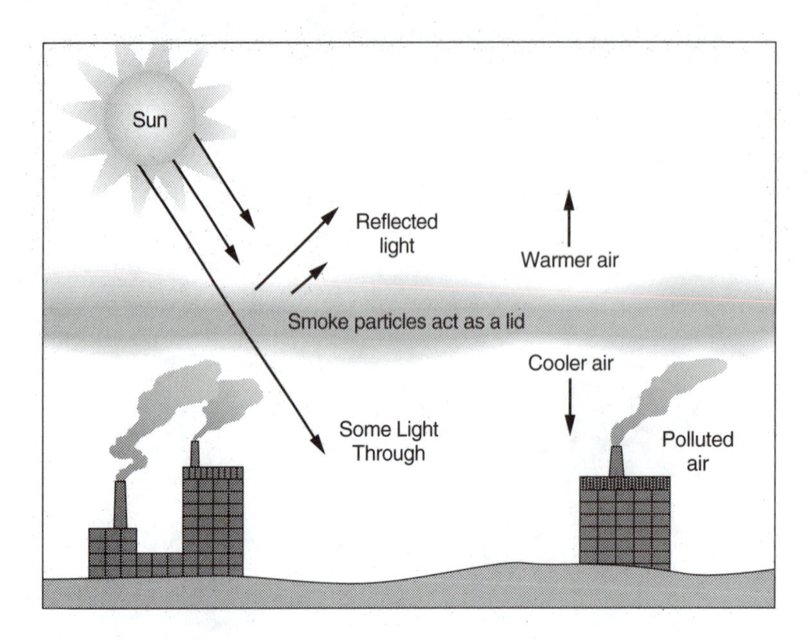

Solid smoke particles, unburned carbon compounds produced by combustion, volcanic dust, sulfur aerosols, desert or farming dust, can rise to the top of an air layer and screen out sunlight, making the underlying layer colder than the upper layer—a temperature inversion.

If there were worldwide increases of particles (including volcanic and agricultural dusts), the earth would cool. One theory about the mass extinction of dinosaurs is that an asteroid or large meteor hit the earth 65 million years ago, causing widespread fires and high-altitude dust bands. High-altitude dust bands from volcanoes also block sunlight and create severe winters. When the volcano

Krakatoa in Indonesia erupted over 100 years ago, a very severe winter occurred in Europe as sunlight was blocked. There is some mineral evidence of a meteor impact in rocks deposited at the so-called K-T (Cretaceous-Tertiary geologic time) boundary. There is a layer of dust and soot containing the metal iridium, common in meteors, and glassy "shock crystals" created by rock impact.

At that time the dinosaurs disappear and the small mammals that had been around for a long time, but in small numbers and size, suddenly, geologically speaking, become dominant and enlarge in size tremendously.

What chemical process results in smoke? Incomplete combustion. Complete combustion with oxygen produces no smoke.

What is "nuclear winter?" Nuclear winter is a cooling effect produced by the atmospheric dust created by the explosions of atomic bombs. Similar to meteor impacts with the Earth that apparently caused mass extinctions of plants and animals, 50 large yield nuclear bombs exploding in cities could produce radical cooling of the atmosphere and collapse of plant populations and the food chains they depend on.

Smog

Unburned hydrocarbons, carbon monoxide, oxygen, and NO_2 interact to produce harmful **ozone (O_3)** and **peroxyacetyl nitrate (PAN)** that are the eye-irritating chemicals in smog, especially in Mexico City, Los Angeles, and New York.

Nitrogen dioxide is formed.

$$O_3 + NO \text{ (nitric oxide)} \rightarrow NO_2 \text{ (nitrogen dioxide)} + O_2$$

Nitric oxide (NO) acts to remove ozone (O_3) from the atmosphere. This mechanism occurs naturally in the atmosphere.

$$NO + C_XH_XO_2 \rightarrow NO_2 + \text{other products}$$

Sunlight UV can break down nitrogen dioxide (NO_2) back into nitrogen oxide (NO).

$$NO_2 + \text{Ultraviolet light} \rightarrow NO + O$$

The single atom (nascent) oxygen (O) formed reacts with oxygen in the air producing ozone (O_3).

$$O + O_2 \rightarrow O_3$$

Nitrogen dioxide (NO_2) can also react with radicals produced from volatile organic compounds in a series of reactions to form peroxyacetyl nitrates (PAN) that sting eyes.

$$NO_2 + C_XH_X \text{ (hydrocarbons)} \rightarrow PAN$$

A clear day and smog in Los Angeles.

Global warming, ozone depletion, and acid deposition may change the environment of the future. Assuming the worst case scenarios for the three, what changes in the health, numbers and types of organisms, and their habitats are in store?

Health or Environmental Effects

1. Decrease in coral reefs.
2. Increase in heat-related deaths.
3. Midlatitude droughts and desertification.
4. Stronger hurricanes with wave-related damage to reef and shoreline communities.
5. Further decrease in phytoplankton of the California Ocean current and in Antarctica with concomitant losses of sea life.
6. Submersion of some islands and coastal areas.
7. Thawing of permafrost, reduction of lichen food for reindeer. Santa can't get south!
8. Reduction in western U.S. snowfall and stream flow resulting.
9. A mistiming of the flowering and emergence patterns of plants and their pollinators, with a decline in populations of both.
10. An increase in insect-vector populations and the diseases they transmit.
11. Decrease in buffalo grass in the prairies accompanied by an increase of invasive species.
12. U-V damage to photosynthetic chemicals and concurrent reduction of the ability of plants to mitigate carbon dioxide increase.
13. Increased skin cancers in humans and animals and leaf damage to plants.
14. Further decreases seen in mountaintop trees and in those northeastern areas of North America and Europe due to acid rain.
15. Further decreases in the populations of amphibians, insect larvae, and fishes in the streams and lakes affected by acid rain.

ENERGY

There are several types of energy which are important for Environmental Sciences.

Atomic energy is stored in atoms, it literally holds the particles of the atomic nucleus together. When atoms are split (as in a uranium or plutonium fission bomb or nuclear reactor) or fused (as in the Sun or laser fusion devices) some mass is converted to radiant energy.

Kinetic energy is the energy of moving atoms or molecules. Kinetic energy provides the energy to collide **reactants A** with **B** to react to make **C product.** Heat increases molecular motion.

Chemical energy is stored in the chemical bonds made by atoms exchanging or sharing electrons.

Potential energy is energy that can be stored and released. A good example is the evaporation of water and its rising against the force of gravity to form clouds. When enough water accumulates in a cloud, it falls releasing its potential energy. When water falls through the turbines of a hydroelectric dam, that gravitational potential energy is converted to electrical energy.

Electrical energy is carried by a current of electrons being pushed along in wires and batteries.

Radiant energy travels in waves. Electromagnetic waves, like ocean waves, rise in potential energy and fall in energy as they pass a set point. Similarly, electromagnetic waves have two characteristics, amplitude or power and wavelength.

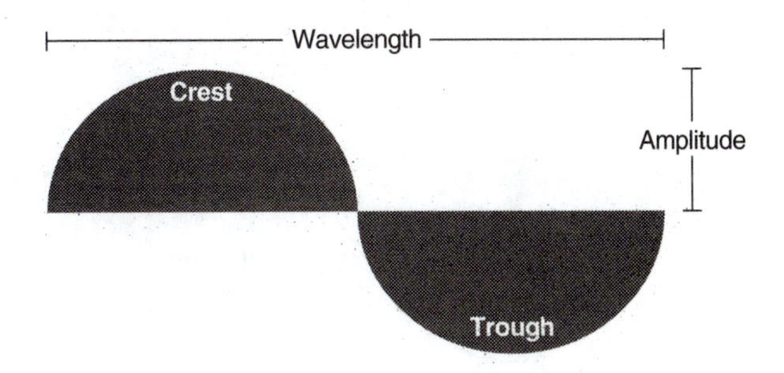

Atoms and Radiation

Radioactive atoms have unstable nuclei that emit radiation as they disintegrate. Radiation includes various forms of energy emitted by a radioisotope or form a nuclear fission or fusion reaction. Radiation consists of two general types: *atomic particles* (electrons, positrons, neutrons or protons), and *electromagnetic* waves, actually particles but their mass is not measurable. The wave forms include gamma rays, X-rays, ultraviolet, visible light spectra, infrared or heat, microwaves, and radio waves.

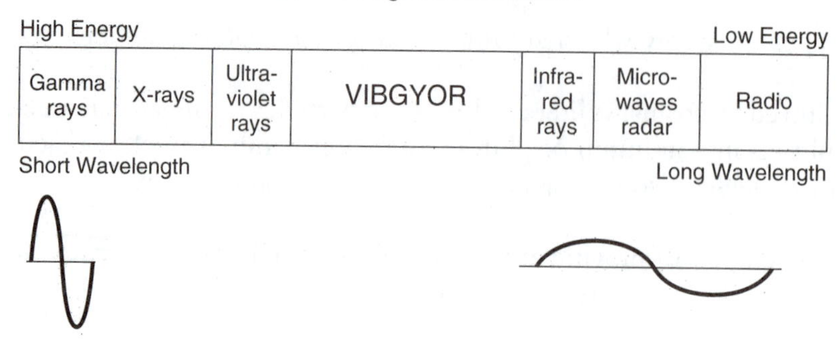

[VIBGYOR = violet, indigo, blue, green, yellow, orange, red]

Gamma rays, X-rays, and ultraviolet rays are mutagenic; that is, they mutate DNA, the chemical in our cells that carries the messages of inheritance that regulate cell growth, reproduction and metabolic activities. These **gamma, X-ray and ultraviolet (UV-C, *shorter wavelength*,** and **UV-B, *longer wavelength***) "rays" are more precisely packets of electromagnetic energy called photons. This **ionizing radiation** knocks off the electrons of gas atoms in the upper atmosphere, specifically the **ionosphere,** producing the northern or southern **"auroras,"** they collide with **ozone** in the **stratosphere** and with the melanin in your skin—there these high energy photons are degraded into low energy **infrared** photons (heat) that are not mutagenic.

The energy of the ultraviolet photon is decreased by the collision with electrons in **ozone** or **melanin** (skin pigment) molecules, or the photosynthetic chemical **chlorophyll** in plants. See the figure following.

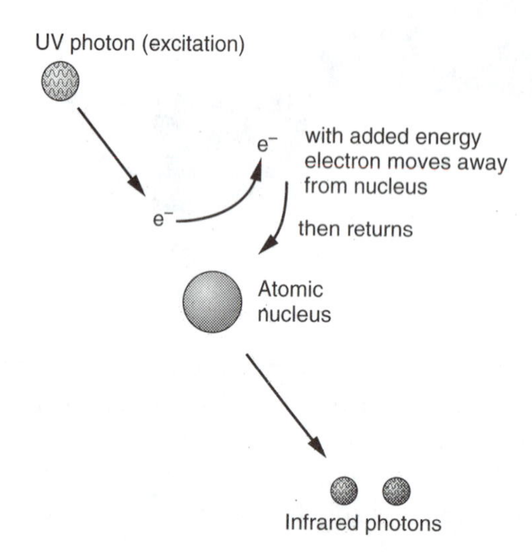

The electron moves away entirely in the ionosphere and in chlorophyll in living cells, or returns to the nucleus in a phosphorescent mineral. In auroras and phosphorescence/fluorescence visible light photons are also produced.

Van Allen Radiation Belts

Radiation, including electrons and protons, is trapped in the magnetosphere of the earth and circulates 500 + mi. above the surface of the earth. This radiation can be seen at the poles (the Aurora Borealis). More distant magnetic belts deflect gamma rays and the "solar wind."

Will skin cancer rates increase as the ozone layer decreases in O_3 concentration?

Who gets more skin cancer due to skin exposure, light-skinned or dark-skinned people?

Who gets more skin cancer? Folks in Michigan or folks in Arizona? Cloud cover absorbs some UV, but who gets more exposure? People living nearer the equator get more exposure than those living closer to the poles generally.

Why wear "sun screens"?

The general relationship of ionizing/mutagenic radiation exposure to mutation-induced disease

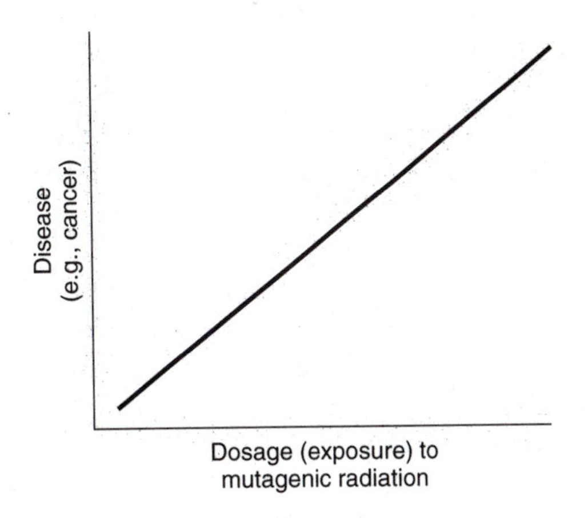

The graphed line was not connected at the bottom because cells have mechanisms for some gene repair.

At very high levels of radiation exposure, **radiation sickness,** characterized by severe nausea, diarrhea, and resulting dehydration occurs as cells lining the digestive tract and other organs are destroyed. Many who immediately survived the atomic bomb explosions in Japan died of this condition some days to a few weeks thereafter. Every time we go out in the sun, some DNA mutations occur in our skin cells due to UV-A and UV-B exposure. We have enzymes that repair the thymine dimer mutations that result. Some children have inherited mutated genes that do not make these enzymes, and they cannot have their skin exposed to sunlight. After exposures to UV, their skin blisters repeatedly and turns cancerous.

Light, Chemicals, Color, and Heat

Chemicals, such as chlorophyll and other photosynthetic pigments, absorb light and direct the energy to the process of photosynthesis that makes food and structural materials for the plants. Why is chlorophyll green? Because it reflects green photons but absorbs strongly in the violet-blue and orange-red areas of VIBGYOR. Reflection and transmittance (light going through a leaf) are opposite to absorbance. A black surface absorbs all visible wavelengths and makes infrared heat. A white surface reflects all wavelengths.

I bet you know this—what happens when sunlight strikes a flat black surface? In Chicago, the city government is experimenting with making the tops of buildings (usually covered with black tar that creates a lot of heat) reflective or planted in thick vegetation. Reflective surfaces reduce surface temperature, but what about temperature higher in the air? When the heat "sinks" of tropical rain forests are replaced by more reflective grassy field surfaces, the overall effect is warming of the air.

Which photon has more energy? A blue or a red one? Blue is correct. See the electromagnetic spectrum above.

More Radiation Risks

When atomic bombs explode in the atmosphere, a dust of "radioactive fallout" is spread downwind of the source of testing. Also materials from leaking bombmaking plants, such as Hanford, WA, have contaminated groundwater. When atoms are split in **"fission"** reactions, smaller atoms are formed. Many of these are "radioisotopes" that emit radiation.

Iodine-131 and persistant I-129 are a common radioisotope found in fallout from nuclear bomb testing. Natural iodine, such as in iodized salt, accumulates in the thyroid gland, where it is incorporated into thyroid hormones.

See the map of I-131 exposure and the map of thyroid cancer distribution. The hot spot in Idaho is partly due to Hanford, WA, groundwater pollution. Otherwise the distribution pattern follows the general jet stream movement of air in a serpentine pattern over the country.

Reference: **http://rex.nci.nih.gov/massmedia/fig1.html**

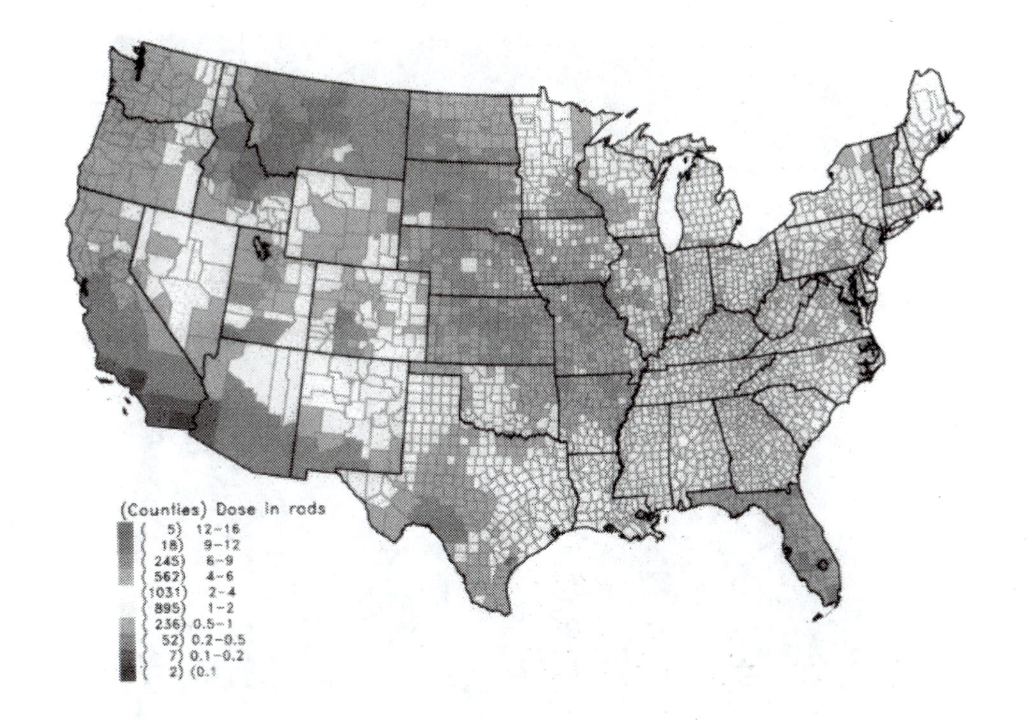

On your periodic chart at www.webelements.com, identify strontium-90 as an isotope. What periodic chart column is it in? It is a relative of calcium found in bones. Cesium (Caesium)-137 is in the same column with sodium and potassium (closer). That means it will be incorporated into the body as sodium and potassium ions are. Due to cesium's similarity to potassium, it will be found inside cells where potassium normally accumulates.

If one eats plants that have soaked up I-129, Sr-90 and Cs-137, does one's radioactivity level increase? Yes. If one acquires radioisotopes then one's level of radioactivity increases.

When the nuclear plant in Chernobyl, USSR (now the Ukraine), exploded, a cloud of radioactive fallout was spread by the wind northwest as far as Sweden.

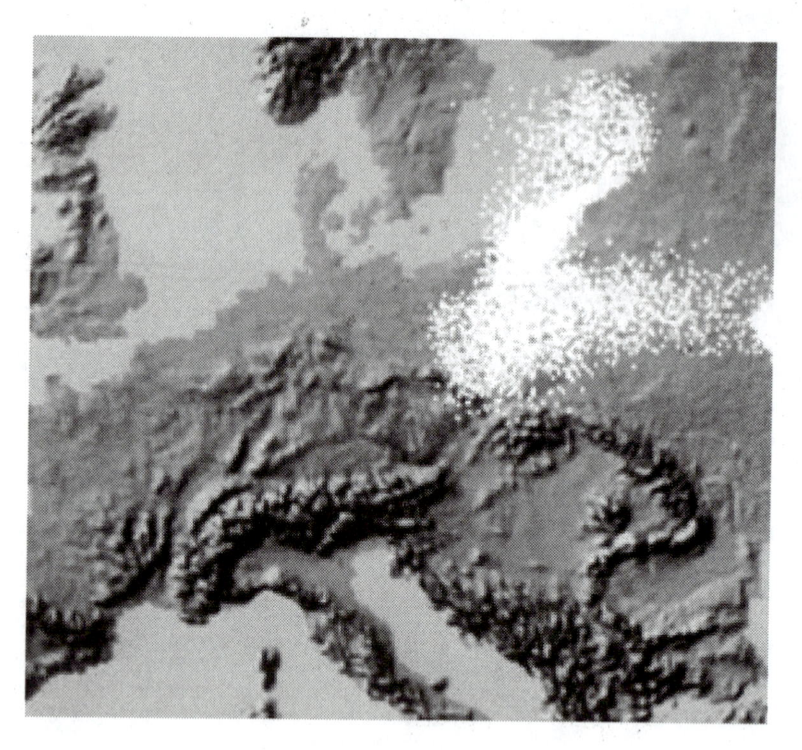

The Chernobyl fallout cloud over Europe.

Many of the "big box" hardware stores sell radon testing kits. There is a radioactive isotope of radon that naturally diffuses from rocks containing uranium-238 and accumulates in basements. It is thought to be another cause of lung cancer, far behind exposure to cigarette smoke, of course.

If food is bathed by gamma rays, as in sterilization/preservation, does that make the food radioactive? Hint: When you are bathed with light radiation, does that make you glow in the dark?

The Effects of the Nagasaki Atomic Bombing on the Human Body

Most directly heat and other radiation from the blast literally cooked the skin off some victims. The death toll within one kilometer from the explosion was 97% among burn cases, 97% among those with external injuries, and 94% among people who were apparently unscathed. Those deaths were due to internal damage done by radiation.

The relative risk of subsequent malignancy (by tumor site) is illustrated below.

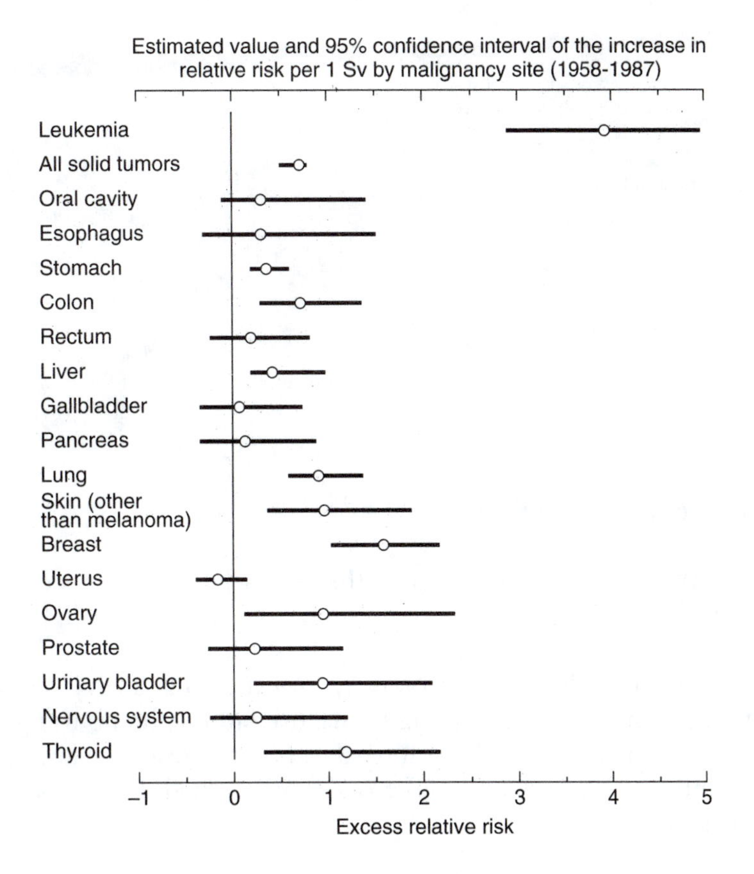

A statistically significant increase in risk was recognized in cancer of the stomach, colon, lung, breast, ovary, urinary bladder, thyroid, and liver, and in skin cancer other than melanoma. (After Thompson et al., RERF TR 5-92)

The first atomic bombs were **fission** bombs. Large uranium atoms are split to form smaller atoms and the neutrons splitting more atoms in a chain reaction.

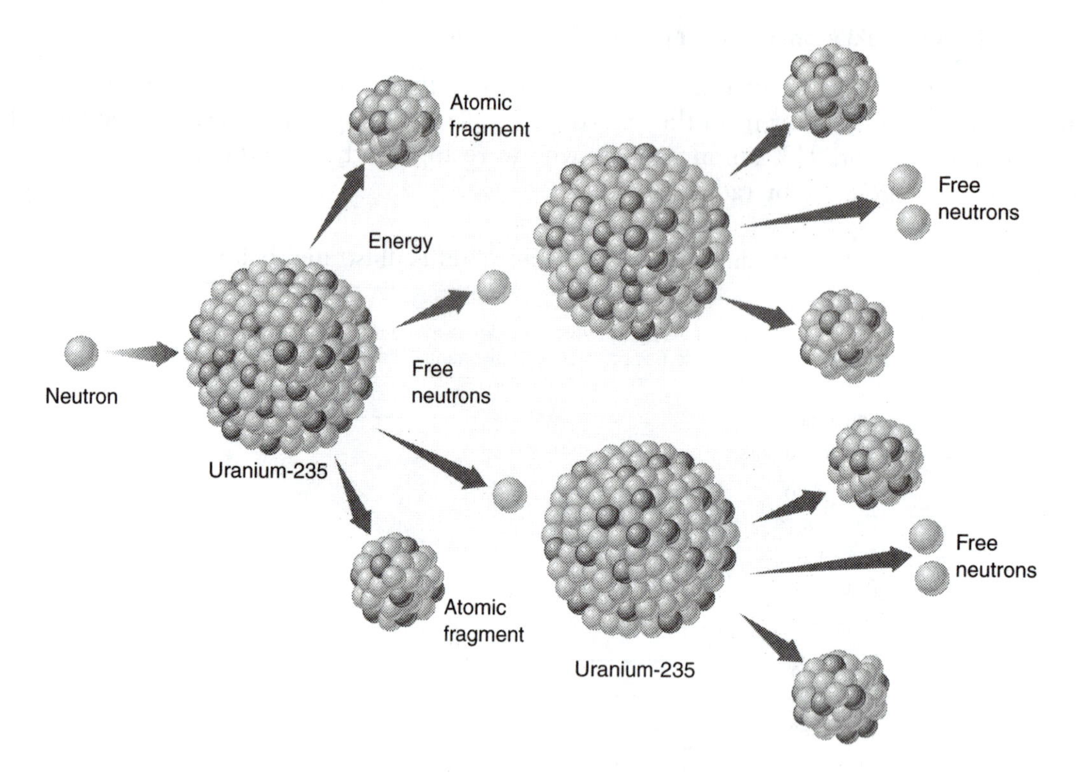

Later bombs used **atomic fusion** like the sun:

$${}^{1}_{3}\text{H (tritium)} + {}^{1}_{2}\text{H (deuterium)} \rightarrow {}^{2}_{4}\text{He (helium)} + \text{large amounts of energy}$$

(the superscript is the atomic number and the subscript the atomic weight of the atoms).

In World War II the Germans built a plant in Norway for extracting "heavy water," deuterium oxide or D_2O. It is thought that they were trying to make an atomic bomb using the deuterium. Because the plant was well concealed, the British tried to bomb it from the air. Failing at this, saboteurs were parachuted into the back country and cross-country skied there to set off bombs that put the plant out of commission.

The North Koreans are operating a breeder reactor that can make the nuclear fuel uranium or plutonium for making fission bombs. The second atomic bomb that was dropped on Nagasaki was a plutonium bomb.

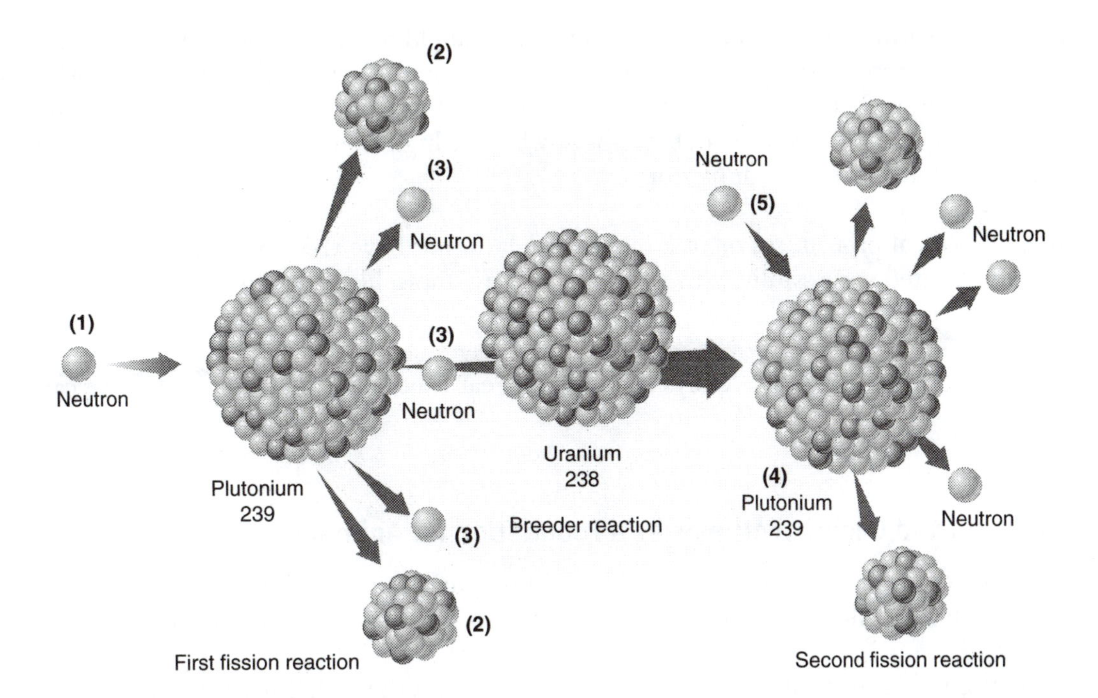

Medical Imaging and Radiation

Many imaging techniques use forms of radiation produced by decaying or excited atomic nuclei. Some forms are more dangerous than others. In general, short wavelength forms have very high energy, mutate the chemical of inheritance (DNA) and cause cancers or the death of cells. In radiation therapy, beams of gamma rays are focused on a tumor to "burn out" the cancer cells. One goal of physicians has been to lower the dosage of medical imaging X-rays or gamma rays that patients are exposed to during their lifetime. Unfortunately, the threat of malpractice litigation causes the ordering of more X-rays and CT scans than are medically necessary.

X-rays are produced when a high-voltage electron current strikes a metal plate in a cathode ray tube. Structures which absorb X-rays, like solid organs and bones, produce white images on the negative. Problems include exposure to mutagenic radiation, above, and the overlapping of two dimensional structures.

CAT or CT scans (computerized axial tomography) apply X-rays that are taken in an arc around the body and fed into a computer. A sectional view of organs can be obtained. Again, X-ray exposure is a problem, particularly to children.

Why can't pregnant women get X-rays? X-ray exposure to a fetus can cause miscarriages and birth defects.

PET (positron emission tomography) scans are produced when a patient is injected with a radioactive substance (sugar) that produces positrons. Positrons are positive electrons usually found in the nucleus but are emitted during the disintegration of certain radioisotopes. When the positrons collide with electrons, they destroy each other and produce low levels of gamma rays that are imaged. The technique is highly useful for diagnosing brain area dysfunction.

MRI (magnetic resonance imaging) uses 1) a magnetic field to cause protons to realign and 2) radio waves that increase the energy of the protons. The radio wave reflections are then "read" by the image monitor. Radio waves are the least damaging form of radiation. They are used to produce the MRI image. There is some debate about whether high-energy magnetic fields are dangerous or not, but most researchers think not.

Are there any medical procedures or treatments which do not have risk? No. Are cancer patients who have been treated successfully with radiation therapy more likely to get tumors of a new source (different cells)? Yes!

Therefore, is it contradictory that X-rays are used to treat cancer but that they can cause cancer? No!

Photosynthesis and Chemosynthesis as a Foundation for Animal Life

CO_2 + H_2O + Sunlight (electromagnetic, VIBGYOR wavelengths) energy $\rightarrow$ $C_6H_{12}O_6$/glucose (used to make cellulose and other solid compounds, and food for the plant) + O_2 gas.

Most people think that most of the mass of a tree comes from the soil. That is not correct. The mass mostly comes from the carbon in the carbon dioxide of the air that is turned into solid carbon compounds in wood.

Sunlight provides the electromagnetic energy that is absorbed by the green chemical chlorophyll and accessory pigments, such as those seen when sugar maple leaves die in the fall. The electromagnetic energy is converted to chemical bond energy in the glucose molecule. Where there is a C-C or a C-H covalent bond, energy is stored in that bond that can be extracted in cellular metabolism.

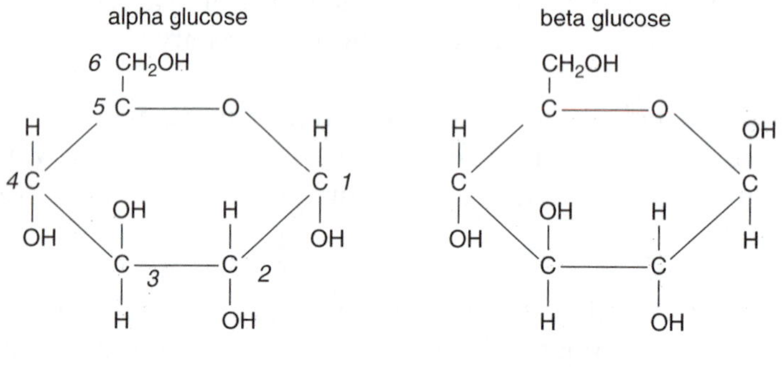

alpha glucose

beta glucose

Energy Transformations

The food energy we acquire in eating originally came from the sun's atomic energy transformed into radiant electromagnetic energy. Plants trap the electromagnetic energy from sunlight and transform it into chemical energy.

Chemosynthesis

However, some bacteria living near volcanic vents in the Pacific and Atlantic Oceans, too deep, at 10,000 feet down, to receive sunlight, nevertheless produce food for a whole community of organisms from tube worms to crabs and fishes. These bacterial **producers** do **chemosynthesis:** H_2S (hydrogen gas from volcanic vents) is oxidized to yield SO_2 and H_2O and the important energy currency of all cells, **ATP.**

Besides using their ATP for cell movement, transport, etc., ATP is used to produce light or **bioluminescence** in some of these sea-bottom dwellers. If life is found on Mars, it will likely be chemosynthetic bacteria, so called **rock eaters** that can derive the energy to make food from the oxidation of chemicals in the rock such as sulfur. On Mars, the chemosynthetic bacteria, if they exist, are probably sulfur or iron bacteria.

Sulfur bacteria thrive at 170 degrees F at the edge of this Yellowstone geyser pond.

"The Grand Balance of Life"

In the biosphere, to maintain stable populations of all organisms,

Photosynthesis and Chemosynthesis = Cellular Respiration

Photosynthesis:

$$CO_2 + H_2O + \text{sunlight energy} \rightarrow \text{glucose } (C_6H_{12}O_6) \text{ and other food molecules} + O_2$$

Electrons are acquired from the water molecule to replace those energized in chlorophyll. Those energized electrons make chemical bonds in food molecules used in the metabolism of the plants and stored in the leaves, stems, seeds, and roots of the plant. Remember that plants do photosynthesis and cellular respiration at the same time.

Chemosynthesis (Two Examples):

a. Iron bacteria: $Fe^{+2} \rightarrow Fe^{+3} + 1\ \mathbf{e-}$

b. Sulfur bacteria: $H_2S + O_2 \rightarrow SO_2$ or SO_4^{-2} (sulfate) + 2 H^+ (protons) + $2\mathbf{e-}$

The high energy electrons acquired are used to make energy containing C-H and C-C covalent bonds in the food of cells.

Cellular Respiration:

glucose ($C_6H_{12}O_6$) and other food molecules + $O_2 \rightarrow CO_2 + H_2O$ + chemical energy (ATP) for cells.

The chemical energy stored in glucose is chemical bond energy which is transferred to ATP. ATP then powers cell activities like movement, growth, and transport. All creatures do this process when they utilize food made by themselves (algae and plants) or food made by other organisms (animals) that consume them.

Debate this statement: "Plants can survive without animals, but animals cannot survive without plants." Plants can live using cellular respiration and photosynthesis. Animals can only do cellular respiration. The food chain must start with photosynthesis or chemosynthesis.

Will homeostasis of carbon dioxide levels be possible if plant and algae populations decrease? No!

Links

To see a good online periodic chart, go to **www.webelements.com**

For another dinosaur extinction theory, see **www.knowledge.co.uk/frontiers/sf080/sf080g12.htm**

For more information on radioactive fallout, thyroid cancer, and radiation risks, see **rex.nci.nih.gov/**

To study the radiation effects of the atomic bombs dropped on Japan, see **www.sdc.med.nagasaki-u.ac.jp/n50/disaster/medical-E.html**

For more on the K-T extinction and the rise of mammals, see **www.wilson.ucsd.edu/education/pchem/airpollution/inversion.html**

For more on global warming and climate change, go to **www.epa.gov/globalwarming/index.html**

What are the potential effects of global warming on cities and plant and animal communities? See **www.epa.gov/globalwarming/index.html**

Catalytic converters, see **www.chemcases.com/converter/converter_19.htm**

Acid rain, go to **www.epa.gov/airmarkets/acidrain/**

See the sections on photochemical smog, effects of air pollution on plants, and how catalytic converters work at **www.msc-smc.ec.gc.ca/cd/factsheets/smog/index_e.cfm**

Magnetic belts deflect gamma rays and the "solar wind." See **ssdoo.gsfc.nasa.gov/education/lectures/fig05.gif**

Ozone hole, go to **www.epa.gov/ozone/science/process.html**

For a discussion on the effects of air pollution on tree communities in Appalachia, see **www.dieoff.com/page47.htm.**

See **www.nwf.org/climate** for the effects of global warming and what we can do about it.